Reinhard Selten (Ed.)

Game Equilibrium Models IV

Social and Political Interaction

With Contributions by

R. Avenhaus, R. Gardner, W. Güth, R. K. Huber
M. Mitzkewitz, A. Okada, B. O'Neill, J. Pool
J. Potters, R. Selten, F. van Winden, D. Wendt
S. Zamir

With 55 Figures

Springer-Verlag Berlin Heidelberg GmbH

Professor Dr. Reinhard Selten
Institut für Gesellschaft- und
Wirtschaftswissenschaften
der Universität Bonn
Wirtschaftstheoretische Abteilung I
Adenauerallee 24-42
D-5300 Bonn 1, FRG

ISBN 978-3-642-08111-8 ISBN 978-3-662-07369-8 (eBook)
DOI 10.1007/978-3-662-07369-8

Originally published by Springer-Verlag Berlin Heidelberg New York Tokyo in 1991
Softcover reprint of the hardcover 1st edition 1991

2142/7130-543210

Preface to the Series "Game Equilibrium Models"

The four volumes of the series "Game Equilibrium Models" are the result of a research year at the Center for Interdisciplinary Research of the University of Bielefeld, Germany. The German name of this center is Zentrum für interdisziplinäre Forschung, but everybody who is familiar with this unique institution refers to it by the official acronym *ZiF*.

In the time from October 1, 1987, to September 30, 1988, the ZiF was the home of the interdisciplinary research group which produced the papers in the four volumes of this series. Participants coming from many parts of the world lived in the guest apartments of the ZiF for the whole time or part of it and worked on a common project. The name of the project was "Game Theory in the Behavioral Sciences". It occurred to me only later that "Game Equilibrium Models" - the title of the series - even more appropriately indicates the unifying theme of the research group.

Among the participants were economists, biologists, mathematicians, political scientists, psychologists and a philosopher. A lively interaction resulted from the mix of disciplines. The common methodological basis of non-cooperative theory was the shared culture which facilitated communication across disciplines. The intense exchange of ideas coming from different fields had a profound influence on the thinking of many among the participants.

It was not easy to find a coherent way to group the papers into the four volumes and to find appropriate titles for the books. These and other difficult decisions have been made by an editorial committee consisting of Wulf Albers, Rudolf Avenhaus, Eric van Damme, Werner Güth, Peter Hammerstein, Ronald Harstad, Franz Weissing, and myself.

In the behalf of the whole research group I want to thank all those who helped to make the research year possible. We owe special thanks to the staff of the ZiF and in particular to Mrs. Lilo Jegerlehner for her technical help in the preparation of the four volumes.

Finally, I want to express my gratitude to all those who assisted me in the organizational and editorial work, especially to Franz Weissing whose efforts were indispensable.

Bielefeld/Bonn, January 1991 Reinhard Selten

Contents

Contributors

Rudolf Avenhaus, Institut für angewandte Systemforschung und Operations Research, Fakultät für Informatik, Universität der Bundeswehr München, Werner-Heisenberg-Weg 39, D-8014 Neubiberg, FRG

Roy Gardner, Department of Economics, Indiana University, Ballantine Hall, Bloomington, Indiana 47405, USA

Werner Güth, Johann Wolfgang Goethe-Universität Frankfurt/M., Fachbereich Wirtschaftswissenschaften, Mertonstr. 17, D-6000 Frankfurt am Main 11, FRG

Reiner Huber, Fakultät für Informatik, Universität der Bundeswehr München, Werner-Heisenberg-Weg 39, D-8014 Neubiberg, FRG

Michael Mitzkewitz, Institut für Gesellschafts- und Wirtschaftswissenschaften der Universität Bonn, Wirtschaftstheoretische Abteilung I, Adenauerallee 24-42, D-5300 Bonn 1, FRG

Akira Okada, Graduate School of Policy Science, Saitama University, Urawa 338, Japan

Barry O'Neill, Northwestern University, Department of Industrial Engineering and Management Sciences, Evanston, Illinois 60201-9990, USA

Jonathan Pool, Department of Political Science, DO-30, University of Washington, Seattle, Washington 98195, USA

Jan Potters, Faculteit der Economische Wetenschappen, Jodenbreestraat 23, KR 3142, NL-1011 NH Amsterdam, The Netherlands

Reinhard Selten, Institut für Gesellschafts- und Wirtschaftswissenschaften der Universität Bonn, Wirtschaftstheoretische Abteilung I, Adenauerallee 24-42, D-5300 Bonn 1, FRG

Dirk Wendt, Institut für Psychologie, Christian-Albrechts-Universität Kiel, Ohlshauserstr. 40-60, D-2300 Kiel, FRG

Frans van Winden, Faculteit der Economische Wetenschappen, Jodenbreestraat 23, KR 3142, NL-1011 NH Amsterdam, The Netherlands

Shmuel Zamir, 6537, Darlington Road, Pittsburgh, Pennsylvania 15217, USA

Introduction to the Series "Game Equilibrium Models"

Game equilibrium models are descriptions of interactive decision situations by games in extensive or normal form. The analysis of such models is based on the equilibrium point concept, often refined by additional requirements like subgame perfectness. The series consists of four volumes:

I: Evolution and Game Dynamics
II: Methods, Morals and Markets
III: Strategic Bargaining
IV: Social and Political Interaction.

The game equilibrium models presented in these books deal with a wide variety of topics. Just one example from each of the volumes may provide an illustration: Egg trading in hermaphrodite fish (*Friedman and Hammerstein* in Volume I), the social organization of irrigation systems (*Weissing and Ostrom* in Volume II), wage bargaining (*Haller* in Volume III), and beheading games in mediaeval literature (*O'Neill* in Volume IV).

Non-cooperative game theory is a useful research tool not only in economics and the social sciences, but also in biology. Game theory has been created as a theory of conflict and cooperation among rational individuals. For a long time strong rationality assumptions seemed to be indispensable at the foundations of game theory. In this respect, biological applications have changed our perspectives. Game equilibrium may be reached as the result of processes of natural selection without any rational deliberation. Therefore, game equilibrium models can contribute to the explanation of behavioral and structural features of animals and plants.

The interpretation of game equilibrium as the result of evolution raises the question of dynamic stability with respect to evolutionary processes. Similar problems also arise in theories of game learning. The first volume contains three papers on game dynamics. Two of them are concerned with the dynamic foundations of evolutionary game theory and the third one explores stability in a model of anticipatory learning. The remaining papers in the first volume present evolutionary game equilibrium models ranging from abstract investigations of phenomena like bluffing or group-based altruism to the examination of concrete systems observed in nature like "competition avoidance in a dragonfly mating system". Not only theoretical clarifications of the foundations of evolutionary game theory and related research can be found in **Evolution and Game Dynamics**, but also exciting new biological applications.

The title of the second volume, **Methods, Morals, and Markets**, points to several areas of research which attract the interest mainly of economists, but also of political scientists, mathematicians and philosophers. The first paper is a sophisticated mathematical contribution which applies new tools to basic questions of non-cooperative game theory. The word "method" mainly refers to this paper, but to some extent also to the next three contributions, which discuss basic conceptual problems in the interpretation of game equilibrium. Two papers relate to the philosophical notion of the social contract and its exploration with the help of game theoretical models. This work in concerned with "morals", a theme which is also touched by a paper on irrigation institutions. The remaining four papers of the second volume explore game equilibrium models of markets; two of these contributions are experimental and compare theoretical solutions with experimental data.

The third volume on **Strategic Bargaining** collects ten papers on game equilibrium models of bargaining. All these papers look at bargaining situations as non-cooperative games. Unlike in cooperative game theory, cooperation is not taken for granted, but explained as an outcome of equilibrium analysis. General models of two-person and n-person bargaining are explored, sometimes enriched by institutional detail like the availability of long-term contracts. Other papers explore bargaining in special contexts like wage negotiations. Two contributions concern spatial games; one of these contributions is experimental.

The exploration of strategic models of bargaining is an active field of research which attracts the attention of many game theorists and economists. The ten papers in the third volume contribute to the progress in this field.

The fourth volume on **Social and Political Interaction** mainly presents game equilibrium models in the area of political science. Three of the papers concern topics in other fields: the distribution of foreign language skills, altruism as a social dilemma (an experimental paper) and beheading games in mediaeval literature. Five contributions to the area of international relations deal with game theoretical models of the balance of power, of alliance formation, and of an issue in armament policy. An investigation of inspection problems like those arising in connection with the non-proliferation treaty also touches the area of international relations. Other papers on problems of political science deal with the game theoretical resolution of the Condorcet paradox by equilibrium selection, the modelling of political pressure exerted by firms on the government and the draft resistance problem.

The main emphasis is on biology in Volume I, on economics in Volumes II and III, and on political science in Volume IV. This is the result of an attempt to group the great variety of papers resulting from a year long interdisciplinary research project in a reasonably coherent way. However, not only biologists, but also economists and a psychologist have contributed to Volume I. Similarly, not only economists and mathematicians, but also political scientists as well as a biologist and a psychologist are among the authors of Volumes II and III. All four volumes are the result of the cooperation of researchers from many disciplines united by a common interest in game equilibrium models within and beyond the borders of their fields.

Bielefeld/Bonn, January 1991 Reinhard Selten

Introduction to Volume IV: "Social and Political Interaction"

This volume contains game equilibrium models focussing on social and political interaction within communities or states or between states, i.e. national and international social and political interaction. In short, the contributions try to illuminate specific aspects of such interaction systems by modelling them as noncooperative games and analysing their equilibria. It should be mentioned that social interaction is not only the topic of this volume, but also the essential prerequisite of its contributions since they resulted from fruitful and stimulating interaction within the research group.

Various aspects of international military policy problems are analysed *by Avenhaus, Güth, and Huber*, by *Avenhaus, Okada, and Zamir*, by *Gardner*, by *Gardner and Güth* and by *Güth* as well as by *Selten*. Till now basic aspects of international policy problems have often been modelled as cooperative games. This is especially inappropriate for international security problems where self-enforcement is highly exceptionable. Of course, very realistic models will mostly be too complex to be analytically tractable. Thus, authors had to balance their desire to incorporate the most important aspects on the one hand and the complexity of the game model on the other. In general, more simple models have been analysed more profoundly whereas the analysis of more complex models is usually less ambitious.

In Part I of their study, *Rudolf Avenhaus*, *Werner Güth and Reiner K. Huber* distinguish between a peaceful and an imperialistic military policy. After analysing the complete information games where the given military intentions of the two antagonistic countries are common knowledge they then proceed with the two cases of asymmetric incomplete information and the case of mutually incomplete information. The sequential game model in Part II allows the two antagonistic countries to restructure their forces from offensive to defensive ones if the first stage of choosing between an offensive or defensive force structure led to the highly unstable situation of mutually offensive force postures. Unlike Part I, in the first decision stage the restructuring phase is sequential and asymmetric in the sense that one party has to initiate the restructuring process.

The crucial assumption underlying the game equilibrium models analysed by Avenhaus, Güth, and Huber is the so-called Defense Efficiency Hypothesis meaning that conventional forces, specifically designed for defensive operations and incapable of any major offensive operations, can exploit intrinsic defense advantages more efficiently than forces designed for all combat modes, i.e. also for offensive operations. Under this hypothesis conventional stability can be brought about by mutual adoption of defensive force structures. This is what is meant by a successful restructuring of armed forces.

Inspection problems arise when an inspector has to control whether an inspectee behaves legally, i.e. according to a formal and well-specified agreement. To decide whether behavior has been legal or illegal and to limit the inspectee's incentive for illegal acitvities the inspector will often rely on a well-defined procedure which is based on observations of random variables and which may or may not be announced to the inspectee. *Rudolf Avenhaus, Akira Okada, and Shmuel Zamir* show that the inspectee will act legally if the monitoring procedure is announced whereas the inspectee will deviate from legal behavior with positive probability if both, inspector and inspectee, have to decide independently. The authors describe this result by saying that "inspector leadership" prevents illegal behavior. Whereas the leadership solution appears unstable in case of complete information this instability disappears if one assumes that the inspector is only incompletely informed about the inspectee's payoff function. The results are applied to material accounting and data verification problems of International Nuclear Material Safeguards. Similar problems will result for any arms control convention which hopefully will be agreed upon in a not too distant future.

Roy Gardner studies a three-stage model of conscription and possible resistance to it. The volunter stage requires a call for volunteers and its possible fulfillment. If the call is fulfilled, the game ends. If not, play continues with the participation stage where potential conscripts can form a resistance movement which then might carry out an agreed upon threat. In the final conscription stage conscripts are called up and the resistance movement carries out the threat. The paper assumes a continuum of players with additive utilities for the personal status (civilian, draftee, volunteer, or resister) and for the actual size of the armed force. It is shown that there is a unique subgame perfect equilibrium point which is characterized by interesting phenomena such as reluctant volunteers, i.e., volunteers who want to avoid the less favorable conscription.

Roy Gardner and Werner Güth analyse three different extensive game models illuminating special aspects of alliance formation. The authors prefer to illuminate these aspects by different models since a game which takes into account all aspects would be extremely complex and difficult to analyse. The first model considers the formation of various alliances from an initial state of nonalignment. The model helps to answer questions like: will nonalignment continue or are bipolarity or balance-of-power configurations more likely to develop? The second type of models allows one to predict arms expenditures, at least qualitatively, in bipolar situations such as that one prevailing in Europe after World War II. The third model considers how an alliance containing one superpower and a potential periphery of smaller countries can develop if the superpower is only incompletely informed about its potential partners. The paper relies partly on the method of equilibrium scenarios which restricts the analysis of the game to the more limited task to check whether certain policies are consistent with equilibrium behavior or not. Throughout the paper the authors try to relate their results to actual political observations.

Problems of forming a Western defense alliance are also investigated by *Werner Güth* who analyses two special game models with the players USA and two stylized

Western European countries, resembling French and German defense policy. In the first model with incomplete information the two types of Western European countries reflect the beliefs of the USA about the true security interests in Western Europe. The second model with complete information concentrates on alliance instability resulting from the fact that one can enjoy an alliance's military protection without being its member. By comparing the results with the actual state of the Western defense alliance NATO, one can derive restrictions for political parameters which often cannot be directly estimated.

Social decision making within a society by majority voting is the topic of *Werner Güth and Reinhard Selten* who investigate the so-called Condorcet Paradox, i.e. the case where three individuals have to choose among three alternatives and where preferences are such that for each alternative there is another one which is preferred by a majority. By applying the theory of equilibrium selection the authors show that one nevertheless can solve the game uniquely and thereby determine a unique public decision. Here it is, of course, essential that voters assign cardinal utilities to alternatives including the status quo which continues to prevail in case of no majority for one of the three proposals. It is an interesting result that the inefficient status quo is only obtained in degenerate cases, i.e. the solution will be Pareto-optimal with probability 1.

Jan Potters, Frans van Winden, and Michael Mitzkewitz use a repeated game in which the government is pressed by a lobby to concede a certain claim. Unlike in a previous study by Potters and van Winden the pressure group can choose its punitive action even if the government concedes. The enlarged strategy space increases significantly the number of equilibria. By applying the stability concept of Kohlberg and Mertens as well as the equilibrium selection theory of Harsanyi and Selten, the authors try to derive more definite results. It is a non-degenerate result that punishment would be chosen after a concession which, in turn, will often induce the government not to concede. A pressure group will often prefer to have no chance of choosing punitive acts after concessions.

By balance of power one often describes a state of affairs where potential agressors are deterred from military operations which might result in the elimination of one of the actors. *Reinhard Selten* has invented a parlor game whose rules are given in ordinary language. It should be mentioned that the research group played this game repeatedly. The players represent sovereign states in a world with 30 provinces but without any geographical structure. Provinces can be won or lost during war which may or may not result in one of the 30 rounds. In each round an alliance and a counteralliance can be built up by formal moves on the board where alliances may change from round to round. To analyse essential aspects of such a game situation, Selten considers a modified version of the game with only 6 provinces and an infinite number of rounds and discusses its solution, especially the reasons for stability and instability of province distributions.

In their study on the distribution of foreign language skills *Reinhard Selten and Jonathan Pool* try to predict some general features of distributions of nonnative language skills by studying a game model with a continuum of players representing the members of the world population. A community in such a world is a continuum of players with positive size. Players must choose which set of nonnative languages they want to learn. Payoffs are the net benefits from their language skills, i.e. the communicative benefit minus the player's learning cost. Selten and Pool assume for each community an importance function assigning a numerical importance to each member of the world population. The communicative benefit is then proportional to the importance-weighted fraction of the world population with which one shares at least one language. The importance function is intended to reflect that economic, geographical and other aspects affect the value of interpersonal communication. Players are distinguished by their learning costs which are influenced by a personal factor and a linguistic factor which depends purely on the available languages. The paper proves the existence of an equilibrium in pure strategies. An example is used to illustrate how such an equilibrium can be analysed.

A social dilemma describes a situation where every player has a dominating strategy and where the mutual choice of dominating strategies is worse than not using it for all members of the society. A normative solution to such a dilemma is to propagate concerns for others, i.e. altruism in the form of being concerned with the well-being of other members in the community. *Dirk Wendt* discusses this approach and tries to estimate empirically such concerns by performing experiments with various scenarios since concerns are expected to be different under different circumstances. The main aim of the study is to assess how much influence the context, i.e. experimental scenario, might have on altruistic concerns. Scenarios were therefore selected which considerably differ in the extent according to which one intuitively expects altruistic behavior to be appropriate.

Social interactions in works of literature is analysed with the help of game theory by *Barry O'Neill*. He takes his point of departure from the Middle English poem "Sir Gawain and the Green Knight" and develops a theory of "daring" as a competition of reputation and especially for the reputation of valuing reputation. Several game theoretical models are constructed in order to throw light on Gawain's decision problem; one of them describes the interaction of two sides of Gawain's personality. O'Neill suggests that the conventional repertoire of literary analysis lacks an adequate method for the analysis of plot and proposes the use of non-cooperative game theory for this purpose. At the end of the paper, the relevance of daring for international relations is discussed. Attempts to gain a reputation for valuing reputation can also be observed here.

Neubiberg/Frankfurt/M., January 1991 Rudolf Avenhaus and Werner Güth

MAJORITY VOTING IN THE CONDORCET PARADOX
AS A PROBLEM OF EQUILIBRIUM SELECTION

by

Werner Güth and Reinhard Selten

Abstract: Voting by simple majority is often viewed as undesirable since it can lead to cyclical majority decisions (Condorcet paradox). In general, there can be no transitive social ordering of alternatives based on majority decisions. Here, we do not follow the welfare theoretic attempt to derive a transitive social ordering but rather consider the situation as a game where agents select among alternatives by majority decisions. Of course, the phenomenon of cyclical majorities entails the fact that such a game has more than just one equilibrium point. But by applying the theory of equilibrium selection, one nevertheless can solve the game uniquely and thereby determine a unique public decision.

To illustrate our approach, we consider the most simple form for the so-called Condorcet paradox with three alternatives and three agents. It is assumed that agents assign cardinal utilities to alternatives including the status quo which results if none of the three proposals is accepted. It is an interesting fact that the set of uniformly perfect equilibrium points depends crucially on cardinal utilities although they always imply the same cyclical majorities. Furthermore, the status quo will only sustain in degenerate cases. In other words: the uniquely determined alternative is Pareto-optimal with probability 1. This indicates that the application of equilibrium selection to majority voting offers new ways to derive mechanisms of social choice. Since agents choose among alternatives and not among preference profiles, etc., such mechanisms are, in our view, much more in line with actual democratic decision processes.

1. Introduction

The Condorcet paradox (Condorcet, 1785) concerns a situation where three individuals have to choose among three alternatives and where preferences are such that for each alternative there is another alternative preferred by a majority. This seems to lead to the conclusion that in such situations majority voting cannot yield a unique social decision. However, recently a theory of equilibrium selection has been developed which for every finite noncooperative game determines one of its equilibrium points as the solution (Harsanyi and Selten, 1988).

Majority voting in the Condorcet paradox can be modeled as a noncooperative game. This will be done in the most obvious way by giving each of the participants the opportunity to vote for one of the three alternatives. Of course, it may happen that no majority is reached. It is assumed that in this case the result is a 'status quo' alternative. We also permit the possibility of voting for the status quo.

The payoffs are von Neumann-Morgenstern utilities attached to the three alternatives and the status quo. We only consider utility specifications which produce the Condorcet paradox.

Various assumptions could be made concerning the relationship of the utilities at the status quo to those at the three alternatives. We want to restrict our attention to a situation where each of the three alternatives is preferred to the status quo by a majority of two voters whereas the third voter has an interest to resist the change. This means that at each of the three alternatives two players receive greater utilities than at the status quo and one receives a smaller utility. Actually, we do not exclude the limiting case where the third player is indifferent between the status quo and his worst alternative.

The noncooperative game model sketched above has infinitely many equilibrium points. However, equilibrium selection theory is not directly applied to the game itself but to its ε-perturbed games where every pure strategy must be chosen with a small minimum probability ε. These ε-perturbations have three strong equilibrium points. Each of these equilibrium points belongs to one of the three alternatives. Those players who prefer the alternative to the status quo vote for it, and the remaining player votes for his most preferred alternative. The possibility of mistakes by the other players gives him an incentive to do this.

The three strong equilibrium points are the initial candidates for a solution. Equilibrium selection prescribes a process of candidate elimination and substitution which, starting from the first candidate set, finally determines the solution of the ε-perturbed game. The solution of the original game is found by letting the perturbance parameter ε go to zero.

For almost all parameter combinations, the solution determined by equilibrium selection is a pure strategy equilibrium point which results in the choice of one of the three alternatives. Only in degenerate cases the solution turns out to be a mixed strategy equilibrium point.

Equilibrium selection together with the simple majority voting model sketched above can be looked upon as a mechanism of social choice. Usually, social choice rules are characterized by axioms which express intuitively desirable properties (Arrow, 1963, Sen, 1970). Here wo do not take this approach. We apply equilibrium selection theory to a natural institutional framework and look at the properties of the implied social choice rule. The results of the analysis show that the social choice rule implied by this mechanism is not unreasonable. It is Pareto-optimal almost everywhere and clearly non-dictatorial. Of course, it would not be correct to say that majority voting conforms to the ideal of social justice. Minorities may suffer too much by majorities which form in order to obtain relatively small advantages. However, social choice mechanisms should not be judged by considerations of social justice alone. Stable institutions cannot be in complete disagreement with power relationships. Therefore, it may be necessary to rely on social choice rules which to some extent permit deviations from the ideals of social justice in favor of more power adequate social decisions.

Noncooperative game models of social decision together with equilibrium selection theory provide an approach to social choice which is quite different from the usual type of social choice theory. Instead of looking for rules satisfying a set of desirable properties, one examines the game theoretical consequences of natural mechanisms similar to those used in practice. If the mechanism turns out to work quite well in theory, one may expect it to be not too bad in practice. Of course, the assumption of complete rationality is a dubious one. But it is defensible as an instrument of theoretical investigation. Our analysis shows that the majority rule as specified by our game model yields reasonable results even if strategic considerations rather than ethical judgements single out the final solution.

One may raise the objection that much of social choice literature is concerned with voting situations where the participants have incomplete information about the utilities of other participants. However, the Condorcet paradox is generally not viewed as a problem of incomplete information but rather as a phenomenon arising from circular majority relations. In principle, it is also possible to apply equilibrium selection theory to voting situations involving incomplete information. This shall not be done here. We restrict our attention to the important limiting case of complete information. The assumption of complete information may not be too unrealistic for many committees whose members know each other quite well.

The next section will introduce basic notations and definitions concerning games in normal form. Equilibrium selection theory makes use of *uniformly perturbed games*. Section 3 is devoted to this topic. The formal definition of the *Condorcet game* is given in Section 4. Equilibrium selection theory will not be explained in full detail. The general structure of equilibrium selection theory is discussed in Section 5. Essential features, which are used in this paper, are introduced where they are needed in later sections. The first step in the application of equilibrium selection theory to Condorcet games replaces the original perturbed games, where every player has four strategies, by a reduced perturbed game, where every player has only two strategies. In Section 6, the reduced perturbed Condorcet game is derived. After a reduced game has been obtained, equilibrium selection theory determines a set of initial candidates. This step in the process of finding the solution is taken in Section 7. Payoff and risk dominance relationships are important for selection among the initial candidates. This topic is discussed in Section 8. The tracing procedure which derives an equilibrium point from prior disequilibrium beliefs is described in Section 9. Equilibrium selection theory first derives solutions for perturbed games, one then examines the consequences of mistake probabilities going to zero. A limit solution is determined in this way for non-degenerate Condorcet games in Section 10.

The properties of the limit solution viewed as a social choice rule are discussed in Section 11. Finally, the last section is devoted to degenerate cases which for the sake of simplicity have been omitted in earlier sections.

2. Games in Normal Form

A game in normal form $G = (\Phi_1,\ldots,\Phi_n;H)$ consists of n nonempty finite sets $\Phi_1,\ldots,\Phi_n$, the pure strategy sets of the players $1,\ldots,n$, and a payoff function H. A pure strategy combination is an n-tuple

$$\varphi = (\varphi_1,\ldots,\varphi_n) \quad \text{with} \quad \varphi_i \in \Phi_i \ . \tag{1}$$

The set of all pure strategy combinations is denoted by Φ. The payoff function H assigns a payoff vector

$$H(\varphi) = (H_1(\varphi),\ldots,H_n(\varphi)) \tag{2}$$

to every pure strategy combination. A mixed strategy of player i is a probability distribution q_i over Φ_i. The probability assigned to a strategy φ_i by q_i is denoted by $q_i(\varphi_i)$. A mixed strategy combination

$$q = (q_1,\ldots,q_n) \tag{3}$$

contains a mixed strategy q_i for every player i. A pure strategy φ_i is identified with that mixed strategy which assigns probability 1 to this pure strategy. The set of all mixed strategies of player i is denoted by Q_i. A mixed strategy combination $q = (q_1,\ldots,q_n)$ determines a probability

$$q(\varphi) = q_1(\varphi_1) \cdot \ldots \cdot q_n(\varphi_n) \tag{4}$$

for every pure strategy combination $\varphi = (\varphi_1,\ldots,\varphi_n)$. The payoff function H is extended to mixed strategy combinations in the usual way:

$$H(q) = \sum_{\varphi\in\Phi} q(\varphi)\ H(\varphi) \ . \tag{5}$$

An i-incomplete mixed strategy combination is an n-1-tuple of mixed strategies

$$q_{-i} = (q_1,\ldots,q_{i-1},q_{i+1},\ldots,q_n) \tag{6}$$

with $q_j \in Q_j$ for $j = 1,\ldots,n$ and $j \neq i$. The set of i-incomplete mixed strategy combinations is denoted by Q_{-i}. In the case that all components of an i-incomplete strategy combination are pure strategies, we speak of an i-incomplete pure strategy combination. The symbol Φ_{-i} is used for the set of all i-incomplete pure strategy combinations. The notation

$$q = q_i \; q_{-i} \tag{7}$$

is used in order to express the fact that q contains q_i and the mixed strategies in q_{-i} as its components.

A mixed strategy r_i is called a best reply to the i-incomplete mixed strategy combination q_{-i} if we have

$$H_i(r_i q_{-i}) = \max_{q_i \in Q_i} H_i(q_i q_{-i}) \; . \tag{8}$$

r_i is a strong best reply if every other strategy s_i of player i yields a lower payoff $H_i(s_i q_{-i})$. It is clear that a strong best reply must be a pure strategy. We also say that q_i is a best reply or a strong best reply to q if it is a (strong) best reply to the i-incomplete strategy combination q_{-i} whose components are in q.

An equilibrium point is a mixed strategy combination $r = (r_1,\dots,r_n)$ with the property that for $i = 1,\dots,n$ strategy r_i is a best reply to r. The equilibrium point r is called strong if for every player $i = 1,\dots,n$ his equilibrium strategy r_i is a strong best reply to r.

Equilibrium selection theory makes use of best replies not only to mixed strategy combinations but also to joint mixtures. A joint mixture $q_{.i}$ is a probability distribution over Q_{-i}. A hybrid mixture $q_i q_{.i}$ consists of a mixed strategy q_i and a joint mixture $q_{.i}$. The symbol $q_{.i}(\varphi_{-i})$ denotes the probability assigned to φ_{-i} by $q_{.i}$. The payoff function H is extended to hybrid mixtures as follows:

$$H_i(q_i q_{.i}) = \sum_{\varphi_i \in \Phi_i} \; \sum_{\varphi_{-i} \in \Phi_{-i}} q_i(\varphi_i) q_{.i}(\varphi_{-i}) H_i(\varphi) . \tag{9}$$

A mixed strategy r_i is a best reply to $q_{.i}$ if we have

$$H_i(r_i q_{.i}) = \max_{q_i \in Q_i} H_i(q_i q_{.i}). \tag{10}$$

In the framework of equilibrium selection theory, joint mixtures $q_{.i}$ express possible beliefs of player i about the behavior of his opponents. Player i may, for example, be in doubt about which of two i-incomplete strategy combinations corresponding to different theories about his opponents' behavior will be played. In such cases, his probability distribution must be described by a joint mixture.

3. Uniform Perturbations

Equilibrium selection theory is not applied directly to the game under consideration but to uniform perturbations of this game. A uniform perturbation can be thought of as a slightly changed game where the players have small probabilities of making mistakes in the execution of their strategy choices. Each strategy has the same probability ε of being executed by mistake. Consider the game $G = (\Phi_1,\dots,\Phi_n;H)$. Let ε be a number with

$$0 < \varepsilon < 1/|\Phi_i| \quad \text{for} \quad i = 1,\dots,n. \tag{11}$$

Here $|\Phi_i|$ denotes the number of elements in Φ_i. The ε-perturbation

$$G_\varepsilon = (\Phi_1,\dots,\Phi_n;H_\varepsilon) \tag{12}$$

has the same pure strategy sets as G, but a different payoff function. For every pure strategy $\varphi_i \in \Phi_i$ we define an ε-extreme mixed strategy $\varphi_{\varepsilon i}$ as follows:

$$\varphi_{\varepsilon i}(\varphi_i) = 1-(|\Phi_i|-1)\varepsilon \tag{13}$$

$$\varphi_{\varepsilon i}(\psi_i) = \varepsilon \quad \text{for} \quad \psi_i \in \Phi_i\backslash\{\varphi_i\}\ . \tag{14}$$

Note that this definition is based on the idea that in the case of a mistake every pure strategy including φ_i is selected with the same probability. For every pure strategy combination $\varphi = (\varphi_1,\dots,\varphi_n)$ the corresponding ε-extreme mixed strategy combination

$$\varphi_\varepsilon = (\varphi_{\varepsilon 1},\dots,\varphi_{\varepsilon n}) \tag{15}$$

is defined as that combination which contains the ε-extreme strategies $\varphi_{\varepsilon i}$ belonging to the components of φ. The payoff function H_ε is defined as follows:

$$H_\varepsilon(\varphi) = H(\varphi_\varepsilon) \tag{16}$$

for every $\varphi \in \Phi$ where φ_ε is the ε-extreme combination corresponding to φ. The i-th component of H_ε is denoted by the symbol $H_{\varepsilon i}$. It is clear that the idea of small mistakes in the execution of strategies leads to this definition.

4. The Condorcet Game

The name Condorcet game refers to the game model which has been informally described in the Introduction. Each of the three players 1, 2, 3 votes for one of the three alternatives A,B,C or the status quo S. The payoffs connected to these alternatives are shown in Table 1. Without loss of generality, a player's von Neumann-Morgenstern utilities for the status quo and his most preferred alternative can be normalized to zero and one, respectively. We use the symbols w_i for player i's utility of his least preferred (<u>w</u>orst) alternative and m_i for his second most preferred (<u>m</u>iddle) alternative.

For every specification of the parameters m_i and w_i with

$$1 > m_i > 0 \geq w_i \qquad \text{for } i = 1, 2, 3 \tag{17}$$

we obtain a Condorcet game $G = (\Phi_1,\Phi_2,\Phi_3;H)$ whose components are as follows:

$$\Phi_i = \{A,B,C,S\} \quad \text{for} \quad i = 1,2,3 \tag{18}$$

	1	2	3
A	1	m_2	w_3
B	w_1	1	m_3
C	m_1	w_2	1
status quo S	0	0	0

$1 > m_i > 0 \geq w_i$

for $i = 1,2,3$

<u>Table 1</u>: Cardinal utilities: The entries show the utility attached to the proposal corresponding to the row by the committee member indicated at the top of the column.

$$H(\varphi) = \begin{cases} (1,\ m_2,\ w_3) & \text{if } \varphi \text{ contains at least two } A\text{'s} \\ (w_1,\ 1,\ m_3) & \text{if } \varphi \text{ contains at least two } B\text{'s} \\ (m_1,\ w_2,\ 1) & \text{if } \varphi \text{ contains at least two } C\text{'s} \\ (0,\ 0,\ 0) & \text{otherwise .} \end{cases} \tag{19}$$

The application of equilibrium selection theory to the Condorcet game will require a distinction of a number of cases. Solutions are different in different regions of the parameter space. In some border cases, the determination of the solution meets special difficulties. The discussion of such degenerate cases is deferred to Section 13.

We call a Condorcet game G non-degenerate if the following condition is satisfied:

$$m_i \neq m_j \quad \text{for} \quad i \neq j \quad \text{and} \quad i,j = 1,2,3. \tag{20}$$

Consider the ε-perturbation G_ε of the Condorcet game G. The payoff function H_ε of G_ε can be described as follows:

$$\begin{aligned} H_\varepsilon(\varphi) &= (1-4\varepsilon)^3 H(\varphi) \\ &+ \varepsilon(1-4\varepsilon)^2 \sum_{j=1}^{3} \sum_{\psi_j \in \Phi_j} H(\psi_j \varphi_{-j}) \\ &+ \varepsilon^2(1-4\varepsilon) \sum_{j=1}^{3} \sum_{\psi_{-j} \in \Phi_{-j}} H(\varphi_j \psi_{-j}) \\ &+ \varepsilon^3 \sum_{\psi \in \Phi} H(\psi) . \end{aligned} \tag{21}$$

The first term on the right-hand side of (21) is due to the event that no player makes a mistake (the intended strategy φ_i can also be selected by mistake). The second term corresponds to the event that only one of the three players makes a mistake. Similarly, the third and the fourth term are connected to mistakes by two or all three players, respectively.

Equilibrium selection theory has the property that positive linear payoff transformations do not influence the solution. It will be convenient to exploit this invariance property in the investigation of uniform perturbations. We apply the following positive linear payoff transformation to H_ε: We divide by $(1-4\varepsilon)^3$ and delete the last term which does not depend on φ. With the help of the definition

$$\bar{\varepsilon} = \frac{\varepsilon}{1-4\varepsilon} \tag{22}$$

this yields the following new payoff function $\hat{H}_{\bar{\varepsilon}}$:

$$\begin{aligned} \hat{H}_{\bar{\varepsilon}}(\varphi) &= H(\varphi) + \bar{\varepsilon} \sum_{j=1}^{3} \sum_{\psi_j \in \Phi_j} H(\psi_j \varphi_{-j}) \\ &+ \bar{\varepsilon}^2 \sum_{j=1}^{3} \sum_{\psi_{-j} \in \Phi_{-j}} H(\varphi_j \psi_{-j}) . \end{aligned} \tag{23}$$

A further simplification of this payoff function is possible. We can omit terms which do not depend on φ_i in player i's payoff function. Obviously, this amounts to a positive linear transformation. In the first double sum of (23), the terms with $j = i$ do not depend on φ_i. In the second double sum of (23), the terms with $j \neq i$ do not depend on φ_i. Moreover, the terms in the second double sum with $j = i$ do not depend on φ_i if in ψ_{-i} both players vote for the same alternative. We obtain a new payoff function $\bar{H}_{\bar{\varepsilon}i}$ for player i if we omit all these terms. In the following formula for $\bar{H}_{\bar{\varepsilon}i}$ the symbol Φ^*_{-i} denotes the set of all i-incomplete pure strategy combinations ψ_{-i} where both players vote for different alternatives

$$\bar{H}_{\bar{\varepsilon}i}(\varphi) = H_i(\varphi) + \bar{\varepsilon} \sum_{\substack{j=1 \\ j\neq i}}^{3} \sum_{\psi_j \in \Phi_j} H_i(\psi_j \varphi_{-j})$$

$$+ \bar{\varepsilon}^2 \sum_{\psi_{-i} \in \Phi^*_{-i}} H_i(\varphi_i \psi_{-i}) \tag{24}$$

for $i = 1,2,3$. We call the game $\bar{G}_{\bar{\varepsilon}} = (\Phi_1,\ldots,\Phi_n;\bar{H}_{\bar{\varepsilon}})$ with the payoff function $\bar{H}_{\bar{\varepsilon}}$ defined in (24) the shortened $\bar{\varepsilon}$-perturbation of G. It will often be notationally convenient to replace $\bar{\varepsilon}$ by ε even if the shortened ε-perturbation corresponds to a perturbance $\varepsilon/(1-4\varepsilon)$ rather than ε.

5. The Structure of Equilibrium Selection Theory

Equilibrium selection theory is not applied to a game G directly but to its ε-perturbation G_ε. For every G_ε a solution $L(G_\varepsilon)$ is determined which is an equilibrium point of G_ε. One then goes to the limit for $\varepsilon \to 0$ in order to find the limit solution $\underset{\neg}{L}(G)$ of G :

$$\underset{\neg}{L}(G) = \lim_{\varepsilon \to 0} L(G_\varepsilon). \tag{25}$$

The limit solution is a perfect equilibrium point of G (Selten, 1975).

One may ask whether a limit solution always exists. Up to now, no rigorous answer has been given to this question even if there are good reasons to conjecture that the limit solution always exists (Harsanyi and Selten, 1977). The existence problem is not really an important conceptual problem. Equilibrium selection theory takes the point of view that small mistake probabilities can never be really avoided in practical situations. The theory is meant to be applied to such game situations only. Unperturbed games are idealizations which should not be considered to be approximations of real situations if no limit solution exists.

The determination of the solution for an ε-perturbation G_ε begins with the process of reduction and decomposition. In our case, reduction and decomposition amounts to the elimination of dominated strategies. As we shall see, two strategies of every player are dominated in the ε-perturbations G_ε of the Condorcet game G, namely the choice of the worst alternative and of the status quo. One obtains the reduced games where each player has two strategies only. All further steps of equilibrium selection are applied to this reduced game.

The next step is the determination of initial candidates. In the case at hand, the initial candidates are the strong equilibrium points (A,A,C), (A,B,B), and (C,B,C) of the reduced perturbed Condorcet game. One has to determine payoff and risk dominance relationships for pairs of initial candidates. A process of candidate substitution and elimination is used in order to generate a sequence of smaller and smaller candidate sets terminating in a set of only one candidate. This candidate is the solution of the game.

In our case, the process of candidate elimination and substitution takes a very simple form. Since dominance relationships are circular, none of the initial candidates can be eliminated. In such cases, it is necessary to determine a *substitute* for the initial candidate set. The substitute is the solution of the game.

The determination of the substitute makes use of the tracing procedure (Harsanyi, 1975). Starting from a prior distribution (a special mixed strategy combination), an equilibrium point is determined by a continuous equilibrating process. In our case, the prior distribution is the *centroid* of this initial candidate set. In the centroid, each player uses each pure strategy with the relative frequency with which it occurs in the initial candidate set. With the exception of some degenerate cases, one of the three initial candidates emerges as the substitute of the initial candidate set and as the solution of the game.

We shall not describe equilibrium selection theory in full detail. Instead of this, attention will be concentrated on those features which are decisive for the case at hand. At first, we shall look only at cases of Condorcet games where the payoffs m_i for the second best alternative are pairwise different. We refer to such cases as non-degenerate Condorcet games. The degenerate cases where at least two of the m_i are equal will be discussed in the last section.

6. The Reduced Perturbed Condorcet Game

The first step in the application of equilibrium selection theory is a procedure of reduction and decomposition which is applied to the ε-perturbations of the original game. This procedure involves a decomposition into *cells*. These cells are substructures similar to subgames of extensive forms. In our case, decomposition into cells is not possible. Therefore, we shall omit the description of decomposition into cells. The procedure of reduction and decomposition also eliminates inferior strategies, duplicate classes and semi-duplicate classes. Duplicate classes and semi-duplicate classes do not occur in ε-perturbed Condorcet games and will not be discussed here.

An inferior strategy of player i is a pure strategy $\varphi_i \in \Phi_i$ with the property that for some pure strategy $\psi_i \in \Phi_i$ the following is true: If φ_i is a best reply to an i-incomplete mixture $q_{.i}$, then ψ_i is a best reply to $q_{.i}$, too, but for at least one $q_{.i} \in Q_{.i}$ the strategy ψ_i is a best reply to $q_{.i}$ whereas φ_i is not a best reply to $q_{.i}$. A special case of an inferior strategy is a strictly dominated strategy. A strategy φ_i is strictly dominated by ψ_i in $G = (\Phi_1,\ldots,\Phi_n;H)$ if we have

$$H_i(\psi_i\varphi_{-i}) > H_i(\varphi_i\varphi_{-i}) \qquad \text{for every } \varphi_{-i} \in \Phi_{-i} \;. \tag{26}$$

The strategy φ_i is strictly dominated in G if it is strictly dominated by at least one pure strategy ψ_i of the same player.

A pure strategy φ_i of player i in G is dominated by a pure strategy $\psi_i \in \Phi_i$ if condition (26) holds with "$\geq$" instead of "$>$" and if for at least one $\varphi_{-i} \in \Phi_{-i}$ the payoff $H_i(\psi_i\varphi_{-i})$ is greater than the payoff $H_i(\varphi_i\varphi_{-i})$. A strategy $\varphi_i \in \Phi_i$ is called dominated if it is dominated by at least one $\psi_i \in \Phi_i$.

It can be seen immediately that a strategy is strictly dominated in every ε-perturbed game G_ε of G if it is dominated in the original game. Table 2 shows the payoffs for player 1's strategies A, B, and S against all possible 1-incomplete pure strategy combinations φ_{-1} in the Condorcet game G. Obviously, A dominates B and S since w_1 is not positive. Therefore, B and S are strictly dominated in the ε-perturbed Condorcet game G_ε. These strategies are removed by the procedure of reduction and decomposition. In the same way, C and S are removed for player 2 and A and S are removed for player 3.

The resulting 2×2×2-game $G_\varepsilon^r = (\Psi_1,\Psi_2,\Psi_3;H_\varepsilon)$ is the reduced ε-perturbed Condorcet game. We have

$$\Psi_1 = \{A,C\} \tag{27}$$

$$\Psi_2 = \{A,B\} \tag{28}$$

$$\Psi_3 = \{B,C\}. \tag{29}$$

This game cannot be further reduced or decomposed.

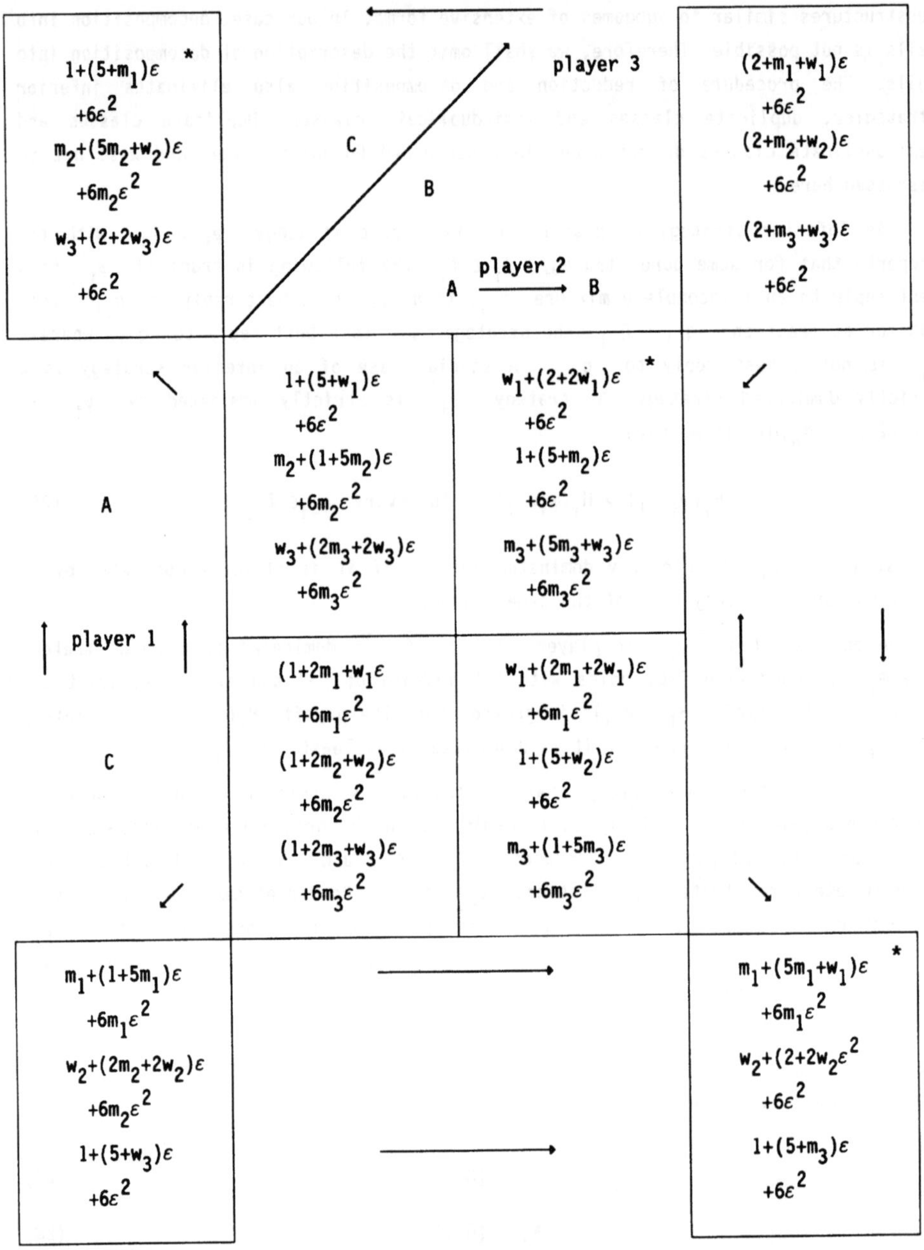

Figure 1: The reduced shortened ε-perturbation $\tilde{G}^r_\varepsilon$ of the Condorcet Game G. Player 1 chooses between the upper four squares (A) and the lower four squares (C); player 2 chooses between the four squares on the left-hand side (A) and the four squares on the right-hand side (B); player 3 chooses between the four inner squares (B) and the four outer squares (C). the payoff entries for players 1,2,3 are given from above to below. Arrows indicate the direction of payoff incentives for sufficiently small ε.

φ_{-1} / φ_1	AA	AB BA	AC CA	AS SA	BB	BC CB	BS SB	CC	CS SC	SS
A	1	1	1	1	w_1	0	0	m_1	0	0
B	1	w_1	0	0	w_1	w_1	w_1	m_1	0	0
S	1	0	0	0	w_1	0	0	m_1	0	0

Table 2: Weak dominance of A over B and S for player 1 in the Condorcet game. - Entries are payoffs for the strategy combinations $\varphi_1\varphi_{-1}$.

Figure 1 shows the reduced game $\bar{G}^r_\varepsilon = (\Psi_1,\Psi_2,\Psi_3;\bar{H}_\varepsilon)$ of the shortened ε-perturbation $\bar{G}_\varepsilon$ of the Condorcet game G. In view of the invariance of equilibrium selection theory with respect to positive linear payoff transformations, further steps of equilibrium selection theory can be applied to $\bar{G}^r_\varepsilon$ instead of G^r_ε. Figure 1 also shows arrows which indicate the direction of payoff incentives for sufficiently small ε. Arrows for 1 and 2 near the inner squares point in the direction of positive payoff differences between the two adjacent squares. The outer arrows relate to the outer squares. Arrows for the third player indicate the sign of his payoff differences between adjacent inner and outer squares. We shall always assume that ε is sufficiently small in the sense that the arrows in Figure 1 describe the sign of the corresponding payoff differences correctly.

7. Initial Candidates

In this section, we take the first step toward the solution of the reduced shortened ε-perturbations $\bar{G}^r_\varepsilon$ of the Condorcet game G. We determine the set of initial candidates. In order to explain the notion of an initial candidate, we first have to introduce the concept of a formation.

Consider a game in normal form $G = (\Phi_1,\ldots,\Phi_n;H)$. Loosely speaking, a formation of G is a smaller game contained in G, which is closed with respect to best replies. More precisely, a formation $G' = (\Psi_1,\ldots,\Psi_n;H')$ is characterized by the following properties:

$$\Psi_i \subseteq \Phi_i \qquad \text{for } i = 1,\ldots,n \tag{30}$$

$$H'(\psi) = H(\psi) \tag{31}$$

for every pure strategy combination ψ in G'. Moreover, closedness with respect to best replies in the following sense must hold for $i = 1,\ldots,n$: For every joint mixture $q_{.i}$ of G such that

$$q_{.i}(\varphi_{-i}) = 0 \qquad \text{for } \varphi_{-i} \notin \Psi_{-i} \tag{32}$$

every pure best reply ψ_i to $q_{.i}$ is in Ψ_i. Here, Ψ_{-i} is the set of i-incomplete pure strategy combinations for G'.

Since according to this definition the game itself is one of its formations, we shall speak of a proper formation if we mean a formation which is different from the game itself. Formations may themselves have proper formations which we call subformations. Obviously, these subformations are also formations of the whole game. A formation is called primitive if it does not have any proper subformation.

An initial candidate is defined as a solution of a primitive formation. We shall not have to discuss the question of how solutions of primitive formations are determined since in the case at hand the primitive formations contain only one equilibrium point. This equilibrium point is the solution of the primitive formation.

The game $\bar{G}^r_\varepsilon$ of Figure 1 has three strong equilibrium points (A,A,C), (A,B,B), and (C,B,C). Each of the three strong equilibrium points "generates" a primitive formation where each player has his equilibrium strategy as his only strategy. As we shall see, these three primitive formations are the only ones of the game in Figure 1.

Obviously, a pure strategy combination differing from one of the three strong equilibrium points by the strategy of only one player cannot belong to a primitive formation since this formation would have to contain one of the three strong equilibrium points. There is only one pure strategy combination (C,A,B) which differs from all three strong equilibrium points by the strategies of two players. If (C,A,B) is in a formation, then (C,A,C) and therefore also (C,B,C) must be in the same formation. This shows that a formation containing (C,A,B) cannot be primitive.

The three strong equilibrium points are the solutions of the primitive formations of the game $\bar{G}^r_\varepsilon$ of Figure 1. The solutions of the primitive formations form the first candidate set Ω_1 :

$$\Omega_1 = \{(A,A,C),(A,B,B),(C,B,C)\} \ . \tag{33}$$

8. Dominance

Equilibrium selection theory makes use of two dominance relationships between equilibrium points, payoff dominance and risk dominance. Let $q = (q_1,\dots,q_n)$ and $r = (r_1,\dots,r_n)$ be two different equilibrium points of a game $G = (\Phi_1,\dots,\Phi_n;H)$. We say that r payoff dominates q if we have

$$H_i(r) > H_i(q) \quad \text{for all } i \text{ with } r_i \neq q_i \,. \tag{34}$$

The definition of payoff dominance does not impose any conditions on players i with $r_i = q_i$ since in a situation where it is known that either q or r is the solution, these players know which strategy to choose. Payoff dominance may serve to coordinate the behavior of the other players.

For sufficiently small ε, the equilibrium point (A,B,B) payoff dominates (A,A,C) in the game of Figure 1. Analogously, for sufficiently small ε the equilibrium point (C,B,C) payoff dominates (A,B,B) and (A,A,C) payoff dominates (C,B,C). We shall always assume ε to be sufficiently small so that this circular pattern of payoff dominance relationships emerges.

In equilibrium selection theory, payoff dominance takes precedence over risk dominance. Therefore, it is not necessary to give a definition of risk dominance for the purposes of this paper. Equilibrium selection theory distinguishes dominance relations of different strength. This is done by a concept of strategic distance which is connected to the number of players who use different strategies in both equilibrium points. In our case, there are always exactly two players who use different strategies in both equilibrium points. Without giving a detailed definition of strategic distance and the strength of dominance relationships, we want to indicate that this is the basic reason for the fact that in our case all dominance relationships are equally strong.

The process of candidate elimination and substitution involves elimination steps and substitution steps. An elimination step removes candidates from a current candidate set. If dominance relationships are circular and equally strong, an elimination step cannot be performed. In such cases, a candidate set has to be replaced by a *substitute*. This is the situation we have to face with respect to the first candidate set Ω_1 of the game in Figure 1.

9. The Tracing Procedure

In this section, we shall explain the tracing procedure to be used in the computation of the substitute for the set of initial candidates. Consider an n-person game $G = (\Phi_1,\dots,\Phi_n;H)$. Let $p = (p_1,\dots,p_n)$ be a mixed strategy combination for G. We think of p as a preliminary theory for the game G which does not yet have the property of an equilibrium point. We refer to p as the prior strategy combination since its role is similar to a prior distribution in the ordinary bayesian analysis. The tracing procedure continuously adapts the preliminary theory until finally an equilibrium point of G is found.

The tracing procedure makes use of a set of auxiliary games $G^t = (\Phi_1,\dots,\Phi_n;H^t)$. For every t with

$$0 \leq t \leq 1 \tag{35}$$

the payoff function H^t of G^t is defined as follows:

$$H_i^t(\varphi) = t\, H_i(\varphi) + (1-t)H_i(\varphi_i p_{-i}) \tag{36}$$

for all pure strategy combinations φ of G and $i = 1,\dots,n$; here p_{-i} is the i-incomplete combination containing components of the prior strategy combination p. The payoff $H_i^t(\varphi)$ is a convex linear combination of the payoff obtained against φ_{-i} and the payoff obtained against p_{-i}. This means that player i trusts the prior strategy combination to an extent expressed by $1-t$. As the parameter t, called the tracing parameter, increases from 0 to 1, less and less trust is put on the prior strategy combination.

A pair (q,t), where q is a mixed strategy combination of G and t is a number with $0 \leq t \leq 1$, will be called an indexed strategy combination. The set of all indexed strategy combinations will be referred to as the tracing space. The equilibrium set E is the set of all indexed strategy combinations (q,t) such that q is an equilibrium point of G^t. Assume that every player i has a uniquely determined best reply ψ_i to p_{-i}. We call $\psi = (\psi_1,\dots,\psi_n)$ the best reply to the prior strategy combination p. Obviously, $(\psi,0)$ belongs to E. Almost always E contains a unique continuous path which connects $(\psi,0)$ with a point $(q^*,1)$. The precise sense of 'almost always' in this statement will not be discussed here (see Harsanyi, 1975). We refer to the path from $(\psi,0)$ to $(q^*,1)$ as the trace generated by p. The equilibrium point q^* of G, reached at the end of the trace, is denoted by $T(G,p)$. We call $T(G,p)$ the result of the tracing procedure applied to p in G.

The explanation given above concerns the simpler version of the tracing procedure, also referred to as the linear tracing procedure. It may happen that there is no uniquely determined best reply to the prior strategy combination or that there is more than one continuous path starting from $(\psi,0)$ in E. In such cases, a more complicated version, the so-called logarithmic tracing procedure, has to be used. The non-degeneracy conditions remove the necessity to consider the logarithmic tracing procedure.

Generally, a trace can have a very complicated form. As one moves along the trace from $(\psi,0)$ to $(q^*,1)$, one may meet segments of the trace where t is decreasing. But this does not happen in the analysis of the non-degenerate Condorcet game. Here, all traces will have a special form called *rectangular*. A rectangular trace can be subdivided into *constant* and *jump* segments. A constant segment is a set of indexed strategy combinations (φ,t) with $t' \leq t \leq t''$ where φ is a pure strategy combination and t' and t'' are the borders of the maximal interval such that (φ,t) is on the trace. In case of $t'' < 1$, the endpoint (φ,t'') of a constant segment is a *reversal point* where φ_i stops to be the only best reply to φ_{-i} in G^t for some player i. In non-degenerate cases, there is only one such player i and he has exactly two pure best replies φ_i and φ_i' to φ_{-i} in $G^{t''}$.

In a rectangular trace, a jump segment follows a reversal point (φ,t''). The jump segment consists of all indexed strategy combinations of the form $(q_i\varphi_{-i},t'')$ where q_i gives positive probabilities to φ_i and φ_i' only:

$$q_i(\rho_i) = 0 \qquad \text{for } \rho_i \neq \varphi_i \text{ and } \rho_i \neq \varphi_i' \quad . \tag{37}$$

In order to show that the set of all indexed strategy combinations $(q_i\varphi_{-i},t'')$ with (37) belongs to the trace, it is sufficient to prove that $\varphi_i'\varphi_{-i}$ is an equilibrium point of $G^{t''}$ with the property that the equilibrium strategies of the players $j \neq i$ are strong best replies to $\varphi_i'\varphi_{-i}$. This can be seen as follows: For every player $j \neq i$, the payoff loss involved in a deviation from $q_i\varphi_{-i}$ to a pure strategy ρ_j is a linear function of the probability $q_i(\varphi_i')$. Therefore, this payoff loss is positive for every $q_i\varphi_{-i}$ with (37) if it is positive at the extreme points $\varphi_i\varphi_{-i}$ and $\varphi_i'\varphi_{-i}$ of the set. Since these payoff losses are positive at the reversal point, it is sufficient to show that they are non-negative at $\varphi_i'\varphi_{-i}$.

In a rectangular trace constant and jump segments alternate until finally the result of the tracing procedure is reached at the end of a constant segment.

If (q,t) is a reversal point, we refer to t as the reversal parameter of the reversal point. It is often analytically convenient to look at the quotient

$$\tau = \frac{t}{1-t} \tag{38}$$

instead of t. We refer to τ as the reversal index of the reversal point (q,t).

10. The Limit Solution of Non-Degenerate Condorcet Games

As we have seen, dominance relationships among the initial candidates fail to narrow down the first candidate set. In such cases, equilibrium selection theory determines a substitute of the first candidate set. The substitute is the result of tracing a special prior distribution, the centroid of the first candidate set. The centroid strategy of a player is a mixed strategy which uses each pure strategy with its average probability in the first candidate set. In our case, the three elements of the first candidate set are (A,A,C), (A,B,B), and (C,B,C). Therefore, the centroid $c = (c_1,c_2,c_3)$ of the first candidate set is as follows:

$$c_1(A) = \frac{2}{3}, \quad c_1(C) = \frac{1}{3} \tag{39}$$

$$c_2(A) = \frac{1}{3}, \quad c_2(B) = \frac{2}{3} \tag{40}$$

$$c_3(B) = \frac{1}{3}, \quad c_3(C) = \frac{2}{3}\,. \tag{41}$$

The first step in the application of the tracing procedure to this centroid is the determination of the indexed best reply combination $(\psi,0)$ to the prior. Define:

$$\Delta_1 = H_{\varepsilon 1}(Ac_{-1}) - H_{\varepsilon 1}(Cc_{-1}) \tag{42}$$

$$\Delta_2 = H_{\varepsilon 2}(Bc_{-2}) - H_{\varepsilon 2}(Ac_{-2}) \tag{43}$$

$$\Delta_3 = H_{\varepsilon 3}(Cc_{-3}) - H_{\varepsilon 3}(Bc_{-3})\,. \tag{44}$$

The payoff difference Δ_i can be described as player i's incentive to vote for his best alternative rather than his second best one if the other players use their centroid strategies. Player i's best reply to the prior is his best alternative if Δ_i is positive and his second best alternative if Δ_i is negative. With the help of Figure 1 we obtain

$$\Delta_i = \frac{1}{3} - \frac{2}{3}m_i + \left(\frac{8}{3} - \frac{10}{3}m_i\right)\varepsilon + 6(1-m_i)\varepsilon^2 \tag{45}$$

for $i = 1,2,3$. It is interesting to see that w_i does not influence Δ_i. Obviously, for $m_i < .5$ Δ_i and sufficiently small ε the amount Δ_i is positive and for $m_i > .5$ and sufficiently small ε the amount Δ_i is negative. It can also be seen that Δ_i is positive for $m_i = .5$. Consequently, the best replies to the centroid are as follows:

$$\psi_1 = \begin{cases} A & \text{for } m_1 \leq .5 \\ C & \text{for } m_1 > .5 \end{cases} \tag{46}$$

$$\psi_2 = \begin{cases} B & \text{for } m_2 \leq .5 \\ A & \text{for } m_2 > .5 \end{cases} \tag{47}$$

$$\psi_3 = \begin{cases} C & \text{for } m_3 \leq .5 \\ B & \text{for } m_3 > .5 \, . \end{cases} \tag{48}$$

Table 3 shows all best reply combinations to the centroid and, with the exception of two cases, indicates the final solution. The easiest cases are those where the best reply to the centroid is one of the three equilibrium points. If this happens, the trace consists of one constant segment and the best reply to the centroid is the final solution. In three other cases the best reply to the centroid shows a majority for one alternative, e.g., (A,A,B), where one member of the majority does not use a best reply to ψ. It is sufficient to look at the case (A,A,B) since the two other cases of this kind can be obtained from this case by cyclical permutations.

We consider the case that (A,A,B) is the best reply to the centroid. As long as the other players stick to their strategies in (A,A,B), player 1 has no incentive to deviate from A. As t increases, player 3 will eventually become indifferent between B and C if the others stick to their best replies to the centroid. However, player 3's incentive to shift to C approaches 0 for $\varepsilon \to 0$. Therefore, player 3's reversal time approaches 1 for $\varepsilon \to 0$. As t increases, player 2 becomes indifferent between A and B for some tracing parameter t_1 which approaches a limit smaller than 1 for $\varepsilon \to 0$.

For sufficiently small ε, $((A,A,B),t_1)$ is a reversal point where a jump segment begins with player 2 shifting from A to B. Since (A,B,B) is an equilibrium point of the auxiliary game $\tilde{G}_\varepsilon^{t_1}$ and since for both players the equilibrium strategies are strong best replies to this equilibrium point, the jump segment does not end before $((A,B,B),t_1)$ is reached. From there, a constant segment leads to the result ((A,B,B,),1) of the tracing procedure.

We still have to investigate the two cases (A,B,C) and (C,A,B) as best replies to the centroid. In both cases, there is no majority for any alternative. We first look at the case (A,B,C). Here, along the line $((A,B,C),t)$ each of the three players i has a parameter value t_i, where he becomes indifferent between his two pure strategies. Assume that there is a smallest such parameter value, say t_j. Then, $((A,B,C),t_j)$ is a reversal point. The limit of the reversal index τ_j, corresponding to t_j, for $\varepsilon \to 0$ is as follows:

$$\lim_{\varepsilon \to 0} \tau_j = \frac{\frac{1}{3} - \frac{2}{3}m_j}{m_j} = \frac{1}{3m_j} - \frac{2}{3} . \qquad (49)$$

This can be seen with the help of Figure 1 and (38) as well as Equations (42) to (44). It follows from (49) that m_j must be greater than the parameters m_i for the other players. In non-degenerate Condorcet games G, we obtain the limit solution $\underline{L}(G)$ described by (50).

Parameter conditions			Best replies to the centroid for players			Solution strategies for players		
m_1	m_2	m_3	1	2	3	1	2	3
$\le \frac{1}{2}$	$\le \frac{1}{2}$	$\le \frac{1}{2}$	A	B	C	see (50)		
$\le \frac{1}{2}$	$\le \frac{1}{2}$	$> \frac{1}{2}$	A	B	B	A	B	B
$\le \frac{1}{2}$	$> \frac{1}{2}$	$\le \frac{1}{2}$	A	A	C	A	A	C
$\le \frac{1}{2}$	$> \frac{1}{2}$	$> \frac{1}{2}$	A	A	B	A	B	B
$> \frac{1}{2}$	$\le \frac{1}{2}$	$\le \frac{1}{2}$	C	B	C	C	B	C
$> \frac{1}{2}$	$\le \frac{1}{2}$	$> \frac{1}{2}$	C	B	B	C	B	C
$> \frac{1}{2}$	$> \frac{1}{2}$	$\le \frac{1}{2}$	C	A	C	A	A	C
$> \frac{1}{2}$	$> \frac{1}{2}$	$> \frac{1}{2}$	C	A	B	see (51)		

Table 3: Best reply to the centroid and solution. In the two extreme cases the solution is determined by:

$$\underline{L}(G) = \begin{cases} (C,B,C) & \text{for } m_1 = \max\limits_{i=1,2,3} m_i \\ (A,A,C) & \text{for } m_2 = \max\limits_{i=1,2,3} m_i \\ (A,B,B) & \text{for } m_3 = \max\limits_{i=1,2,3} m_i \end{cases} \qquad (50)$$

if $m_i \leq \frac{1}{2}$ for $i = 1,2,3$ and

$$\underset{\sim}{L}(G) = \begin{cases} (A,B,B) & \text{for } m_1 = \min\limits_{i=1,2,3} m_i \\ (C,B,C) & \text{for } m_2 = \min\limits_{i=1,2,3} m_i \\ (A,A,C) & \text{for } m_3 = \min\limits_{i=1,2,3} m_i \end{cases} \tag{51}$$

if $m_i > \frac{1}{2}$ for $i = 1,2,3$.

We now consider the case where (C,A,B) is the best reply to the centroid. Here, too, along the line $((C,A,B),t)$ each player i has a point t_i, where he is indifferent between his two pure strategies. Assume that t_j is the smallest of the three numbers t_i. Then $((C,A,B),t_j)$ is a reversal point where player j shifts to his other pure strategy. In the same way as before, it can be seen that for $\varepsilon \to 0$ the reversal index τ_j, corresponding to t_j, approaches the following limit:

$$\lim_{\varepsilon \to 0} \tau_j = \frac{\frac{2}{3} m_j - \frac{1}{3}}{1} = \frac{2}{3} m_j - \frac{1}{3} . \tag{52}$$

Consequently, player j is the player whose parameter m_j is the smallest of the m_i. Consider the case $j = 1$ (the cases $j = 2$ and $j = 3$ are analogous). In this case, $((C,A,B),t_1)$ is the first reversal point. A jump segment begins there. Player 1 shifts from C to A. The strategies A and B of players 2 and 3 continue to be strong best replies in the auxiliary game $\bar{G}_\varepsilon^{t_1}$ as player 1's probability for A is increased. Player 2's incentive to choose A is strengthened and player 3's incentive to switch to C is weakened. The jump segment ends with $((A,A,B),t_1)$. A new constant segment begins there. (A,A,B) is not an equilibrium point of $\bar{G}_\varepsilon$. Therefore, a second reversal point must be reached. The situation is analogous to the case where (A,A,B) is the best reply to the centroid. For the same reasons as in this simpler case, a reversal point, where player 2 becomes indifferent between A and B, is reached here, too. From there, a jump segment leads to (A,B,B), which becomes the final solution. Therefore, the limit solution $\underset{\sim}{L}(G)$ of G is (A,B,B) if m_1 is the smallest m_i. The other cases in (51) are obtained by cyclical permutation.

The formulae (50) and (51) do not only cover the two cases of best replies ψ to the centroid without a majority. Let $\bar{m}$ be the median of m_1, m_2, and m_3. With the help of Table 3 it can be seen that (50) holds for $\bar{m} \leq .5$ and (51) holds for $\bar{m} > .5$:

$$\underline{L}(G) = \begin{cases} (C,B,C) & \text{for } m_1 = \max\limits_{i=1,2,3} m_i \\ (A,A,C) & \text{for } m_2 = \max\limits_{i=1,2,3} m_i \\ (A,B,B) & \text{for } m_3 = \max\limits_{i=1,2,3} m_i \end{cases} \tag{53}$$

if $\bar{m} \leq .5$ and

$$\underline{L}(G) = \begin{cases} (A,B,B) & \text{for } m_1 = \min\limits_{i=1,2,3} m_i \\ (C,B,C) & \text{for } m_2 = \min\limits_{i=1,2,3} m_i \\ (A,A,C) & \text{for } m_3 = \min\limits_{i=1,2,3} m_i \end{cases} \tag{54}$$

if $\bar{m} > .5$. With these equations we have obtained a complete overview of the limit solutions of all non-degenerate Condorcet games G.

11. The Structure of the Limit Solution

Equations (53) and (54) show that two different principles emerge from the application of equilibrium selection theory to non-degenerate Condorcet games G. In the following, we shall try to give an intuitive interpretation of these two principles and of the case distinction based on the median of the m_i.

If the median $\bar{m}$ of m_1, m_2, and m_3 is smaller than or equal to .5, then that alternative is selected which is connected with the highest m_i. The amount $1-m_i$ can be interpreted as player i's proportional sacrifice relative to his best alternative if he agrees to his second best one. The *maximization rule* of Equation (53) minimizes this proportional sacrifice. The rule applies to a case where the proportional sacrifices of both other players would be relatively great, namely greater than .5.

The *minimization rule* of Equation (54) applies if there are at least two players whose proportional sacrifices are smaller than .5. A player with a proportional sacrifice of less than .5 has a relatively strong motivation to agree to a compromise at his second best alternative. If at least two players are in this situation, a majority is formed against that player who faces the greatest proportional sacrifice. He has the relatively strongest motivation to aim for his best alternative.

Of course, these motivational tendencies caused by differences of proportional sacrifices are disequilibrium phenomena. Equilibrium selection theory transforms a situation of strategic uncertainty into an equilibrium. The reasons why one equilibrium point is selected and not another one are related to a hypothetical situation of strategic uncertainty where the solution is not yet known. It is reasonable to assume that in this hypothetical situation the player whose proportional sacrifice is higher will have a stronger inclination to aim at his most favored alternative: The *maximization rule* of (53) and the *minimization rule* of (54) are both compatible with this idea. The *maximization rule* applies where at least two players strongly aim at their best alternatives since they face relatively great proportional sacrifices. The player with the least proportional sacrifice yields first and thereby determines the solution. In fact, this interpretation is almost a description of what happens along the trace in the determination of the solution if for every player his most favored alternative is his best reply to the centroid of the first candidate set.

The *minimization rule* applies to a situation where at least two players i have proportional sacrifices $1-m_i$ which are smaller than $.5$. Players facing relatively small proportional sacrifices may be inclined not to aim at their best alternative in order to increase their chance of avoiding their least preferred alternative. Suppose that initially all players aim for their second most favored alternative as in the case of the best reply (C,A,B) to the centroid of the first candidate set. The player with the highest proportional sacrifice is the first one to be attracted by the possibility to form a majority for his most favored alternative. However, if he yields to this inclination, a new disequilibrium is created since the second most preferred alternative of the player outside the majority is the best one for his partner. This is the reason why finally his partner and the outside player form a majority against him.

It is interesting that the parameters w_i do not have any influence on the alternative selected by the solution. Intuitively, one might expect that a player who has more to lose if a majority is formed against him will try harder to avoid this event. However, our analysis suggests that this is not a valid idea. As far as incentives for or against the selection of strategies are concerned, only payoff differences matter. It can be seen with the help of Figure 1 that the differences of the payoffs obtained for both strategies of a player for fixed strategies of the other players do not depend on his parameter w_i. This is due to the fact that player i receives w_i only in the event that a majority is formed against him either intentionally or as a consequence of low-probability mistakes. If both others vote for his worst alternative, his payoff is independent of his own choice. Therefore, the value of w_i is without strategic significance as long as it is non-positive. Matters would be different for values of positive w_i since then it would not always be possible to reduce the original game by the elimination of dominant strategies to the 2×2×2-game of Figure 1.

We now turn our attention to questions of comparative statics with respect to the parameters m_i. In particular, we want to examine how the alternative selected by the solution changes if m_1 is increased. It is sufficient to look at m_1 since the situation is analogous for m_2 and m_3. For three selected values of m_1, Figure 2 shows regions in the (m_2,m_3)-plane where the three alternatives are selected by the solution. For other values of m_1 with $m_1 < .5$ the situation is similar to the case $m_1 = .3$ with the only difference that the square where C is selected is smaller or larger according to the value of m_1. A similar statement can be made for $m_1 > .5$.

The limiting case $m_1 = .5$ is not really different from $m_1 <.5$. As m_1 is increased from 0 to .5, the region for the alternatives in the (m_2,m_3)-plane change continuously. However, at $m_1 = .5$ a further increase results in a discontinuous change which involves a sudden increase of the region for C and a sudden decrease of the region for B.

Player 1 receives his payoff m_1 if alternative C is realized. It can be seen with the help of Figure 2 that every pair (m_2,m_3), where C is selected for a particular m_1, remains in the region for C if m_1 is increased. An increase of m_1 always increases the region for C. This means that the solution has a payoff monotonicity with respect to the payoff parameters m_i. If the equilibrium point where m_i appears as a payoff is selected for the parameter combination (m_1,m_2,m_3), then it will also be selected for a changed parameter combination where m_i is increased and nothing else is changed.

The fact that the equilibrium point which yields m_i as a payoff is strengthened by an increase of m_i does not necessarily mean that the situation of player i is improved by an increase of m_i. As an example, we may look at the parameter combinations (.3,.4,.3) and (.5,.4,.3). Figure 2 shows that A is selected in case of the first parameter constellation and C in case of the second one. The transition decreases the payoff of player 1 from 1 to m_1.

We can also look at the effects of increasing the payoff for the most favored alternative from 1 to a value $x > 1$. In order to explore these effects, we have to renormalize player i's payoff in such a way that his payoff for his most favored alternative becomes 1 again. This means that all payoffs of player i are divided by x. Therefore, an increase of the payoff for the most favored alternative is equivalent to a decrease of the payoff m_i for his second most preferred alternative. From an intuitive point of view, this is not unreasonable. What really matters is the proportional sacrifice which increases with the payoff for the most preferred alternative.

In non-degenerate Condorcet games, equilibrium selection theory always determines a Pareto-optimal outcome since all three pure strategy equilibrium points are Pareto-optimal. Deviations from Pareto-optimality are possible in degenerate cases only. In this sense, we can say that equilibrium selection theory applied to Condorcet games is an almost always Pareto-optimal social choice mechanism. Clearly, this mechanism is also non-dictatorial, neutral, and anonymous. Names of players and alternatives do not influence the solution.

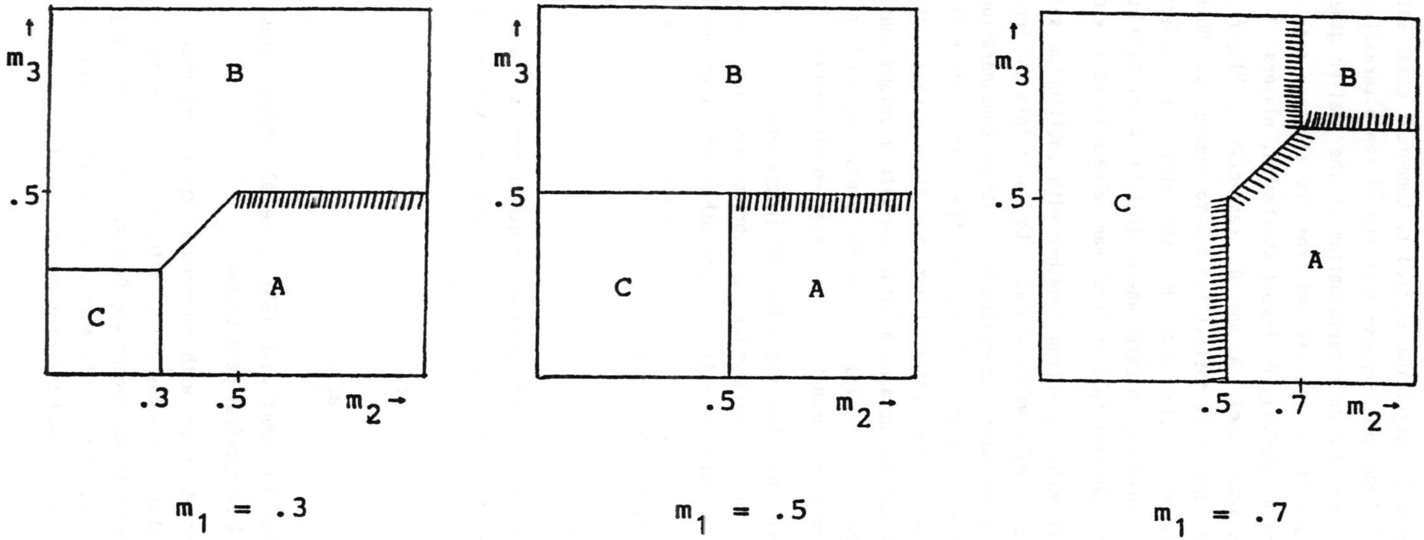

Figure 2: Solution regicns in the (m_2,m_3)-plane for selected values of m_1. - Shades adjacent to a borderline segment indicate that the points on the borderline (with the possible exception of the segment endpoints) belong to the region where the shades are. For the results on degenerate border points, see Section 12. There, it is also shown that border points on segments without shades belong to one of the adjacents regions (with the exception of border points where three regions meet). At border points where three regions meet the limit solution is completely mixed (see Table 4).

One may ask whether equilibrium selection applied to Condorcet games satisfies the principle of irrelevant alternatives. Suppose that one of the alternatives, say alternative C, is deleted. Consider an ε-perturbation of the smaller game which results from the Condorcet game if C is removed from the strategy sets of all players. After the elimination of strictly dominated strategies, players 1 and 3 have only one strategy in this game, namely A and B, respectively. Player 2 has the two strategies A and B and it is optimal for him to choose B. Obviously, (A,B,B) is the only equilibrium point and B the alternative selected by equilibrium selection theory. Therefore, in cases where (A,A,C) would have been the solution in the presence of C the principle of irrelevant alternatives is violated.

It is not surprising that majority voting together with equilibrium selection theory yields a social decision rule which violates the principle of irrelevant alternatives. Equilibrium selection theory is guided by strategic considerations. The strategic situation is radically changed if one of the three alternatives is eliminated. Before the elimination the perturbed game has three strong equilibrium points, but after the elimination one obtains a situation with a unique equilibrium point. The presence or absence of a third alternative profoundly influences the strategic reasons for and against the support of one of the two alternatives.

Our discussion has shown that the violation of independence of irrelevant alternatives is not a feature of the particular equilibrium selection theory used here. What has been said above would be valid for any other method of equilibrium selection which always yields one of the three pure strategy equilibrium points as the solution of a non-degenerate Condorcet game.

12. Degenerate Condorcet Games

Up to now, we have only looked at Condorcet games whose parameters m_i are pairwise different. With (53) and (54) we have obtained a complete overview of these non-degenerate cases. We shall now turn our attention to the degenerate cases.

Table 4 is similar to Table 3. It contains degenerate cases with

$$m_2 = m_3 = m \ . \qquad (55)$$

We can restrict our attention to games with (55) since all other cases can be obtained from those in Table 4 by cyclical permutations.

We first look at the easiest cases with rectangular traces. We shall indicate the results without giving detailed reasons since in these four cases the situation is the same as in the corresponding non-degenerate Condorcet games. For $m \leq .5$ and $m_1 > .5$, the best reply to the centroid is the equilibrium point (C,B,C). The trace consists of only one constant segment and the solution is (C,B,C).

We now consider the case $m < m_1 \leq .5$. Here, the best reply to the centroid is (A,B,C). As we have seen in Section 10, in such cases the player with the maximal m_1 is the first one to reach a reversal point. In view of $m < m_1$, this is player 1. At this reversal point ((A,B,C),t*) a jump segment begins which leads to (C,B,C). From there, a constant segment leads to the solution (C,B,C). In the case $m > .5 \geq m_1$, the trace also consists of a constant segment ((A,A,B),t) followed by a jump segment and, finally, by a constant segment ((A,B,B),t). The solution is (A,B,B).

In the case $.5 < m_1 < m$, the trace begins with a constant segment ((C,A,B),t). Since m_1 is the smallest of the m_i, player 1 is the first one to become indifferent between his two pure strategies for a reversal parameter t*. A jump segment leads to ((A,A,B),t*). There, a constant segment ((A,A,B),t) begins. Since (A,A,B) is not an equilibrium point of the perturbed Condorcet game, a further reversal point is reached where player 2 becomes indifferent between his two pure strategies. A jump segment follows which leads to the starting point of the final constant segment ((A,B,B),t). The solution is (A,B,B).

In the two cases with $m_1 = m_2 = m_3$, the reduced shortened Condorcet game is fully symmetric as far as the tracing procedure is concerned. It can be seen immediately that for fixed strategy choices of the other players the payoff differences between the two pure strategies in the reduced shortened ε-perturbation, represented by Figure 1, do not depend on the parameters w_i. As far as these payoff differences are concerned, the game is fully symmetric for $m_1 = m_2 = m_3$. The result of the logarithmic tracing procedure depends only on these payoff differences and the prior which in our case is symmetric, too. Therefore, the result of the logarithmic tracing procedure must reflect the symmetry of the game. There is only one equilibrium point which satisfies this condition, namely the uniquely determined completely mixed equilibrium point. It can be seen easily that the game of Figure 1 has, in fact, a uniquely determined, completely mixed equilibrium point. An easy computation shows that for $\varepsilon \to 0$ this equilibrium point approaches the following strategy combination (q_1,q_2,q_3):

$$q_1(A) = q_2(B) = q_3(C) = \frac{1}{1+m} \quad . \tag{56}$$

This strategy combination is the limit solution for all cases with $m_1 = m_2 = m_3 = m$.

We now turn our attention to the case $m_1 > m > .5$. This case is more difficult than those being considered up to now. The best reply to the centroid is (C,A,B). A constant segment ((C,A,B),t) extends up to a point ((C,A,B),t*), where both players 2 and 3 become indifferent between their two pure strategies. We shall avoid to discuss the question of how the trace continues from here. There may be several linear traces but, as we shall see, they all must come to the same result.

Parameter conditions		Best replies to the centroid			Limit solution strategies		
		Players			Players		
$m = m_2 = m_3$	m_1	1	2	3	1	2	3
$m \leq \frac{1}{2}$	$m_1 < m \leq \frac{1}{2}$	A	B	C	three linear traces*)		
	$m_1 = m$	A	B	C	completely mixed+)		
	$m < m_1 \leq \frac{1}{2}$	A	B	C	C	B	C
	$m_1 > \frac{1}{2}$	C	B	C	C	B	C
$m \leq \frac{1}{2}$	$m_1 \leq \frac{1}{2}$	A	A	B	A	B	B
	$\frac{1}{2} < m_1 < m$	C	A	B	A	B	B
	$m_1 > m$	C	A	B	completely mixed+)		
	$m_1 > m$	C	A	B	A	A	C

*) possible results:
(A,A,C)
(A,B,B)
(A,A,q_3) with $q_3(B) = m$

+) $q_1(A) = q_2(B) = q_3(C) = \frac{1}{1+m}$

Table 4: Best reply to the centroid and solution for degenerate cases with $m_1 = m_2 = m$.

It can be seen easily that there is one rectangular trace. In this trace, a jump segment leads from $((C,A,B),t^*)$ to $((C,A,C),t^*)$. A constant segment $((C,A,C),t)$ follows, which ends in a reversal point $((C,A,C),\bar{t})$, where player 1 becomes indifferent. A jump segment leads to $((A,A,C),\bar{t})$. A constant segment follows. At the end of the trace, the equilibrium point (A,A,C) is reached.

We shall argue that all other linear traces which may exist must eventually emerge into the rectangular trace described above. For this purpose, we first show that for $t \leq t^*$ player 1's strategy C is a strong best reply to every 1-incomplete strategy combination in the auxiliary game G_ε^t. It can be seen easily that player 1's incentive to choose C in an auxiliary game G_ε^t with $t \le t^*$ is smallest if player 2 chooses A and player 3 chooses B: The choice of A by player 2 diminishes the incentive to choose C in favor of A. If player 3 did choose C with positive probability, he would strengthen player 1's incentive to choose C. The fact that player 1 was not the first one to become indifferent along the first constant segment implies that, up to $t = t^*$, C is the strong best reply to all 1-incomplete strategy combinations.

As has been pointed out above, only payoff differences matter. Consider the 2×2-game which results if in the auxiliary game the strategy of player 1 is fixed at C. If in this game an appropriate constant is subtracted from player 2's payoffs in each column and player 3's payoffs in each row, one obtains the *payoff difference game* shown in Figure 3. As far as the tracing procedure is concerned, one can replace the 2×2-game with the original payoffs by this payoff difference game.

For $t < t^*$, player 2 receives positive payoffs if he uses A and 0-payoffs if he uses B. Similarly, player 3 receives positive payoffs for B and 0-payoffs for C. Therefore, (A,B) is the only equilibrium point of this game for $t < t^*$.

2 \ 3	B	C
A	$-t[1+(4-2m)\varepsilon+6(1-m)\varepsilon^2]$ $-(1-t)\Delta$ $-t[1+(4-2m)\varepsilon+6(1-m)\varepsilon^2]$ $-(1-t)\Delta$	$-t(1-m)[2\varepsilon+6\varepsilon^2]$ $-(1-t)\Delta$ 0
B	0 $-t(1-m)[1+4\varepsilon+6\varepsilon^2]$ $-(1-t)\Delta$	0 0

Figure 3: The payoff difference game for players 2 and 3 if in the auxiliary game player 1 is fixed at C. - Player 2's payoffs are shown above and player 3's payoffs are shown below. Δ is the common value of Δ_2 and Δ_3 (see (45)).

We can conclude that for $t < t^*$ (C,A,B) is the only equilibrium point of the auxiliary game G_ε^t.

For $t = t^* + h$ with h positive and sufficiently small, player 1 continues to have C as his only strong best reply to all 1-incomplete strategy combinations. Therefore, it can be seen with the help of Figure 3 that for sufficiently small h the auxiliary game G_ε^t has exactly three equilibrium points, namely (C,A,C), (C,B,B) and a mixed equilibrium point $(C,q_2^t,q_3^t,)$. Three possible continuations of the trace correspond to these three equilibrium points.

Let $\tilde{t}$ be that value of t for which player 3's payoff at (B,B) in Figure 3 is zero. We shall now show that for $t^* \leq t \leq \tilde{t}$ player 1's strategy C is a best reply to the 1-incomplete strategy combinations (B,B) and (q_2^t,q_3^t). As usual, this statement is asserted for sufficiently small ε only. If both other players use B, then player 1's payoff incentive in G_ε is in the order of ε. Therefore, the incentive supplied by the prior has the effect that player 1's strategy C remains a strong best reply.

Define

$$x^t = \lim_{\varepsilon \to 0} q_2^t(A) \tag{57}$$

$$y^t = \lim_{\varepsilon \to 0} q_3^t(B) \ . \tag{58}$$

With the help of Figure 3 we obtain

$$x^t = -\frac{t(1-m)+(1-t)\Delta}{tm} \tag{59}$$

$$y^t = -\frac{(1-t)\Delta}{t} \ . \tag{60}$$

This yields

$$H_1^t(C,q_2^t,q_3^t) - H_1^t(A,q_2^t,q_3^t) = \frac{t(1-m)+(1-t)\Delta}{m} + m_1(t-(1-t)\Delta) + (1-t)(\tfrac{2}{3} m_1 - \tfrac{1}{3}) . \tag{61}$$

Here, H_1^t is player 1's payoff in the auxiliary game G_0^t for the limiting case $\varepsilon = 0$. The right-hand side of (61) is linear in t. Therefore, it is positive for $t^* \leq t \leq \tilde{t}$ if it is positive for $t = t^*$ and $t = \tilde{t}$. We have already shown that player 1's strategy C is a strong best reply to all 1-incomplete strategy combinations in auxiliary games G_ε^t with $t \le t^*$. The arguments used there do not lose their validity for the limiting case $\varepsilon = 0$. Therefore, the right-hand side of (61) is positive for $t = t^*$. For $t = \tilde{t}$, the first term on the right-hand side of (61) vanishes. It can be seen without difficulty that the other two terms are positive. In view of the fact that the trace cannot return to tracing parameters t smaller than t^* since there the auxiliary game G_ε^t has a unique equilibrium point, we can conclude that, between t^* and $\tilde{t}$, the trace must follow one of the three paths connected to the equilibrium points (C,A,C), (C,B,B) and (C,q_2^t,q_3^t).

At $t = \tilde{t}$, x^t assumes the value 0 and y^t the value 1-m. The path, formed by the indexed combinations $((C,q_2^t,q_3^t),t)$, can be continued from there by a jump segment leading to $((C,B,B),\tilde{t})$. From there, one can go back along a backward constant segment to $((C,B,B),t^*)$ and by a jump segment to $((C,A,B),t^*)$. It is interesting to see that here we find a path with a loop.

For $t > \tilde{t}$, the game of Figure 3 has only one equilibrium point, namely (A,C). Therefore, the paths corresponding to the equilibrium points (C,B,B) and (C,q_2^t,q_3^t) cannot be continued in a forward direction beyond $\tilde{t}$. It follows that the rectangular trace is the only possibility of continuation beyond $\tilde{t}$. We can conclude that (A,A,C) is the solution in the case $m_1 > m > .5$.

We now turn our attention to the remaining case $m_1 < m \le .5$. This case is the most difficult one. In order to find the solution, one would have to apply the logarithmic tracing procedure. Even if, in principle, the logarithmic tracing procedure determines a unique path, there seems to be no easily accessible way to determine it in practice. Therefore, we must confess that we are not able to determine the solution for this degenerate case. However, we can say that either (A,A,C) or (A,B,B) or a mixed equilibrium point (A,A,q_3) is the solution of the game. In the following, we shall indicate why these are the only possibilities for the solution.

The best reply to the centroid is (A,B,C). A constant segment ((A,B,C),t) extends up to a point $((A,B,C),t^*)$, where both players 2 and 3 become indifferent between their two pure strategies. We first show that from there the trace cannot be continued in a backward direction. This is due to the fact that in all auxiliary games G_ε^t with $t \le t^*$ player 1's strategy A is a strong best reply to every 1-incomplete strategy combination. It can be seen easily that the best replies B and C of players 2 and 3 to the centroid minimize player 1's incentive to choose A over all 1-incomplete strategy combinations in G_ε^t. At t^*

player 1's strategy A is still a strong best reply to the 1-incomplete strategy combination (B,C). Therefore, the assertion holds.

2 \ 3	B	C
A	0 0	0 $t(1-m)[2\varepsilon+6\varepsilon^2]$ $+(1-t)\Delta$
B	$t(1-m)(1+4\varepsilon+6\varepsilon^2)$ $+(1-t)\Delta$ 0	$t[-m+(2-4m)\varepsilon+6(1-m)\varepsilon^2]$ $+(1-t)\Delta$ $t[-m+(2-4m)\varepsilon+6(1-m)\varepsilon^2]$ $+(1-t)\Delta$

Figure 4: The payoff difference game for players 2 and 3, if in the auxiliary game player 1 is fixed at A. - Player 2's payoffs are shown above and player 3's are shown below. Δ is the common value of Δ_2 and Δ_3 (see (45)).

Figure 4 is analogous to Figure 3. For $t = t^*$ the payoffs for (B,C) in Figure 4 vanish and become negative for $t > t^*$. The other two nonzero payoffs are always positive. It can be seen immediately that from the reversal point at $t = t^*$two jump segments lead to $((A,A,C),t^*)$ and $((A,B,B),t^*)$. From there one can continue by constant segments which finally lead to $((A,A,C),1)$ and $((A,B,B),1)$, respectively.

A third possibility of continuation beyond t^* is a segment of the form $((A,q_2^t,q_3^t),t)$ where q_2^t and q_3^t are genuinely mixed strategies. At least for sufficiently small $t - t^* > 0$ the trace can be continued in this way since at t^* player 1's strategy A is a strong best reply to all 1-incomplete strategy combinations. For every $t > t^*$, let (q_2^t,q_3^t) be the uniquely determined, genuinely mixed equilibrium point of the game ofFigure 4. Moreover, let x^t be the limit of $q_2^t(A)$ and y^t be the limit of $q_3^t(B)$ for $\varepsilon \to 0$. With the help of Figure 4, it can be seen that the following equations hold:

$$x^t = 1 - \frac{(1-t)\Delta}{tm} \tag{62}$$

$$y^t = m - \frac{1-t}{t}\Delta . \tag{63}$$

We shall show that for $t^* \le t \le 1$ player 1's strategy A is a strong best reply to the 1-incomplete strategy combination (q_2^t,q_3^t) if ε is sufficiently small. As before, let H_1^t be player 1's payoff in the auxiliary game G_0^t for the limiting case $\varepsilon = 0$. With the help of (62) and (63) we obtain

$$H_1^t(A,q_2^t,q_3^t) - H_1^t(C,q_2^t,q_3^t) = (1-t)(\tfrac{1}{3} - \tfrac{2}{3}m_1)$$

$$+ t(1-m_1(1-m)) - (1-t)\frac{\Delta}{m}(1-m(1-m_1)). \tag{64}$$

The right-hand side of (64) is positive for $t = t^*$ and $t = 1$ and it is linear in t. Therefore, it is positive for all $t^* \le t \le 1$ if ε is sufficiently small. This shows that for sufficiently small ε the forward continuation $((A,q_2^t,q_3^t),t)$ of the trace extends until $t = 1$ without meeting a reversal point. For $t = 1$, we obtain $x^t = 1$ and $y^t = m$. Therefore, for $\varepsilon \to 0$ the endpoint of the mixed path converges to the equilibrium point (A,A,q_3) with

$$q_3(B) = m . \tag{65}$$

We have shown that there are exactly three linear traces, whose endpoints are (A,A,C), (A,B,B) and a mixed equilibrium point, whose limit for $\varepsilon \to 0$ is (A,A,q_3) with q_3 according to(65).

The discussion of the linear tracing procedure in degenerate cases is now complete. On the basis of the results obtained, we may ask the question of whether our conclusions on the properties of majority voting in Condorcet games together with the equilibrium selection theory as a social choice mechanism also extend to the degenerate cases.

We can now strengthen our assertions on efficiency. The discussion of degenerate cases shows that the only case where Pareto-optimality of the solution may not be obtained is the case $m_1 = m_2 = m_3$, where the solution is completely mixed.

Another conclusion, which can be strengthened, concerns the influence of an increase of m_1. It can now be seen that an increase of m_1 always strengthens the alternative C in the following sense: If for fixed m_2 and m_3 alternative C is realized by the solution for a particular value of m_1, then the same is true for every m_1' greater than m_1. What has been said on the influence of an increase of one of the m_i-values on the solution in Section 11 also holds with the inclusion of degenerate Condorcet games. This can be seen with the help of Figure 2.

References:

Arrow, K.J. (1963). Social Choice and Individual Values. 2nd ed., New York: John Wiley.

Condorcet, M.J. Marquis de (1785). Essai sur l'application de l'analyse à la plurité des voix. Paris.

Harsanyi, J.C. (1975). The Tracing Procedure: A Bayesian approach to defining a solution for n-person non-cooperative games. Int. J. Game Th., **4**: 61-95.

Harsanyi, J.C. and R. Selten (1977). Simple and Iterated Limits of Algebraic Functions. Working Paper CP-370 Center for Research in Management, Berkeley: University of California.

Harsanyi, J.C. and R. Selten (1988). A General Theory of Equilibrium Selection in Games. To appear.

Selten, R. (1975). Reexamination of the Perfectness Concept for Equilibrium Points in Extensive Games. Int. J. Game Th., **2**: 141-201.

Sen, A. (1970). Collective Choice and Social Welfare. San Francisco: Holden-Day.

DOES CONCESSION ALWAYS PREVENT PRESSURE?

by

Jan Potters, Frans van Winden and Michael Mitzkewitz*

1. Introduction

In Potters and Van Winden (1989) a repeated game model is used to analyze whether and under which conditions a government would be put under pressure (punished) by a pressure group for not conceding to a certain claim. In the model it is *assumed* that the punitive (aggressive) action is not used by the pressure group if the government concedes to the claim. The same kind of assumption is used in similar models. In the models of Selten (1978), Kreps and Wilson (1982) and Milgrom and Roberts (1982), for example, the monopolist cannot employ an aggressive marketing strategy if the entrant stays out, in Calvert (1987) the political leader cannot punish if the follower obeys, in Alt et al. (1988) the hegemon cannot punish if the ally obeys. In the present paper it will be examined to what extent the outcome of the game changes if this assumption is dropped, that is (in our terminology), if the pressure group is *allowed* to use the punitive strategy even if the government takes the action which is preferred by the group.

It appears that extending the strategy space in this way makes the repeated game much more complex, even in a two-period context. The number of equilibria increases significantly, and some of them are intuitively implausible. In order to reduce the set of equilibria we will employ two completely different concepts which refine the set of Nash-equilibria. First, we will use a concept which restricts "out-of-equilibrium" beliefs of a sequential equilibrium. Following Van Damme (1987, Ch.10) we will call this concept (Kohlberg-Mertens) admissibility. Secondly, we apply the selection theory of Harsanyi and Selten (1988). An additional motivation for the present paper is to see the differences and agreements beteen the two theories. We will start off with sequential equilibria, since for our purposes - the study of pressure, as defined in Potters and

* We are grateful to Eric van Damme, Werner Güth, Akira Okada and Reinhard Selten for helpful discussions. We alone are responsible for any errors and the views expressed.

Van Winden (1989) - it is useful that the beliefs of the players and the payoffs in any single period of the repeated game are explicitly modelled.

It will be seen that, for a fairly large set of parameter values, the use of a punitive action after a concession is part of the equilibrium strategies. However, often it need not come into force because the government anticipates on that strategy and does not concede. As a consequence, the pressure group is often worse off with the extended strategy space. Also the comparative statics of the extended game are different for some admissible equilibria. Hence, the analysis indicates that the *assumption*, mentioned above, of no punishment after a concession may have an important impact on the outcome of the game.

In Section 2 the "standard" game without the possibility to punish a conceding opponent will shortly be discussed. Section 3 presents the "extended" game and the equilibria of the one period game. Section 4 and 5 present the equilibria in case of a repetition of the extended game, using the admissibility concept and Harsanyi-Selten's selection theory, respectively. Section 6 concludes.

2. The standard game

2.1 Structure of the game

In the sequel we will use an interpretation of the model which refers to a context of political pressure (cf. Potters and Van Winden (1989)). Other interpretations, such as those in the papers mentioned in the introduction, are also possible.

Assume that a firm f, at a certain point in time, asks a government g for a subsidy. The firm claims that it can invest profitably only with the subsidy. If the government does not give the subsidy (N) the firm must decide whether to punish (D, not invest) or not (I, invest). If the government subsidizes (S) it is assumed that the firm does not punish (invests). Furthermore, it is assumed that g has incomplete information about f's payoffs. The firm can be of two possible types (in the sense of Harsanyi (1967-8)). With probability 1-P the firm is of the "weak" type (type 1), which prefers to invest (not to punish) even if there is no subsidy. With probability P the firm is of the "strong" type (type 2), which is better off not to invest if there is no subsidy (e.g., because of better opportunities elsewhere). The government prefers not to subsidize if the firm would still invest. Table 1 illustrates this situation.

		payoffs to g	payoffs to f
g concedes: S	f does not punish: I	0	1
g does not concede: N	f does not punish: I	a	0
	f punishes: D	a-1	b-d

with $0 < a,\ b < 1$

and $d = \begin{cases} 1 \text{ with probability } 1-P \text{ (f is type 1, "weak")} \\ 0 \text{ with probability } P \text{ (f is type 2, "strong")} \end{cases}$

TABLE 1. The standard game

2.2 Equilibrium analysis

It is a matter of simple calculation to show that for all $P \neq a$ there is a unique sequential (and perfect) equilibrium. The government will concede (S) if $P > a$, and will not concede (N) if $P < a$. A firm of type 1 will always invest and a firm of type 2 will invest only if g plays subsidy. The situation becomes more interesting if this game is repeated a number of times because then a weak type of firm can try to mimic the strong type by punishing the government if it does not concede, thereby exerting pressure and building a "reputation" for being strong.

The repeated game has many sequential equilibria, some of which are intuitively "implausible". In a two-fold repetition, for instance, the following is a sequential equilibrium. The government has a high initial belief $P_1 > a$ that the firm is of type 2 but does not give a subsidy in the first period (x_1 = N), and plays x_2 = S if and only if the firm invests (y_1 = I). Both types of firm respond to this strategy by investing in the first period (if $b < \frac{1}{2}$). This equilibrium rests on an implausible adjustment of the belief P_2 after a deviation from the equilibrium. It is implausible because y_1 = D after x_1 = N should not decrease g's belief P_2 that the firm is of type 2, and therefore g should play x_2 = S after (x_1,y_1) = (N,D). Such equilibria, however, do not pass the test of *admissibility*. This concept requires an equilibrium to be invariant with respect to elimination of never weak best responses. Translated to our signalling game, it requires (cf. Van Damme (1987, Ch.10)) that beliefs after a deviation from the equilibrium are concentrated on the type which is "most easily" induced to make that deviation voluntary. Admissibility is a necessary, but not suffient condition for *stability* in the sense of Kohlberg Mertens' (1986). Furthermore, in our game it leads to the same equilibria as the concept of *universal divinity* by Banks and Sobel (1987).

The following notation will be useful to describe the sequential equilibrium of the repeated game. Let $x_t \in \{S,N\}$ denote g's action in period t and $y_t \in \{I,D\}$ f's action. P_t is g's belief in period t that f is of type 2. Let $\sigma_t(P_t)$ denote the probability (mixed strategy) that g plays S in period t when g's current belief is P_t. Finally, let $\rho_t^k(P_t,x_t)$ indicate the probability that f type k (= 1,2) plays I in period t when g plays x_t (= S,N) and g's belief is P_t. Beliefs are common knowledge and updated according to Bayes' rule. The following proposition is a special case of Proposition 3 in Potters and Van Winden (1989) (see also Van Damme (1987, p.297) and Kreps and Wilson (1982)).

Proposition 1

The unique admissible sequential equilibrium strategies are:

$$\sigma_t(P_t) = \begin{cases} 0 & \text{if } 0 \le P_t < a^{T-t+1} \\ 1-b & \text{if } P_t = a^{T-t+1} \\ 1 & \text{if } a^{T-t+1} < P_t \le 1 \end{cases}$$

$$\rho_t^1(P_t,x_t) = \begin{cases} 1 & \text{if } x_t = S \\ 1 - P_t(1-P_t)^{-1}(a^{-(T-t)}-1) & \text{if } x_t = N \text{ and } P_t \le a^{T-t} \\ 0 & \text{if } x_t = N \text{ and } P_t > a^{T-t} \end{cases}$$

$$\rho_t^2(P_t,x_t) = \begin{cases} 1 & \text{if } x_t = S \\ 0 & \text{if } x_t = N \end{cases} \quad \text{for all } P_t$$

Uniqueness, here and in the sequel, refers to the equilibrium *outcome* (the probability distribution over end-nodes in the extensive form of the game). Off the equilibrium path the equilibrium *strategies* need not be unique.

This - very interesting - equilibrium is extensively analyzed and discussed in Potters and Van Winden (1989). In the present paper it will (for T = 2) mainly be used for comparison with the equilibria of the extended game. The following observations are important in that respect.

2.3 Some observations

Ex ante (i.e., in expected value terms) it is never profitable to *both* players to make binding agreements to a certain course of action. The firm would (only) gain from a binding agreement to play $\rho_t^k(P_t,N) = 0$ if $P_1 < a$. The government, however, would always lose from such a commitment.

According to the definition of pressure in Potters and Van Winden (1989) - which says that pressure involves a costly action aimed at influencing the government's beliefs - the firm can only exert pressure if it is of type 1 in this equilibrium. The firm will then play $y_t = D$ if $x_t = N$ (with a positive probability). This action is costly in the short run for the firm of type 1 ($b-1 < 0$) but not for type 2 ($b > 0$).

The government concedes in a certain period t if $P_t > a^{T-t+1}$. So, the probability of concession is non-decreasing in P_t (the probability that the firm "needs" the subsidy), and non-increasing in a (the cost to g of concession). These result are intuitively very plausible.

3. The extended game

We will now extend the game of Section 2 in the sense that the firm will be allowed to punish (not invest) even if the government concedes (subsidizes). It seems reasonable to assume that the payoff (G(x,y)) to the government in case of subsidy and no investment is less than in case of no subsidy and no investment. Supplying the subsidy but still being punished is the worst alternative to g. So the payoff structure to g is:

$$G(N,I) > G(S,I) > G(N,D) > G(S,D) \qquad (3.1)$$

A further reasonable assumption seems to be that the firm gains from a subsidy even if it does not invest: $F(S,D) > F(N,D)$, where $F(x,y)$ denotes the payoff to the firm. Again, it is supposed that the government has incomplete information about the firms' payoffs. With probability 1-P the firm (type 1, with payoff F^1) prefers to invest even if there is no subsidy:

$$F^1(S,I) > F^1(N,I) > F^1(S,D) > F^1(N,D) \qquad (3.2)$$

With probability P the firm (type 2, with payoff F^2) prefers to invest if and only if there is a subsidy:

$$F^2(S,I) > F^2(S,D) > F^2(N,D) > F^2(N,I) \qquad (3.3)$$

A parametrization of the payoffs is given in Table 2. [This parametrization can be generalized without qualitatively affecting the results, provided that payoffs satisfy the constraints (3.1)-(3.3)].

g	f	$G(x,y)$	$F^k(x,y)$
g concedes: S	f does not punish: I	0	1
	f punishes: D	-1	c-d
g does not concede: N	f does not punish: I	a	0
	f punishes: D	a-1	b-d

where $0 < a < 1$, $0 < b < c < 1$

$$\text{and } d = \begin{cases} 1 \text{ with probability } 1-P \text{ (f is type 1, "weak")} \\ 0 \text{ with probability } P \text{ (f is type 2, "strong")} \end{cases}$$

TABLE 2. The extended game

It is easily seen that the unique sequential equilibrium of the one-period extended game is identical to that of the standard game. In a one-period analysis the *assumption* that y = I if x = S has no effect on the equilibrium strategies or the expected equilibrium payoffs. Hence, the sequential equilibrium strategies in the second period of the repeated game are as follows.

$$\sigma_2(P_2) = \begin{cases} 0 & \text{if } P_2 < a \\ m & \text{if } P_2 = a \\ 1 & \text{if } P_2 > a, \text{ where } m \in [0,1] \end{cases} \tag{3.4}$$

$$\rho_2^1(P_2,x_2) = 1, \text{ for all } P_2 \text{ and } x_2 \in \{S,N\} \tag{3.5}$$

$$\rho_2^2(P_2,x_2) = \begin{cases} 0 \text{ if } x_2 = N \\ 1 \text{ if } x_2 = S \end{cases} \tag{3.6}$$

The next two sections present the equilibrium analysis of the two-period extended game.

4. Admissible equilibria of the two-period extended game

It appears that the analysis of the repeated play of the extended game is far more complex than that of the standard game. A two-fold repetition of the stage game of Table 2, however, will suffice to illustrate our arguments. In this case the normal form of the game contains 2^{25} pure strategy combinations, and if the game is reduced such that players are fixed at dominant strategies we are still left with 2^8 pure strategy combinations. There are many sequential equilibria and even more Nash equilibria, some of which are intuitively very implausible. To give a complete list of all (sequential) equilibria does not seem very useful.

We will concentrate on admissible equilibria. It will be shown that in some cases the concept of admissibility leads to a unique equilibrium, but that in other cases this concept is not strong enough to select a unique equilibrium (outcome).

It is useful for the equilibrium analysis to make a distinction between two sets of the government's initial belief P_1 that the firm is of type 2. We will deal separately with the case $P_1 < a$ and the case $P_1 > a$. [The degenerate case $P_1 = a$ will not be discussed. In the sequel we will disregard all such "knife-edge" cases.]

4.1 A low initial reputation: $P_1 < a$

It is easy to check that the strategies of Proposition 1 for the case T=2 still form a sequential equilibrium of the extended game. However, the following proposition shows that these strategies, at least for certain values of the initial belief P_1, no longer are admissible.

Proposition 2

For $P_1 < a$, admissibility requires that

$$\rho_1^1(P_1,x_1{=}S) = 1 \text{ and } \rho_1^2(P_1,x_1{=}S) = 0$$

[The proof is in Appendix A]

This proposition shows the importance of allowing f to punish ($y_1 = D$) even if g concedes ($x_1 = S$). The firm of type 2 will not invest even if the government gives the subsidy. This may seem remarkable in view of the short run cost of this strategy to f: as can be seen in Table 2, also the firm of type 2 prefers to play I if g plays S. This short run cost, however, is outweighed by the longer run benefit of this strategy. Suppose that type 2 would play $\rho_1^2(P_1,x_1) = 1$ if $x_1 = S$. Then Bayesian updating of beliefs after $x_1 = S$, $y_1 = I$ requires:

$$P_2 = \text{Prob}\{f \text{ is type } 2 \mid x_1 = S,\ y_1 = I\} = \frac{P_1\rho_1^2}{(1-P_1)\rho_1^1 + P_1\rho_1^2} = P_1 < a$$

$P_2 < a$ would induce g to play $x_2 = N$ in period 2 and type 2 to play $y_2 = D$. The strategy ρ_1^2 in the proposition, however, leads to $P_2 = 1$ with Bayesian updating after $x_1 = S$, $y_1 = D$, and then g will play $x_2 = S$. The firm of type 2 takes a costly action in order to influence g's beliefs. According to the definition in Potters and Van Winden (1989) the firm *exerts pressure* on the government. Since $F^2(S,D) + F^2(S,I) > F^2(S,I) + F^2(N,D)$ the latter strategy is rational for the firm of type 2 but not for type 1: $F^1(S,D) + F^1(S,I) < F^1(S,I) + F^1(N,D)$. This fact causes that the strategies in proposition 1 are no longer admissible (although they still form a sequential equilibrium).

Proposition 3

For $P_1 < a$ there is a unique admissible equilibrium.

$$\sigma_1(P_1) = \begin{cases} 0 \text{ if } P_1 < \alpha(a) & \text{or } a < \alpha(a) \\ 1 \text{ if } \alpha(a) < P_1 < a, & \text{where } \alpha(a) := a^2/(1-a^2) \end{cases}$$

$$\rho_1^1(P_1,x_1) = \begin{cases} 1 & \text{if } x_1 = S \\ 1-P_1(1-P_1)^{-1}(a^{-1}-1) & \text{if } x_1 = N \end{cases}$$

$$\rho_1^2(P_1,x_1) = 0 \text{ for } x_1 \in \{S,N\}$$

The strategies in the second period, are as in equations (3.4)-(3.6) with $m = 1-b$.

A first observation that can be made is, that the government supplies the subsidy in the first period only if $\alpha(a) < P_1 < a$. [Note that this condition can only hold if $a < (\sqrt{5}-1)/2$.] It is only in this case that the firm of type 2 will (be able to) exert pressure. In the other cases the threat, of no investment even with a subsidy, deters the government from supplying the subsidy. In these latter cases, however, the firm of type 1 will exert pressure by not investing (with a positive probability). So, whatever the government's action (concession or no concession), pressure cannot be avoided!

A simple comparison of the expected payoffs of the players of the equilibria in Proposition 3 and Proposition 1 (for $P_1 < a$) reveals the following.

Proposition 4

The government is worse off in the extended game if $a^2 < P_1 < a$, and never better off. The firm of type 1 is worse off in the extended game if $a^2 < P_1 < \min\{a,\alpha(a)\}$ and never better off. The firm of type 2 is worse off if $a^2 < P_1 < \min\{a,\alpha(a)\}$ and better off if $\alpha(a) < P_1 < a$.

Hence, for $a^2 < P_1 < \min\{a,\alpha(a)\}$ a binding agreement not to play the punitive action after a concession, would be beneficial to both players. Contrary to the standard game, in the extended game there is scope for mutually benefical binding agreements. Note that, for $a^2 < P_1 < \min\{a,\alpha(a)\}$, the firm would not reveal its type if it expressed a wish to enter such an agreement.

4.2 A high initial reputation: $P_1 > a$

Also in this case the strategy by the firm of type 2 not to invest with a positive probability if there is a subsidy is part of two admissible equilibria (denoted E^2 and E^3). A major difference with the case $P_1 < a$, however, is that now there is also a admissible equilibrium (denoted E^1) in which both types of firms invest if the subsidy is being supplied. Moreover, this equilibium E^1 payoff dominates the other admissible equilibria.

Proposition 5

For $P_1 > a$ there are three admissible equilibria, denoted E^1 and E^2 and E^3.

E^1:

$$\sigma_1(P_1) = 1, \; \rho_1^k(P_1,x_1) = \begin{cases} 0 \text{ if } x_1 = N \\ 1 \text{ if } x_1 = S \end{cases} \qquad \text{for } k = 1, 2$$

E^2:

$$\sigma_1(P_1) = \begin{cases} 1 \text{ if } a < P_1 < \beta(a) \\ 0 \text{ if } \beta(a) < P_1 < 1, \text{ where } \beta(a) = a^2-a+1 \end{cases}$$

$$\rho_1^1(P_1,x_1) = \begin{cases} 0 & \text{if } x_1 = N \\ 1 & \text{if } x_1 = S \end{cases} \qquad \rho_1^2(P_1,x_1) = \begin{cases} 0 & \text{if } x_1 = N \\ \dfrac{1-P_1}{P_1} \cdot \dfrac{a}{1-a} & \text{if } x_1 = S \end{cases}$$

E^3:

$$\sigma_1(P_1) = \begin{cases} 1 \text{ if } a < P_1 < \gamma(a) \\ 0 \text{ if } \gamma(a) < P_1 < 1 \text{ or } \gamma(a) < a, \text{ where } \gamma(a) = 1/(1+a) \end{cases}$$

$$\rho_1^1(P_1,x_1) = \begin{cases} 0 \text{ if } x_1 = N \\ 1 \text{ if } x_1 = S \end{cases} \qquad \rho_1^2(P_1,x_1) = 0 \quad \text{for } x_1 = N, S$$

The second period strategies are given by (3.4)-(3.6), where for equilibrium E^2 g must play $m = (c-b)/(1-b)$ if it is indifferent ($P_2=a$).

Proposition 6

Equilibrium E^1 payoff dominates equilibrium E^2, which in turn (weakly) payoff dominates equilibrium E^3.

[The proof is by simple comparison of expected payoffs.]

The outcome of equilibrium E^1 appears to be identical to the outcome of the equilibrium of the standard game in Proposition 1. Contrary to the case $P_1 < a$ (see Proposition 2), this equilibrium is admissible now. The reason for this is that the higher value of P_1 is sufficient to ensure that the government will again supply a subsidy in period 2, even if P_2

remains unchanged. Therefore, there is no reason for the firm to exert pressure in order to raise P_2.

It is remarkable that for the equilibria E^2 and E^3 the comparative statics results of the standard game need no longer hold. A higher belief P_1 that the firm "needs" the subsidy may decrease the "probability" σ_1 that the government supplies the subsidy in the first period. This, of course, is due to the fact that type 2 does not invest (with a positive probability) if the government supplies the subsidy, which is even worse to g then not supplying the subsidy and being punished.

The players could again benefit from a binding agreement always to play the strategy $\rho_1^k(P_1,S) = 1$. However, as long as the possibility to make binding agreements is not explicitly included in the rules of the game, equilibria E^2 and E^3 cannot be excluded. Although we have not checked all the details yet, we conjecture that not only the outcome of equilibrium E^1 but also that of E^3 is stable in the sense of Kohlberg and Mertens (1986), whereas the equilibrium outcome of E^2 is not. In the next Section we will see that the strategy $\rho_1^2(P_1,S)$ which supports equilibria E^2 and E^3 is eliminated in Harsanyi-Selten's selection theory.

5. The Harsanyi-Selten solution of the two-period extended game

Because the theory of equilibrium selection by Harsanyi and Selten (1988) considers *uniformly perfect equilibria* and not sequential equilibria, we have to change some of the terminology used in the previous sections. In order to facilitate comparison of the results we will use similar notations.

Assume that the firm's agents in the second period are fixed at their dominant choices. Because we are only interested in non-degenerate cases, the avoidance of perturbations in the last stage of the game does not change the results. The truncated game has nine agents. We call government's first-period agent g_1, and $\sigma_1(P)$ is his probability (mixed strategy) to play S. The four agents of the firm in the first period are denoted by f_1^{kx}, where $k\in\{1,2\}$ indicates f's type, and $x\in\{S,N\}$ is g_1's action. The local mixed strategy of f_1^{kx} is $\rho_1^k(P,x)$ and represents his probability to play I. The four second-period agents of the government are called g_2^{xy}, where $x\in\{S,N\}$ is again g_1's action and $y\in\{I,D\}$ is the firm's action in the first period. $\sigma_2(P,x,y)$ is the mixed strategy of g_2^{xy}, the probability to choose S. Harsanyi-Selten's theory does not require posterior beliefs to be specified. Therefore, the argument P, also in the strategies for the second period, refers to the *initial* belief (which is denoted P_1 in the preceding sections).

In Appendix B we will briefly describe those elements of the Harsanyi-Selten theory which are used to calculate the solution of the two-period extended game. Appendix B also includes the proof of the following proposition.

Proposition 7

The Harsanyi-Selten theory prescribes the following local strategies for the agents of the two-period extended game.

$$\sigma_1(P) = \begin{cases} 0 & \text{if } P < \min\{a,\alpha(a)\} \\ 1 & \text{if } P > \min\{a,\alpha(a)\} \end{cases}$$

$$\sigma_2(P,S,I) = \begin{cases} 0 \text{ if } P < a \\ 1 \text{ if } P > a \end{cases}$$

$$\sigma_2(P,S,D) = 1$$

$$\sigma_2(P,N,I) = \begin{cases} 0 \text{ if } P < a \\ b \text{ if } P > a \end{cases}$$

$$\sigma_2(P,N,D) = \begin{cases} 1-b \text{ if } P < a \\ 1 \quad \text{if } P > a \end{cases}$$

$$\rho_1^1(P,S) = 1$$

$$\rho_1^2(P,S) = \begin{cases} 0 & \text{if } P < a \\ 1 & \text{if } P > a \end{cases}$$

$$\rho_1^1(P,N) = \begin{cases} 1-P(1-P)^{-1}(a^{-1}-1) & \text{if } P < a \\ 0 & \text{if } P > a \end{cases}$$

$$\rho_1^2(P,N) = 0$$

A low initial reputation: $P < a$

A comparison with the results of Propositions 2 and 3 shows that the Harsanyi-Selten theory selects exactly the unique admissible equilibrium.

It should be noted, however, that it is not always the case that a unique admissible (or stable) equilibrium and Harsanyi-Selten's solution coincide (Figure 10.9 in Harsanyi-Selten (1988,p. 351) provides a counterexample.) But, perhaps, the results of these - completely differently constructed - theories more often coincide than one might a priori expect.

A high initial reputation: P > a

The equilibrium selected by Harsanyi-Selten is close to equilibrium E^1 of Proposition 5. Differences only occur *off* the equilibrium paths (consider the strategy of agent g_2^{NI}).

Proposition 6 shows that the E^1-outcome payoff-dominates the outcomes E^2 and E^3. Although payoff-dominance plays a crucial role in Harsanyi-Selten's theory, it is not true that always a payoff-dominant equilibrium of the *whole* game is selected if such an equilibrium exists. The solution of the *whole* game is obtained by comparison of *solely* government's solution payoffs for the "concession cell" and the "non-concession cell" (see Appendix B, Lemma 3). However, the solution of the "concession cell" (see Appendix B, Lemma 1) is determined by payoff-dominance considerations in the case P > a, which indeed eliminates the strategies which sustain equilibria E^2 and E^3.

Comparative statics

If we neglect the degenerate case P = a, the Harsanyi-Selten solution reveals the following properties:

(a) for each agent of the government the probability to supply the subsidy is a nondecreasing function of P, given a, and a nonincreasing function of a, given P (recall that a indicates the cost to g of concession);

(b) for each agent of the firm's weak type (type 1) the probability to invest is a nonincreasing function of P, given a, and a nondecreasing function of a, given P;

(c) for each agent of the firm's strong type (type 2) the probability to invest is a nondecreasing function of P, given a, and a nonincreasing function of a, given P.

These - intuitively appealing - comparative statics results, are of course analoguous to those obtained for the unique admissible equilibrium in the case $P_1 < a$, and the payoff dominant admissible equilibrium E^1 in the case $P_1 > a$.

6. Conclusions

The use of the punitive strategy after a concession by the opponent, may be part of the (unique) equilibrium strategies for the strong type of firm. The short run costs of this strategy may be outweighed by a longer run benefit, because of the effect on the government's beliefs. However, this (remarkable) strategy need not come into force because of anticipation by the government - conceding but still being punished is the worst alternative to the government. Consequently, both the firm and the government can - for a significant range of parameter values - benefit from a commitment by the firm, not to use the punitive action if the government concedes. This fact, which is not visible in the standard model, may provide a rationale for government regulation (e.g., subsidies being coupled with certain conditions).

The probability that pressure is exerted increases in the extended model. Both types of the firm may now exert pressure (i.e., send a costly signal). Moreover, for certain parameter sets pressure cannot be avoided. If the government concedes the strong type of firm will exert pressure, and, if it does not concede the weak type will exert pressure (with a positive probability).

We have also shown that in the extended game the admissibility concept does not always select a unique equilibrium outcome, contrary to the standard game. For some parameter values payoff dominated equilibria survive the admissibility test. Furthermore, the (three) different admissible equilibria mutually disagree on some comparative statics results. Harsanyi-Selten's theory, by definition, selects a unique equilibrium. On the equilibrium path this equilibrium appears to be identical to the admissible equilibrium, when the latter is unique, and identical to the payoff dominant admissible equilibrium, if admissibility does not yield uniqueness (the comparative statics of these equilibria are intuitively plausible). Hence, in our - fairly complex - game there seems to be more agreement between the two theories then one might perhaps a priori expect.

APPENDIX A

First some additional notation will be introduced. $w_t^k(P_t)$, (for $t = 1,2$ and $k = g,1,2$) denotes the total expected payoffs of playing the repeated extended game in period t, for player g, f type 1, and f type 2, respectively. $P_2(P_1,x_1,y_1)$ denotes the updated (posterior) beliefs of g in period 2 as a function of the initial beliefs and the actions in period 1. Using the strategies (3.4)-(3.6) the values of $w_2^k(P_2)$, $k = g,1,2$, can easily be computed. We will use these values in the proofs of Propositions 2 and 4, without presenting them explicitly here.

Proof of Proposition 2

We will prove that $\rho_1^1(P_1,S) = 1$, $\rho_1^2(P_1,S) = 0$ for all $P_1 < a$ by contradiction.

A) Assume that $\rho_1^1(P_1,S) < 1$. This can only be an equilibrium (i.e., optimal) strategy if the total expected payoffs of playing $y_1 = I$ do not exceed those of playing $y_1 = D$: $1 + w_2^1(P_2(P_1,S,I)) \leq c-1 + w_2^1(P_2(P_1,S,D))$ or equivalently: $w_2^1(P_2(P_1,S,D)) - w_2^1(P_2(P_1,S,I)) \geq 2 - c$ (> 1). This, however, gives a contradiction since the left-hand side can never exceed 1. Hence $\rho_1^1(P_1,S) = 1$.

B) Assume $\rho_1^1(P_1,S) = 1$, $\rho_1^2(P_1,S) = 1$. Consistency with Bayes' rule then implies: $P_2(P_1,S,I) = P_1$. $P_2(P_1,S,D)$ cannot be determined by Bayes' rule since $(x_1=S,y_1=D)$ is a "zero-probability event". For the strategies to be optimal it is required that: $1 + w_2^1(P_2(P_1,S,I)) \geq c-1 + w_2^1(P_2(P_1,S,D))$ and $1 + w_2^2(P_2(P_1,S,I)) \geq c + w_2^2(P_2(P_1,S,D))$. Since $P_2(P_1,S,I) = P_1 < a$ these inequalities amount to, respectively: $w_2^1(P_2(P_1,S,D)) \leq 2 - c$ and $w_2^2(P_2(P_1,S,D)) \leq 1 + b - c$. Since $0 < b < c < 1$, the former inequality holds for any $P_2(P_1,S,D) \in [0,1]$ whereas the latter does not. Thus, only type 2 can - for certain values of $P_2(P_1,S,D)$ - be induced to play $y_1 = D$ after $x_1=S$ voluntarily. Therefore, admissibility requires that the beliefs after the "zero-probability event" $(x_1=D,y_1=I)$ are concentrated on type 2: $P_2(P_1,S,D) = 1$. But then $w_2^2(P_2(P_1,S,D)) = 1$ which means that $\rho_1^2(P_1,S) = 1$ cannot be optimal.

C) Assume $\rho_1^1(P_1,S) = 1$, $0 < \rho_1^2(P_1,S) < 1$. Consistency implies: $P_2(P_1,S,D) = 1$ and $P_2(P_1,S,I) < P_1$. This means that $1 + w_2^2(P_2(P_1,S,I)) = 1 + b < c + w_2^2(P_2(P_1,S,D)) = c + 1$, so $\rho_1^2(P_1,S) > 0$ cannot be optimal.

Combining A, B, and C establishes the claim.

<u>Proof of Proposition 3</u>

First, by condradiction it will be proved under A, B and C that $\rho_1^1(P_1,N) < 1$ and $\rho_1^2(P_1,N) = 0$.

A) Assume $\rho_1^1(P_1,N) = 1$ and $\rho_1^2(P_1,N) < 1$. Consistency with Bayes' rule implies that: $P_2(P_1,N,D) = 1$ and $P_2(P_1,N,S) < P_1$. Then type 2 prefers to play $y_1 = N$ since $b + w_2^2(P_2(P_1,N,D)) = b + 1 > w_2^2(P_2(P_1,N,I))$. Hence, $\rho_1^2(P_1,N) = 0$. Constistency implies $P_2(P_1,N,I) = 0$. But then $y_1 = I$ cannot be optimal for type 1, since $w_2^1(P_2(P_1,N,I)) = 0 < b-1 + w_2^1(P_2(P_1,N,D))$.

B) Assume $\rho_1^1(P_1,N) = 1$ and $\rho_1^2(P_1,N) = 1$. Consistency implies $P_2(P_1,N,I)) = P_1$ and $P_2(P_1,N,D))$ is undetermined. For these strategies to be optimal, it must hold that,

$$w_2^1(P_2(P_1,N,I)) \geq b-1 + w_2^1(P_2(P_1,N,D)) \text{ and} \tag{A.1}$$

$$w_2^2(P_2(P_1,N,I)) \geq b + w_2^2(P_2(P_1,N,D)) \tag{A.2}$$

The first inequality holds for all $P_2(P_1,N,D))$ whereas the second can only hold if $P_2(P_1,N,D) \geq a$. So, only type 2 can be induced to play $y_1 = D$ voluntary. Therefore, admissibility requires that the posterior beliefs are concentrated on this type: $P_2(P_1,N,D) = 1$. But then inequality (A.2) can never hold.

C) Combining A and B yields $\rho_1^1(P_1,N) < 1$. Type 1 never strictly prefers to play $y_1 = N$ after $x_1 = D$, that is, it must always hold that $w_2^1(P_2(P_1,N,I)) \leq b-1 + w_2^1(P_2(P_1,N,D))$. For this it is necessary that $P_2(P_1,N,D) > P_2(P_1,N,I)$. But then type 2 strictly prefers to play $y_1 = N$ after $x_1 = N$: $\rho_1^2(P_1,N) = 0$. This establishes the claim and also that $P_2(P_1,N,I) = 0$.

D) $\rho_1^1(P_1,N) = 0$ cannot be optimal if $P_1 < a$, since then $P_2(P_1,N,D) = P_1$ and thus the strict inequality in (A.1) would hold. Now, assume $0 < \rho_1^1(P_1,N) < 1$. Then (A.1) must hold with equality, or, equivalently $w_2^1(P_2(P_1,N,D)) = 1-b$. This can only hold if $\sigma_2(P_2) = 1-b$, which requires that g is indifferent in the second period: $P_2(P_1,N,D) = a$. To justify this posterior belief, Bayes' rule requires that

$$P_2(P_1,N,D) = \frac{P_1\rho_1^2(P_1,N)}{(1-P_1)\rho_1^1(P_1,N)+P_1\rho_1^2(P_1,N)} = a \quad \text{or} \quad \rho_1^1(P_1,N) = 1 - \frac{P_1(1-a)}{(1-P_1)a}$$

Since, of course, $\rho_1^1(P_1,N) \leq 1$, this can only hold if $P_1 < a$. If $P_1 > a$, $\rho_1^1(P_1,N) = \rho_1^2(P_1,N) = 0$ are the only admissible strategies.

E) Given the unique admissible equilibrium strategies $\rho_1^k(P_1,x_1{=}N)$, for k = 1, 2, and $P_1 < a$, the optimal response $\sigma_1(P_1)$ to these strategies for player g follow by simple calculation. The optimal strategy for player g is $\sigma_1(P_1) = 1$ if and only if $a^2/(1-a^2) < P_1 < a$. This condition, however, can never hold if $a^2/(1-a^2) > a$, that is, if $a > (\sqrt{5}-1)/2$.

<u>Proof of Proposition 5</u>

The unique admissible equilibrium strategies $\rho_1^k(P_1,x_1{=}N) = 0$, k = 1, 2, for $P_1 > a$ follow from part D of the proof of Proposition 3. Note also that $\rho_1^1(P_1,x_1{=}S) = 1$ is a dominant strategy for type 1 (see part A of the proof of Proposition 2). We will now show that (A) $\rho_1^2(P_1,S) = 1$, (B) $\rho_1^2(P_1,S) = [(1-P_1)a]/[P_1(1-a)]$, and (C) $\rho_1^2(P_1,S) = 0$ are part of the admissible sequential equilibria E^1, E^2, and E^3, respectively.

A) Suppose $\rho_1^2(P_1,S) = 1$ is an equilibrium strategy. Consistency with Bayes' rule requires: $P_2(P_1,S,I) = P_1$. (Since $P_1 > a$ this implies $\sigma_2(P_2(P_1,S,I)) = 1$.) $P_2(P_1,I,D)$ cannot be updated according to Bayes' rule, but, whatever its value neither type of player f can be induced to play $y_1 = D$ after $x_1 = S$: $1 + w_2^1(P_2(P_1,S,I)) = 2 > c-1 + w_2^1(P_2(P_1,S,D))$ and $1 + w_2^2(P_2(P_1,S,I)) = 2 > c + w_2^2(P_2(P_1,S,D))$. Since any posterior belief $P_2(P_1,I,D)$ supports the equilibrium it is admissible.

B) Suppose $\rho_1^2(P_1,S) = [(1-P_1)a]/[P_1(1-a)]$. Consistency with Bayes' rule implies: $P_2(P_1,S,D) = 1$, and $P_2(P_1,S,I) = P_1\rho_1^2(P_1,S)/[P_1\rho_1^2(P_1,S)+(1-P_1)\rho_1^1(P_1,S)] = a$. This makes g indifferent in period 2 after $(x_1,y_1) =$

(S,I). To justify type 2's randomization after x_1=S, $\sigma_2(a)$ = m must be such that type 2's expected payoffs of playing y_1 = D and y_1 = I are equal: c+1 = m2 + (1-m)(b+1). Solving for m yields m = (c-b)/(1-b). There are no zero-probability events in this sequential equilibrium, so it is admissible.

C) Suppose $\rho_1^2(P_1,S) = 0$. Consistency requires $P_2(P_1,S,I) = 0$ and $P_2(P_1,I,D) = 1$. Given these beliefs playing y_1 = I after x_1 = S is optimal for type 1: $1 + w_2^1(P_2(P_1,S,I)) = 1 > c-1 + w_2^1(P_2(P_1,S,D)) = c$, and playing y_1 = D after x_1 = S is optimal for type 2: $1 + w_2^2(P_2(P_1,S,I)) = 1 + b < c + w_2^2(P_2(P_1,S,D)) = 1 + c$. Since there can be no zero-probability events in this sequential equilibrium it also is admissible.

D) The optimal strategies $\sigma_1(P_1)$ in reponse to the strategies $\rho_1^k(P_1,x_1)$ in equilibria E^1, E^2, and E^3 follow by simple calculation. [In equilibrium E^3 the optimal strategy for g is $\sigma_1(P_1) = 1$ if $a < P_1 < 1/(1+a)$. This condition, however, can never hold if $a > 1/(1+a)$, that is, if $a > (\sqrt{5}-1)/2$.]

APPENDIX B

Before proving Proposition 7, some of the concepts and procedures from Harsanyi-Selten's selection theory (HS), used in the proof, will be presented.

Uniformly perfect equilibria. Perturb a game such that each agent has to play each of his choices with probability not smaller than ε. The limit equilibria of this perturbed game as ε goes to zero are the uniformly perfect equilibria of the original game. HS requires the solution to be uniformly perfect.

Cell-consistency. A group of agents forms a cell, if for each of these agents the strategic situation only depends on the other members of the group. In the two-period extended game the four agents f_1^{1S}, f_1^{2S}, g_2^{SI}, and g_2^{SD}, which come into play in case g_1 chooses S, form a cell which we call the "concession cell". The "non-concession cell" is formed by the agents f_1^{1N}, f_1^{2N}, g_2^{NI}, and g_2^{ND}. HS has the property that the solution of the game and the solution of a cell always prescribe the same local strategies to all agents in the cell.

This cell consistency enables us to proceed as follows. First, the solution of the concession cell (Lemma 1) and of the non-concession cell (Lemma 2) are calculated. The eight involved agents are fixed at their solution strategies and the optimal behavior of agent g_1 is determined afterwards (Lemma 3).

Elimination of inferior choices. If an agent has a weakly dominated choice, he is fixed at his other choice. In our game each agent has only two possible choices. This means that the dominating choice is played with probability $1-\varepsilon$. To meet HS's terminology we will use the term "inferior choice" instead of "weakly dominated choice". In our game there is no difference between the two concepts (see HS, p. 118).
Solution of 2x2-games. If a 2x2-game has two pure strategy equilibria and one of them is *payoff dominant*, this is the one selected by HS. If no payoff dominance exists, let (q_1,q_2) be the mixed strategy equilibrium, with q_i player i's probability to play the strategy belonging to the, say, first pure equilibrium. If $q_1+q_2 < 1$ then the first pure equilibrium, if $q_1+q_2 > 1$ then the other pure equilibrium, and if $q_1+q_2 = 1$ then the mixed equilibrium is selected as the solution of the game. This is not an ad hoc rule, but the consequence of an axiomatic approach (see HS, p. 86). Also the *risk dominance* comparison between the two pure equilibria using the *tracing procedure* yields the same result.

Now we are well-prepared to calculate the Harsanyi-Selten solution of the two-period extended game.

Lemma 1
The HS solution for the concession cell is:

$$\rho_1^1(P,S) = 1, \quad \rho_1^2(P,S) = \begin{cases} 0 \text{ if } P < a \\ 1 \text{ if } P > a \end{cases}$$

$$\sigma_2(P,S,D) = 1, \; \sigma_2(P,S,I) = \begin{cases} 0 \text{ if } P < a \\ 1 \text{ if } P > a \end{cases}$$

Proof
Strategies in the perturbed game are indicated by a lower index ε.
Agent f_1^{1S} has the inferior choice D since this choice yields him at most c, whereas I yields at least 1. So he is fixed at $\rho_{1\varepsilon}^1(P,S) = 1-\varepsilon$ in the perturbed game. Now consider agent g_2^{SD}. He has the best reply S if:

$$-\varepsilon(1-P)-P(1-\rho_{1\varepsilon}^2(P,S)) \geq \varepsilon(1-P)(a-1)+P(1-\rho_{1\varepsilon}^2(P,S))(a-2)$$

or, equivalently, if:

$$1-\rho_{1\varepsilon}^2(P,S) \geq \frac{(1-P)a}{P(1-a)}\,\varepsilon \qquad \text{(B.1)}$$

Suppose P > a. Then the right-hand side of (B.1) is smaller than ε and the left-hand sight is at least ε. Thus, if P>a (B.1) always holds. Consequently, g_2^{SD} has the inferior choice N and is fixed at $\sigma_{2\varepsilon}(P,S,D) = 1-\varepsilon$. The remaining 2x2-game between f_1^{2S} and g_2^{SI} has two pure strategy equilibria (I,S) and (D,N). The first equilibrium yields the (unperturbed) payoffs 2 for f and 0 for g. The second yields 1+cP for f and a-P-aP for g. Since P>a, the first equilibrium payoff dominates the second one. Thus, $\rho^2_{1\varepsilon}(P,S) = \sigma_{2\varepsilon}(P,S,I) = 1-\varepsilon$ is obtained.

Suppose P < a. Now (B.1) does not always hold. But consider agent g_2^{SI}. He has the best reply N if: $(1-\varepsilon)(1-P)a + P\rho^2_{1\varepsilon}(P,S) \geq 0$, or, equivalently, if:

$$\rho^2_{1\varepsilon}(P,S) \leq \frac{(1-P)a}{P(1-a)}(1-\varepsilon) \qquad (B.2)$$

Since P<a, the right-hand side of (B.2) is always greater than $1-\varepsilon$ and the left-hand side is at most $1-\varepsilon$. This implies (B.2). Hence, agent g_2^{SI} is fixed at $\sigma_{1\varepsilon}(P,S,I) = \varepsilon$. The remaining 2x2-game between f_1^{2S} and g_2^{SD} has two pure equilibria, namely (I,N) and (D,S). (The first of which is only strong in the perturbed game.) Because there is no payoff dominance between them, we consider the mixed equilibrium of the perturbed game, which is given by $\rho^2_{1\varepsilon}(P,S) = 1 - a(1-P)[P(1-a)]^{-1}\varepsilon$ and $\sigma_{2\varepsilon}(P,S,D) = (1-c)(1-b)^{-1} + \varepsilon$. For sufficiently small ε, the sum of the probabilities that f_1^{2S} chooses D and that g_2^{SD} chooses S is strictly less than 1. According to the properties mentioned in the introduction of this appendix (D,S) is selected as the solution of the 2x2-game. We obtain $\rho^2_{1\varepsilon}(P,S) = \varepsilon$ and $\sigma_{2\varepsilon}(P,S,D) = 1-\varepsilon$, if P<a.

By letting ε go to zero the claim of the Lemma follows immediately.

Lemma 2

The HS solution for the non-concession cell is as follows:

$$\rho^1_1(P,N) = \begin{cases} 1 - P(1-P)^{-1}(a^{-1}-1) & \text{if } P < a \\ 0 & \text{if } P > a \end{cases}$$

$$\rho^2_1(P,N) = 0$$

$$\sigma_2(P,N,I) = \begin{cases} 0 & \text{if } P < a \\ b & \text{if } P > a \end{cases}$$

$$\sigma_2(P,N,D) = \begin{cases} 1-b & \text{if } P < a \\ 1 & \text{if } P > a \end{cases}$$

Proof

We start with the determination of the best replies.

Agent f_1^{1N} has the best reply I if:

$$\sigma_{2\varepsilon}(P,N,D) \leq \sigma_{2\varepsilon}(P,N,I) + 1 - b \qquad (B.3)$$

Agent f_1^{2N} has the best reply I if:

$$\sigma_{2\varepsilon}(P,N,D) \leq \sigma_{2\varepsilon}(P,N,I) - b/(1-b) \qquad (B.4)$$

Agent g_2^{NI} has the best reply S if:

$$\rho^2_{1\varepsilon}(P,N) \geq (1-P)a[P(1-a)]^{-1}\rho^1_{1\varepsilon}(P,N) \qquad (B.5)$$

Ag nt g_2^{ND} has the best reply S if:

$$\rho^2_{1\varepsilon}(P,N) \le (1-P)a[P(1-a)]^{-1}\rho^1_{1\varepsilon}(P,N) + (P-a)[P(1-a)]^{-1} \qquad (B.6)$$

Suppose $b \ge 1/2$. In that case (B.4) never holds, so agent f_1^{2N} has the inferior choice I and is fixed at $\rho^2_{1\varepsilon}(P,N)=\varepsilon$. Consider the subcase P>a. After fixing f_1^{2N}, (B.6) always holds for sufficiently small ε. Consequently, g_2^{NI} is fixed at $\sigma_{2\varepsilon}(P,N,D)=1-\varepsilon$. The remaining 2x2-game between f_1^{1N} and g_2^{NI} has a unique equilibrium given by:

$\rho^1_{1\varepsilon}(P,N) = P(1-a)[(1-P)a]^{-1}\varepsilon$ and $\sigma_{2\varepsilon}(P,N,I) = b-\varepsilon$.

Now, consider the subcase $b\ge1/2$ and $P<a$. After fixing f_1^{2N}, (B.5) could never hold, so agent g_2^{NI} has the inferior choice S and is fixed at $\sigma_{2\varepsilon}(P,N,I)=\varepsilon$. The remaining 2x2-game between agents f_1^{1N} and g_2^{ND} has the following unique equilibrium:

$\rho^1_{1\varepsilon}(P,N) = (a-P)[a(1-P)]^{-1} + P(1-a)[a(1-P)]^{-1}\varepsilon$ and $\sigma_{2\varepsilon}(P,N,D) = 1-b+\varepsilon$.

Letting ε go to zero, the claim follows immediately for the case $b \ge 1/2$.

Now consider the more difficult case $b < 1/2$ in which no inferior choices exist. Figure B1 illustrates (B.3) and (B.4) if $b<1/2$. The letters above and below the indifference curves indicate the best reply regions for the respective choices of the agent associated with this curve. The inner square corresponds to the ε-perturbed game. Similarly, Figures B2 and B3 illustrate (B.5) and (B.6), but here we have to distinguish between P>a and P<a.

Since there are no intersections of the indifference curves, we can conclude that *both* agents of a player cannot be indifferent simultaneously. From the fact that no indifference curve intersects with one of the corners of the inner square (perturbed game) it follows that it is impossible that one agent is indifferent while none of the agents of the other players is indifferent, because then they would have to play a "corner" strategy combination. Consequently, in equilibrium points of the perturbed game either no agent is indifferent (all four agents play ε-extreme strategies) or exactly two agents (one of each player) are indifferent. With the help of this conclusion we can list all strategy combinations for g's agents, the best replies against these for f's agents (with the help of Figure B1), and the best replies against these best replies for g's agents (with the help of Figures B2 and B3). If the "last" best replies are equal to the initial strategy combination, an equilibrium of the perturbed game is found. Tables B1 and B2 present the lists. The symbol "i" is used for a completely mixed strategy of an indifferent agent, chosen appropriately to make one agent of the other player indifferent too.

$(\sigma_{2\varepsilon}(\cdot,I),\sigma_{2\varepsilon}(\cdot,D))$		$(\rho^1_{1\varepsilon}(\cdot),\rho^2_{1\varepsilon}(\cdot))$		$(\sigma_{2\varepsilon}(\cdot,I),\sigma_{2\varepsilon}(\cdot,D))$	
$(1-\varepsilon,1-\varepsilon)$	→	$(1-\varepsilon,\ \varepsilon)$	→	$(\varepsilon,1-\varepsilon)$	no equilibrium
$(1-\varepsilon,\ i)$	→	$(1-\varepsilon,\ i)$	→	$(i,1-\varepsilon)$	no equilibrium
$(1-\varepsilon,\ \varepsilon)$	→	$(1-\varepsilon,1-\varepsilon)$	→	$(1-\varepsilon,1-\varepsilon)$	no equilibrium
$(i,1-\varepsilon)$	→	$(i,\ \varepsilon)$	→	$(i,1-\varepsilon)$	EQUILIBRIUM
$(i,\ \varepsilon)$	→	$(1-\varepsilon,\ i)$	→	$(i,1-\varepsilon)$	no equilibrium
$(\varepsilon,1-\varepsilon)$	→	$(\varepsilon,\ \varepsilon)$	→	$(1-\varepsilon,1-\varepsilon)$	no equilibrium
$(\varepsilon,\ i)$	→	$(i,\ \varepsilon)$	→	$(i,1-\varepsilon)$	no equilibrium
$(\varepsilon,\ \varepsilon)$	→	$(1-\varepsilon,\ \varepsilon)$	→	$(\varepsilon,1-\varepsilon)$	no equilibrium

Table B1. Equilibrium analysis of the perturbed non-concession cell if $b < 1/2$ and $P > a$ (see Figures B1 and B2).

Table B1 demonstrates that if $b<1/2$ and $P>a$ the non-concession cell has a unique uniformly perfect equilibrium with $\sigma_{2\varepsilon}(P,N,D) = 1-\varepsilon$ and $\rho^2_{1\varepsilon}(P,N) = \varepsilon$. The specific values for the mixed strategies of agents f_1^{1N} and g_2^{NI} are obtained from (B.3) and (B.5): $\rho^1_{1\varepsilon}(P,N) = P(1-a)[(1-P)a]^{-1}\varepsilon$ and $\sigma_{2\varepsilon}(P,N,I) = b-\varepsilon$. This unique uniformly perfect equilibrium (hence, the solution) is indicated in Figures B1 and B2 by black points.

$(\sigma_{2\varepsilon}(\cdot,I),\sigma_{2\varepsilon}(\cdot,D))$		$(\rho^1_{1\varepsilon}(\cdot),\rho^2_{1\varepsilon}(\cdot))$		$(\sigma_{2\varepsilon}(\cdot,I),\sigma_{2\varepsilon}(\cdot,D))$	
$(1-\varepsilon,1-\varepsilon)$	→	$(1-\varepsilon,\ \varepsilon)$	→	$(\varepsilon,1-\varepsilon)$	no equilibrium
$(1-\varepsilon,\ i)$	→	$(1-\varepsilon,\ i)$	→	$(\varepsilon,\ i)$	no equilibrium
$(1-\varepsilon,\ \varepsilon)$	→	$(1-\varepsilon,1-\varepsilon)$	→	$(\varepsilon,\ \varepsilon)$	no equilibrium
$(i,1-\varepsilon)$	→	$(i,\ \varepsilon)$	→	$(\varepsilon,\ i)$	no equilibrium
$(i,\ \varepsilon)$	→	$(1-\varepsilon,\ i)$	→	$(\varepsilon,\ i)$	no equilibrium
$(\varepsilon,1-\varepsilon)$	→	$(\varepsilon,\ \varepsilon)$	→	$(\varepsilon,\ \varepsilon)$	no equilibrium
$(\varepsilon,\ i)$	→	$(i,\ \varepsilon)$	→	$(\varepsilon,\ i)$	EQUILIBRIUM
$(\varepsilon,\ \varepsilon)$	→	$(1-\varepsilon,\ \varepsilon)$	→	$(\varepsilon,1-\varepsilon)$	no equilibrium

Table B2. Equilibrium analysis of the perturbed non-concession cell if $b < 1/2$ and $P < a$ (see Figures B1 and B3).

Likewise, there is a unique uniformly perfect equilibrium if $P<a$, as Table B2 shows. Here $\sigma_{2\varepsilon}(P,N,I) = \varepsilon$ and $\rho^2_{1\varepsilon}(P,N) = \varepsilon$. (B.3) and (B.6) yield: $\rho^1_{1\varepsilon}(P,N) = (a-P)[a(1-P)]^{-1}+P(1-a)[a(1-P)]^{-1}\varepsilon$ and $\sigma_{2\varepsilon}(P,N,D) = 1-b+\varepsilon$. Figures B1 and B3 indicate this equilibrium by black squares.

Hence, there is no difference between the solutions strategies for the cases $b \geq 1/2$ and $b < 1/2$. Let ε go to zero and the claim of the Lemma follows immediately. □

<u>Lemma 3</u>

The HS solution prescribes the following strategy for agent g_1

$$\sigma_1(P) = \begin{cases} 0 \text{ if } P < \min\{a,\alpha(a)\} \\ 1 \text{ if } P > \min\{a,\alpha(a)\} \end{cases}$$

Proof

Easy calculation shows that g's solution payoffs in the concession cell are 0 if $P > a$ and $a-P(1+a)$ if $P < a$. In the non-concession cell g obtains $a-1$ if $P > a$ and $2a-P(1+a^{-1})$ if $P < a$. Consequently, agent g_1 chooses S if $P > a$. If $P < a$, agent g_1 prefers S only if $P > \alpha(a)$, as simple computation shows.

Proof of Proposition 7

The proposition summarizes the results of Lemmata 1, 2, and 3.

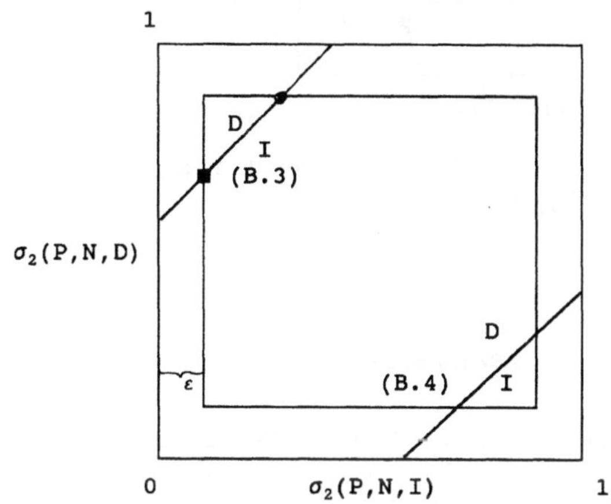

Figure B1. Example (b=1/3) for the best reply structures of agents f_1^{1N} and f_1^{2N} if $b < \frac{1}{2}$.

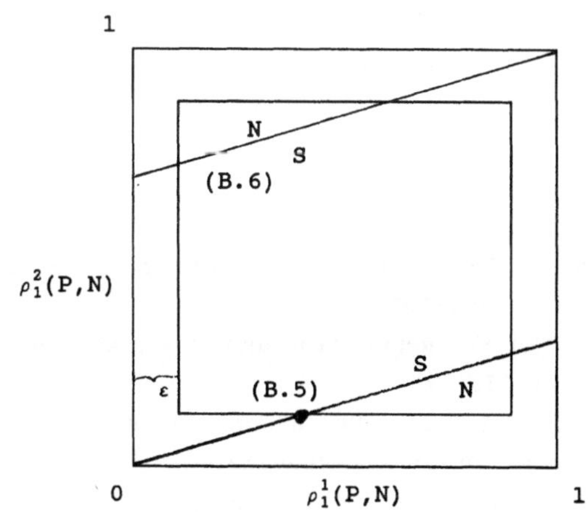

Figure B2. Example (P=2/3,a=1/3) for the best reply structures of agents g_2^{NI} and g_2^{ND} if $P > a$.

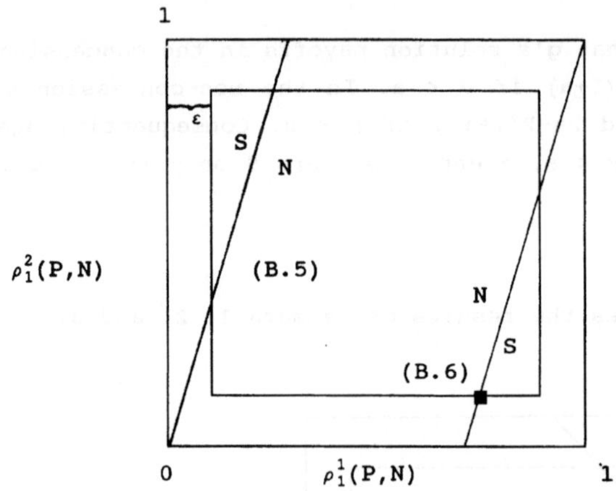

Figure B3. Example (P=1/3,a=2/3) for the best reply structures of agents g_2^{NI} and g_2^{ND} if P < a.

REFERENCES

Alt, J., Calvert, R., and Humes, B. (1988). Reputation and hegemonic stability: a game-theoretic analysis. *American Political Science Review* 82: 445-466.

Banks, J., and Sobel, J. (1987). Equilibrium selection in signalling games. *Econometrica* 55: 647-661.

Calvert, R. (1987). Reputation and legislative leadership. *Public Choice* 55: 81-119.

Harsanyi, J. (1967-8). Games of incomplete information played by Bayesian players, Part I, II and III. *Management Science* 14: 159-182, 320-334, 486-502.

Harsanyi, J., and Selten, R. (1988). *A general theory of equilibrium selection in games*. Cambridge: MIT press.

Kohlberg, E., and Mertens, J.-F. (1986). On the strategic stability of equilibria. *Econometrica* 54: 1003-1038.

Kreps, D., and Wilson, R. (1982a). Reputation and imperfect information. *Journal of Economic Theory* 27: 253-279.

Milgrom, P., and Roberts, J. (1982). Predation, reputation and entry deterrence. *Journal of Economic Theory* 27: 280-312.

Potters, J., and Van Winden, F. (1989). Modelling political pressure as transmission of information. Research Memorandum 8905, University of Amsterdam.

Selten, R. (1978). The chain store paradox. *Theory and Decision* 9: 127-159.
Van Damme, E. (1987). *Stability and perfection of Nash equilibria*. Berlin: Springer Verlag.

The Distribution of Foreign Language Skills as a Game Equilibrium

Reinhard Selten

Jonathan Pool

1. Introduction

The birth, death, growth, and shrinkage of languages over millennia has given us a world containing about three thousand living languages, whose speakers number from 1 up to several hundred million. Our current knowledge of what causes a language to gain more speakers than it loses or lose more speakers than it gains is limited to a few generalizations about bivariate, more–less effects (see Dressler, 1982; Laponce, 1984; Lieberson, 1982). One important generalization is that the children of two native speakers of the same language tend to acquire that native language unless outside the home the language is rarely used or is despised. In addition, persons who spend a few years or more in a milieu (e.g., neighborhood, school, or workplace) where a language other than their native language is the main language tend to add the other language to their repertoire. Persons tend to learn a language through deliberate study (in contrast with immersion in its milieu) when the language is spoken by many persons, has widely distributed speakers, has wealthy and powerful speakers, and has a prestigious literature, art, and history. Languages tend to lose speakers through death, of course, but also through forgetting by their native and nonnative speakers. Forgetting tends to take place among persons who are not in contact with other speakers of the language or whose rewards for using the language are small or negative. There is little evidence as to whether the difficulty of a language or its effectiveness as an instrument of thought and communication influences its acquisition of new speakers or its loss of former speakers.

As Vaillancourt (1985, p. 18) points out, almost all attempts to model aspects of the distribution of language skills in a population have started with the assumption that this distribution is fixed. Given a population distribution of language skills and some mechanism whereby costs of production or benefits of consumption depend on this distribution, one can derive predictions about the production and consumption of linguistically specialized products (Hočevar, 1975; Vaillancourt, 1985a), the earnings of persons with different repertoires of language skills (Sabourin, 1985; Vaillancourt & Lacroix, 1985), and the distributions of language skills within labor markets and firms (Breton & Mieskowski, 1975; Sabourin, 1985).

In the short term, this assumption of a fixed language-skill distribution in a population is realistic, but over a period of years the distribution of language skills changes, partly as a consequence of choices that persons make about which languages to impart to their children, which languages to learn, and how thoroughly to learn languages. Grenier (1985) and Lang (1986) have attempted to model such choices.

Grenier's (1985) model of a two-language economy assumes that each person's earnings level is a function of (1) the person's native language, (2) whether the person is bilingual, and (3) other attributes of the person. A person is assumed to be bilingual if and only if the present value of the increase in earnings for that person resulting from being bilingual exceeds the present value of the cost to that person of becoming bilingual. Using some additional assumptions about the form of the earnings function, the model estimates what the earnings of any member of a population would be if that member's monolingual/bilingual status were reversed and all the other attributes of the person were held constant. The Grenier model uses population data to estimate the parameters of the prediction equation, and for any set of population data it derives parameters. Thus, the model does not make falsifiable predictions of the distribution of language skills in a population. It does not predict that any distribution of language skills is by itself impossible, and it does not predict that any distribution of language skills is incompatible with some distribution of other attributes.

Lang's (1986) first model of a two-language labor market in equilibrium assumes that all employers have the same native language. Employers choose whether to hire workers with the same native language as theirs or the other native language. In the latter case, either these workers must be bilingual or the employer must become bilingual. The wage of bilingual workers is assumed to compensate them exactly for the cost of becoming bilingual. In a more elaborate model, Lang assumes that most but not all employers have the same native language, and that employers hire not only workers but also supervisors. The workers are assumed to be monolingual, and the supervisor in any firm must be bilingual if the employer, the supervisor, and the workers do not all have the same native language. Supervisors' wages compensate them exactly for their supervisorial training, and bilingual supervisors are further compensated exactly for the cost of learning the second language. Lang finds in both models that only native speakers of the language that has few or no employers become bilingual. Their workers are paid less, while their supervisors are paid more, than native speakers of the language that has most or all of the employers. One peculiarity of Lang's models is their treatment of the linguistic distribution of employers as fixed. The overrepresentation of employers with one of the two native languages, combined with the assumed need for employers and their employees to understand one another, creates linguistically differentiated wage rates. The lower wages paid to workers of one native language makes it possible for the employers having that native language to make higher profits. Lang's models do not assume that such an advantage leads to the elimination of the disproportional representation of language groups among employers and

Figure 1. Partition of hypothetical population into native-language communities

a consequent elimination of the wage disparity. In addition, Lang's assumption of uniform learning costs precludes any prediction of an equilibrium in which at least some persons with each native language learn the other language. Yet we know that reciprocal language learning takes place, and it would be interesting to speculate about the assumptions that would predict an equilibrium that exhibits it.

In this paper we do not restrict ourselves to earnings as a mechanism or to firms as a milieu of the incentive to learn languages. We also abandon the two-language restriction imposed by both Grenier and Lang. Our aim is to predict some general features of distributions of nonnative language skills. We shall impose no limit on the number of languages in existence. We shall permit languages with native speakers and languages without native speakers to exist. The most important restriction from others' models that we keep intact is the treatment of the distribution of **native** languages as fixed.

Our model assumes that there is a positive number of languages in the world. At least one of these languages is natural, i.e. a language that has native speakers. There may also be any number of auxiliary languages, i.e. languages with no native speakers. Since everyone is assumed to have one and only one native language, the world population can be partitioned into communities, each community consisting of the native speakers of a particular language. We represent the members of the world population as a continuum of size 1. We represent the ith community as a continuum of size a_i. We choose to identify each member of the ith community with a number s such that $i \leq s \leq i + a_i$. Geometrically, the world population can be represented as a set of n line segments, where n is the number of communities. The ith community is represented as the line segment from i to $i + a_i$. Figure 1 gives an example of such a geometric representation for a world of four communities. The sum of the lengths of the four line segments is 1.

We model this continuum as a continuum of players of a noncooperative normal-form game. Each player chooses what set of additional (non-native) languages to learn. We consider only pure strategies. Thus, if there are m languages in the world, each player (represented as a point on one of the line segments) has a set of 2^{m-1} pure strategies among which to choose. A player in figure 1, for example, has eight possible pure strategies if there are no auxiliary languages, sixteen if there is one auxiliary language, and so forth.

The players' payoffs are the net benefits they derive from learning the languages that they learn. The positive term in a payoff is the player's communicative benefit. The negative term is the player's learning cost. A person's payoff is the difference between

		2 + *t*							
		None		1		3		1 & 3	
1 + *s*	None	0	0	$b_{12}(t)$	$b_{21}(s)$	0	0	$b_{12}(t)$	$b_{21}(s)$
	2	$b_{12}(t)$	$b_{21}(s)$	$b_{12}(t)$	$b_{21}(s)$	$b_{12}(t)$	$b_{21}(s)$	$b_{12}(t)$	$b_{21}(s)$
	3	0	0	$b_{12}(t)$	$b_{21}(s)$	$b_{12}(t)$	$b_{21}(s)$	$b_{12}(t)$	$b_{21}(s)$
	2 & 3	$b_{12}(t)$	$b_{21}(s)$	$b_{12}(t)$	$b_{21}(s)$	$b_{12}(t)$	$b_{21}(s)$	$b_{12}(t)$	$b_{21}(s)$

Figure 2. Weight of two players in one another's communicative benefits

these: the communicative benefit minus the learning cost.

A player's communicative benefit is proportional to the importance-weighted fraction of the world population that shares at least one language with the player. Any language community has an importance function that assigns some importance to each member of the world population. To compute a player's communicative benefit, we use the importance function of the player's community to give a weight to everyone in the world with whom the person shares knowledge of at least one language. We give a weight of 0 to everyone with whom the person does not have a common language. The integral of the world population, thus weighted, is the player's communicative benefit.

To see how the communicative benefit is defined, consider a world with 2 native languages and one auxiliary language. Let $1 + s$ be one of the players in community 1, and $2 + t$ be one of the players in community 2. Suppose that community 1's importance function assigns an importance of $b_{12}(t)$ to $2 + t$, while community 2's importance function assigns an importance of $b_{21}(s)$ to $1 + s$. Then figure 2 shows the weight that each of these two players has in the other's communicative benefit, depending on the pair of strategies that they adopt. The rows are labeled with the four possible strategies for player $1 + s$ (learn no additional language, learn language 2, etc.), and the columns are labeled with the four possible strategies for player $2 + t$. The left entry in each cell is the weight of player 2 + t in player $1 + s$'s communicative benefit, and the right entry in each cell is the weight of player $1 + s$ in player $2 + t$'s communicative benefit.

The importance function is intended to reflect the fact that economic, geographical, and other attributes of a person affect the value of being able to communicate with the person. When a person has unequal importance to several other persons, a typical reason is differences in geographical proximity. Since language communities tend to be geographically concentrated, we allow importance functions to differ among language communities but not within them. With this definition of communicative benefit, any two players who belong to the same community and adopt the same strategy must get the same communicative benefit.

A player's learning cost is the product of two factors: a linguistic factor and a personal factor. The linguistic factor depends only on a person's languages: native and additional. Any two players who have the same native language and learn the same set of additional languages share the same linguistic factor. By assumption, the linguistic factor is always nonnegative, and it is 0 for any player who learns no additional language. These assumptions about the linguistic factor reflect three facts: (1) some languages are more difficult to learn than others, (2) the effort required in learning a combination of languages is not necessarily the sum of the efforts of learning the languages in the combination, and (3) the difficulty of learning a combination of languages depends on the native language of the learner.

The other component of a player's learning cost is a personal factor, which reflects the differences among language learners in aptitudes and motivations. The personal factor is a nonnegative number assigned to a player. For simplicity of analysis, it is assumed that within any language community no player has the same personal factor as any other player. Thus, any two players with the same native language who learn no additional language have the same learning cost (0), but if they learn the same additional language or languages they must have different learning costs. We have already assumed that their communicative benefit is identical. Therefore, with unequal learning costs they must have different payoffs.

Our first substantively interesting result describes a feature of the players' payoffs. The payoff to any player is a function of the strategy of that player and the strategies of all players who are members of other communities. So, a player's payoff is independent of the strategies of the other members of that player's own community. The reason is that a player's communicative benefit depends on the subset of the world population that shares at least one language with the player. But every member of the player's community shares a language with the player (their native language). This fact cannot be changed by any strategy that another member of this community adopts. Thus, the only way in which a player can affect the payoff of another player in the same community is indirectly: by affecting the choices of players in other communities. For example, in our model a native speaker of Japanese does not directly affect the payoff of another native speaker of Japanese. But, if one native speaker of Japanese learns English, this choice may reduce the incentive of a native speaker of English or some other language to learn Japanese, and the reduced learning of Japanese that results from this incentive reduction may affect the payoffs of other native speakers of Japanese.

Our analysis proceeds by examining the features of best replies and group best replies. A best reply is a strategy that maximizes the adopting player's payoff, given the combination of strategies adopted by all other players. The combination of the strategies adopted by all the members of a community is called a group strategy. If every player in a community adopts a best reply to the combination of all other players' strategies, then we say that the community's group strategy is a group best reply. Referring to group strategies and group best replies is only an analytical device, since the players in our game are individuals, not communities.

Our next result describes a regularity in best replies. Consider the members of one language community. For any combination of the other communities' members' strategies, each member of this community has at least one strategy that is a best reply. A strategy that is a best reply for one member may not be a best reply for some other member. We show, however, that whenever a strategy is a best reply for two members it is also a best reply for all members whose personal factors of their learning costs lie between those of the two members. In terms of a geometric representation, we can arrange the members of a community in order of increasing personal factors on the line segment that represents their community. If we do this, then any strategy that is a best reply at more than one point in a community is a best reply everywhere in some interval within the line segment, and nowhere else in the community. This result leads to an empirical prediction that, among the native speakers of any language, those who learn the same set of foreign languages will tend to be more similar in the extent of their (dis)inclination and (in)ability to learn languages than is true for any more loosely defined set of language learners. For example, the variance in scores on a language aptitude test should be greater among all those native speakers of German who have learned English and any one other foreign language than among the subsets of native German speakers who have learned English and French, who have learned English and Russian, etc.

We build on this result to show that any group best reply exhibits a negative association between the two factors of the learning cost. When a community adopts a group best reply, for any two members in the community whose learning costs have unequal linguistic factors, the one with the higher linguistic factor (the member who learns the more expensive set of languages) has the lower personal factor (finds it easier to learn languages). Suppose we arrange the members of a community from left to right in order of increasing personal factors, and suppose the community adopts a group best reply which involves more than one strategy. As we move from left to right, we find personal factors everywhere rising, while at each point where the linguistic factor changes it declines. An example of this pattern for a world with two natural languages and one auxiliary language is illustrated in figure 3.

As we move from left to right in a community that adopts a group best reply, we must also find communicative benefit decreasing whenever there is an decrease in the linguistic factor. If this relationship did not exist, then the community would not be adopting a group best reply, because some member of the community could get a higher payoff by changing to a strategy with a lower linguistic factor.

As the linguistic factor decreases moving from left to right, not only do communicative benefits decrease, but so do payoffs. It might be imagined that this is not always true.

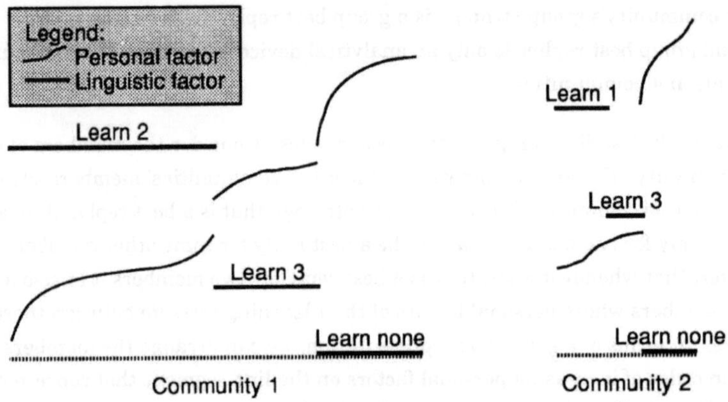

Figure 3. Illustration of group best replies in a two-community, three-language world

Under some conditions, the drop in the linguistic factor of the learning cost might be steep enough to counteract both the drop in the communicative benefit and the rise in the personal factor. But we show that there are no conditions that allow payoffs to rise or remain constant as the linguistic factor decreases. This result can be interpreted as saying that, in a community adopting a group best reply, those who learn more difficult sets of languages get higher profits, despite the difficulty of the languages they learn.

Having determined some features of best replies, we move on to show that it is possible for every player to adopt a best reply at the same time. In other words, at least one equilibrium point exists. There are several steps in proving this result.

First, we arbitrarily resolve any ties that may exist in the linguistic costs of the possible strategies for the players in any community. In other words, we label the possible strategies in a community with the numbers 1 to 2^{m-1}, in order of decreasing linguistic cost. Whenever we encounter two or more strategies having the same linguistic cost, we number them in any order.

The next step is to construct a new game, whose players are the communities. In this new, or aggregated, game, we require the members of any community to adopt individual strategies that are numbered in the same order as the members' personal costs are ordered. Thus, if any two members of a community adopt different individual strategies, the member with the higher personal factor must adopt the higher-numbered strategy. This requirement limits the set of possible group strategies for the communities playing the aggregated game. A community must be partitioned into 2^{m-1} intervals, such that the kth interval is the locus of members adopting the individual strategy numbered k. We limit the communities' strategy sets even further by defining a group strategy only as the set of upper bounds of these 2^{m-1} intervals. To see that this definition restricts the set of strategies, consider a point x in a community and suppose that the members to the immediate left of x adopt strategy k and the members to the immediate right of x adopt strategy $k + w$. Then the member precisely at point x can adopt any of the strategies k, k + 1, ..., $k + w$. But it makes no difference to the group strategy what the individual strategy of this border member is, since the upper bounds of the intervals are unaffected. A community i's strategy in the aggregated game is thus a nondecreasing set of 2^{m-1} border points, each located between i and $i + a_i$ and the last of them being equal to $i + a_i$.

The payoffs to the community players in the aggregated game are the respective integrals of the payoffs to their individual members in the original game. Each combination of strategies in the aggregated game induces a unique payoff to each community, despite the fact that the group strategies may leave the individual strategies of their border members incompletely specified. The reason is that the border members are only an infinitesimal fraction of all the members. The payoffs of the 2^{m-1} border members of a community (or of any finite number of its members) do not affect the integral of the payoffs of all the members of the community.

The last step in proving the existence of an equilibrium in the original game involves establishing relationships between the original and the aggregated games. We first show that whenever there is an equilibrium point in an aggregated game there is also an equilibrium point in the original game to which the aggregated game corresponds. We then show that for each aggregated game there is always at least one equilibrium point. But for each original game at least one corresponding aggregated game exists. These findings imply that there is always at least one equilibrium point in any original game.

In this paper we say nothing about the number and nature of the equilibria whose existence we prove. As an aid to intuitions, however, we picture one equilibrium in figure 4. The example is a world with three language communities, whose shares of the population are 0.5, 0.35, and 0.15, respectively. One auxiliary language exists, implying that there are four languages. For each player, three of the languages are open to choice, so each player has eight available strategies. The importance weight function of each community assigns identical weights to all players; thus, everyone is equally important for everyone else as a potential communication partner. The linguistic factors of the learning costs are additive functions of the languages contained in the learned sets of languages, with each natural language contributing 1.8 and the auxiliary language contributing 0.6 to the linguistic factor, regardless of the learner's native language. Finally, the personal factors of the learning cost are distributed uniformly in each community from 0 to 1, shown as diagonal dashed lines in the figure.

In the equilibrium shown for this world in figure 4, only three of the eight possible strategies in each language community are chosen. These are (1) to learn both nonnative natural languages, (2) to learn the nonnative natural language with the larger number of

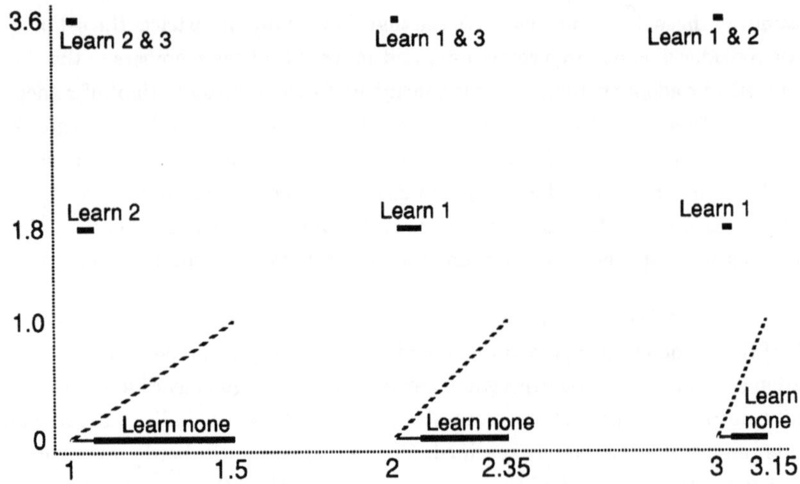

Figure 4. Illustration of equilibrium in a three-community, four-language world

native speakers, or (3) to learn nothing. The players are arranged in order of increasing personal factors of learning cost, with this personal factor plotted by dashed lines. When the players are so arranged, each community is partitioned into three convex sets of players, all players in the ith set choosing the ith strategy. The boundaries between the adjacent sets of players are, in distance from the community's origin, 0.0292 and 0.0729 in community 1, 0.0205 and 0.0875 in community 2, and 0.0219 and 0.0448 in community 3. The sets of players choosing these strategies are plotted at heights equal to the linguistic factors of their learning costs.

In this example, the proportion of each community that learns two nonnative languages is largest in the smallest community and smallest in the largest community. The same is true for the proportion that learns at least one nonnative language. We also notice that no player learns the auxiliary language. These results match well-known tendencies: those who learn nonnative languages tend to choose the most widely spoken ones; members of small language communities are more likely to learn foreign languages than members of large communities; and few learn auxiliary languages, despite their relatively low learning costs.

Another insight we may derive from our results concerns the phenomenon of apparently deliberate foreign-language incompetence. It has been claimed that the native speakers of certain languages (such as English and French) succeed in getting the native speakers of other languages to learn their languages. The strategy allegedly used to attain this effect is the intentional neglect of foreign-language learning and the ostentatious exhibition of low language-learning aptitudes. In the terms of our model, this claim can be rephrased into a suggestion that the native speakers of a language benefit from having high personal factors and hence high learning costs. The predictions from our model do not support this idea. Our predictions are compatible with the possibility that a player's payoff can be an increasing function of the learning costs of the player's fellow native speakers. But payoffs to a member of a community cannot, in any equilibrium, be an increasing function of the member's learning cost. Thus, even if the language community with the highest payoffs is the one whose members have the greatest average difficulty learning foreign languages, it is the easy learners within this group who have the highest payoffs in their group.

2. The model

The model describes a world with m languages, 1, ..., m. Every individual has one and only one of these languages as his *native language*. He is a *native speaker* of that language. Every language other than an individual's native language is a *foreign language* for that individual. Some languages may be foreign to all individuals. Languages with native speakers will be called *natural* languages, and the others will be referred to as *auxiliary* languages. Let languages 1,..., n, with $n \leq m$, be the natural languages. Let M be the set $\{1, ..., m\}$ of all languages, and let N be the set $\{1, ..., n\}$ of all natural languages.

Players and strategies. The set of all individuals with native language i is the *language community* i or, for short, *community* i. Each language community i is described as a continuum of individuals represented as points in an interval $[i, i + a_i]$, with $a_i > 0$. The measure of all individuals is normalized to 1:

(1) $$\sum_{i=1}^{n} a_i = 1.$$

We call a_i the *size* of language community i. Our model may be looked upon as a game with a continuum of players. The players are the members of the language communities 1, ..., n. The pure strategies are possible sets C of languages to be learned. Of course, a member of language community i always learns language i and has no choice in this respect. The set of all pure strategies C of a member $i + s$ of community i is as follows:

(2) $$M_i = \{C \mid i \in C \subseteq M\}.$$

A *group strategy* of community i is a Borel measurable function which assigns a pure strategy $\varphi_i(s)$ to every member $i + s$ of community i:

(3) $$\varphi_i\colon\ [0, a_i] \to M_i.$$

It is notationally convenient to define φ_i on $[0, a_i]$ rather than on $[i, i + a_i]$. The set of all group strategies of community i is denoted by Φ_i. A *strategy combination*

(4) $\varphi = (\varphi_1, \ldots, \varphi_n)$

is an n-tuple of group strategies $\varphi_i \in \Phi_i$. The set of all strategy combinations is denoted by Φ.

An *i-incomplete* strategy combination φ_{-i} is an $(n-1)$-tuple of group strategies $\varphi_j \in \Phi_j$ with $j \neq i$:

(5) $\varphi_{-i} = (\varphi_1, \ldots, \varphi_{i-1}, \varphi_{i+1}, \ldots, \varphi_n)$.

The set of all i-incomplete strategy combinations is denoted by Φ_{-i}. The i-incomplete strategy combination φ_{-i} whose components agree with the corresponding components of a strategy combination φ is called the i-incomplete combination *in* φ.

Learning costs. There are two influences on a player's payoff: "communicative benefit" and "learning cost". We first turn our attention to learning cost. Consider a member $i + s$ of community i who chooses C. His *learning cost* $K_{i+s}(C)$ is modeled as follows:

(6) $K_{i+s}(C) = f_i(s)g_i(C)$,

where f_i is a bounded monotonically increasing nonnegative function defined on $[0, a_i]$ and g_i is a function defined on M_i with

(7) $g_i(C) \geq 0$ for all $C \in M_i$

and

(8) $g_i(\{i\}) = 0$.

Learning costs are the product of a *personal factor* $f_i(s)$ and a *linguistic factor* $g_i(C)$. The personal factor summarizes influences like aptitude. We think of the members of community i as arranged according to their personal factors. Therefore the monotonicity assumption on f_i is not very restrictive.

Communicative benefit. We now turn our attention to the communicative benefit of language learning. Every other individual $j + t$ with whom a player $i + s$ has a common language provides a communication opportunity, which contributes to the communicative benefit of player $i + s$. The importance of the opportunity to communicate with $j + t$ may depend on various characteristics of $j + t$ such as income or geographical distance. Therefore, in the computation of communicative benefit, communicative opportunities are weighted by an importance factor $b_{ij}(t)$. In order to express the communicative benefit we make use of the following notation. For any two subsets C and D of M define:

(9) $$\delta(C, D) = \begin{cases} 1 \text{ for } C \cap D \neq \varnothing \\ 0 \text{ for } C \cap D = \varnothing \end{cases}.$$

Let $\varphi = (\varphi_1, \ldots, \varphi_n)$ be a strategy combination. The *communicative benefit* $U_i(C, \varphi)$ obtained by a member of community i from C if φ is played by all other individuals is defined as follows:

$$(10) \qquad U_i(C, \varphi) = \sum_{j=1}^{n} \int_0^{a_j} b_{ij}(t)\delta(C, \varphi_j(t))dt.$$

It is assumed that for $i = 1, \ldots, n$ and $j = 1, \ldots, n$ the *importance weight functions*

$$(11) \qquad b_{ij}\colon [0, a_j] \to \mathbb{R}_+$$

are bounded, nonnegative and piecewise continuous with finitely many discontinuities. This assumption together with the measurability of the φ_j guarantees the existence of the integrals on the right-hand side of (10).

Since all $C \in M_i$ contain i, we have:

$$(12) \qquad U_i(C, \varphi) = \int_0^{a_i} b_{ii}(t)dt + \sum_{\substack{j=1 \\ j \neq i}}^{n} \int_0^{a_j} b_{ij}(t)\delta(C, \varphi_j(t))dt.$$

This shows that *U*\s\do6(*i*)(*C*, φ) does not depend on φ\s\do6(*i*), but only on the *i*-incomplete combination φ_{-i} in φ. Accordingly, we shall also use the notation $U_i(C, \varphi_{-i})$:

$$(13) \qquad U_i(C, \varphi_{-i}) = U_i(C, \varphi) \text{ with } \varphi_{-i} \text{ in } \varphi.$$

Payoffs. We now define the *payoff* $H_{i+s}(C, \varphi_{-i})$ to a member $i + s$ of community i obtained against an i-incomplete strategy combination $\varphi_{-i} \in \Phi_{-i}$ for $C \in M_i$:

$$(14) \qquad H_{i+s}(C, \varphi_{-i}) = U_i(C, \varphi_{-i}) - f_i(s)g_i(C)$$

for $i = 1, \ldots, n$ and $s \in [0, a_i]$. The *payoff function* H assigns $H_{i+s}(C, \varphi_{-i})$ to every triple $(i + s, C, \varphi_{-i})$, where $i + s$ is a member of community i, where C is one of his pure strategies, and where φ_{-i} is an i-incomplete strategy combination. The pure strategy sets M_i of the members of the language communities together with H constitute a game in normal form with a continuum of players. The measurability requirement imposes a restriction on the region where payoffs are defined, but apart from this a game in normal form with a continuum of players is a straightforward modification of the usual concept of a game in normal form.

Summary of assumptions. Let $a = (a_1, \ldots, a_n)$ be the vector of the a_i and let $f = (f_1, \ldots, f_n)$ and $g = (g_1, \ldots, g_n)$ be the vectors of the functions f_i and g_i, respectively. Moreover, let b be the $n \times n$ matrix of the functions b_{ij} with $i = 1, \ldots, n$ and $j = 1, \ldots, n$. The sextuple (m, n, a, b, f, g) summarizes the elements which have to be fixed in a specification of the model. The notation $\Gamma(m, n, a, b, f, g)$ indicates the normal form game resulting from the specification (m, n, a, b, f, g). In the following we shall give a summary of the assumptions made about the elements to be specified.

1. m, the number of languages, and n, the number of natural languages, are positive integers with $m \geq n$.

2. The components a_i of a, the sizes of language communities, are positive and sum to 1.

3. For $i = 1, \ldots, n$ and $j = 1, \ldots, n$, the importance weight functions b_{ij} are bounded, nonnegative, and piecewise continuous with finitely many discontinuities.

4. For $i = 1, \ldots, n$, the personal factor $f_i(s)$ of the learning cost is a bounded monotonically increasing nonnegative function.

5. For $i = 1, \ldots, n$, the linguistic factor g_i of the learning cost has the properties $g_i(\{i\}) = 0$ and $g_i(C) \geq 0$ for all $C \in M_i$.

3. Equilibrium points and the structure of group best replies

The model will be analyzed as a noncooperative game. In this section we shall prepare for the investigation of equilibrium points in pure strategies. Definitions and statements will refer to an arbitrarily fixed game $\Gamma(m, n, a, b, f, g)$.

Best replies. Loosely speaking, a best reply is a strategy which maximizes the payoff of a player given the behavior of all others. The payoff that $i + s$ obtains by a pure strategy $C \in M_i$ depends only on $\varphi_i(s)$ and φ_{-i}. Therefore, player $i + s$ does not have to know $\varphi_i(t)$ for $t \neq s$ in order to determine his best reply. Let φ_{-i} be an i-incomplete strategy combination. Formally, $D \in M_i$ is a *best reply* of $i + s$ to φ_{-i} if we have:

$$H_{i+s}(D, \varphi_{-i}) = \max_{C \in M_i} H_{i+s}(C, \varphi_{-i}). \tag{15}$$

We say that D is a *best reply* to $\varphi = (\varphi_1, \ldots, \varphi_n)$ if D is a best reply to the i-incomplete strategy combination φ_{-i} in φ.

A group strategy $\psi_i \in \Phi_i$ is a *group best reply* or simply a *best reply* to an i-incomplete combination φ_{-i} if for every member $i + s$ of community i his strategy $\psi_i(s)$ is a best reply to φ_{-i}. We say that ψ_i is a *group best reply* or simply a *best reply* to $\varphi = (\varphi_1, \ldots, \varphi_n)$ if ψ_i is a best reply to the i-incomplete strategy combination φ_{-i} in φ. A strategy combination $\psi = (\psi_1, \ldots, \psi_n)$ is a *best reply combination* for $\varphi = (\varphi_1, \ldots, \varphi_n)$ or simply a *best reply* to φ if for $i = 1, \ldots, n$ the group strategy ψ_i is a best reply to φ.

A group strategy ψ_i is an *almost best reply* to an i-incomplete strategy combination φ_{-i}

or a strategy combination φ if for $0 \le t \le a_i$ with the exception of finitely many values of t the strategy $\psi_i(t)$ is a best reply of player $i + t$ to φ_{-i} or φ, respectively.

Equilibrium points. Since mixed strategies are not considered in this paper, the term "equilibrium point" will always refer to an equilibrium point in pure strategies. A strategy combination $\varphi = (\varphi_1, \ldots, \varphi_n)$ is an equilibrium point if φ is a best reply to itself or, in other words, if for $i = 1, \ldots, n$ the group strategy φ_i is a best reply to φ_{-i} in φ. Analogously, an almost-equilibrium point is a strategy combination which is an almost best reply to itself.

Lemma 1. Let $\varphi = (\varphi_1, \ldots, \varphi_n)$ be an almost-equilibrium point and let $\psi = (\psi_1, \ldots, \psi_n)$ be a strategy combination such that, for $i = 1, \ldots, n$ and $0 \le t \le a_i$, wherever $\varphi_i(t)$ is a best reply of $i + t$ to φ the strategy $\psi_i(t)$ agrees with $\varphi_i(t)$, and wherever $\varphi_i(t)$ is **not** a best reply of $i + t$ to φ the strategy $\psi_i(t)$ **is** a best reply of $i + t$ to φ. Then ψ is an equilibrium point.

Proof. The transition from φ to ψ involves only a finite number of changes. Therefore, we have

(16) $$H_{i+t}(C, \psi_{-i}) = H_{i+t}(C, \varphi_{-i})$$

for $i = 1, \ldots, n$, for $0 \le t \le a_i$, and for every $C \in M_i$, with φ_{-i} in φ and ψ_{-i} in ψ. Consequently, the assertion holds.

Comment. Lemma 1 shows how an almost-equilibrium point can be changed into an equilibrium point by the adjustment of the strategies of finitely many players. In order to prove the existence of equilibrium points and to explore their structural properties, it is sufficient to investigate almost-equilibrium points.

We shall now derive a useful result on the set of the members of a language community i for whom a given strategy C is a best reply to an i-incomplete strategy combination φ_{-i}. Then we shall prove a theorem which exhibits an interesting structural feature of group best replies: those who learn more difficult sets of languages have lower personal learning costs.

Lemma 2. Let φ_{-i} be an i-incomplete strategy combination and let $C \in M_i$ be a strategy for a member of language community i. Moreover, let I be the set of all t with $0 \le t \le a_i$ such that C is a best reply of $i + t$ to φ_{-i}. If the set I contains at least two points, then I is a closed, open, or half-open interval.

Proof. Assume that s and t with $s \ne t$ belong to I. For every $D \in M_i$ with $D \ne C$ we have:

(17) $$U_i(C, \varphi_{-i}) - f_i(s)g_i(C) \ge U_i(D, \varphi_{-i}) - f_i(s)g_i(D);$$

(18) $$U_i(C, \varphi_{-i}) - f_i(t)g_i(C) \ge U_i(D, \varphi_{-i}) - f_i(t)g_i(D).$$

For every t' with $s < t' < t$, the personal cost factor $f_i(t')$ is a convex linear combination of $f_i(s)$ and $f_i(t)$; in other words, for some α with $0 < \alpha < 1$ the following is true:

(19) $$f_i(t') = (1-\alpha)f_i(s) + \alpha f_i(t).$$

The sum of (17) multiplied by $(1-\alpha)$ and (18) multiplied by α yields

(20) $$U_i(C, \varphi_{-i}) - f_i(t')g_i(C) \geq U_i(D, \varphi_{-i}) - f_i(t')g_i(D).$$

Consequently, C is a best reply of $i + t'$ to φ_{-i} for every t' with $s < t' < t$. It follows that the assertion of the lemma holds.

Theorem 1 (on group best replies). Let φ_{-i} be an i-incomplete strategy combination and let φ_i be a group best reply to φ_{-i}. Moreover, let $i + s$ and $i + t$ be two members of community i with

(21) $$g_i(\varphi_i(s)) > g_i(\varphi_i(t)).$$

Then we have:

(22) $$s < t.$$

Proof. The following inequalities are consequences of the fact that $\varphi_i(s)$ and $\varphi_i(t)$ are best replies to φ_{-i}:

(23) $$U_i(\varphi_i(s), \varphi_{-i}) - f_i(s)g_i(\varphi_i(s)) \geq U_i(\varphi_i(t), \varphi_{-i}) - f_i(s)g_i(\varphi_i(t));$$

(24) $$U_i(\varphi_i(s), \varphi_{-i}) - f_i(t)g_i(\varphi_i(s)) \leq U_i(\varphi_i(t), \varphi_{-i}) - f_i(t)g_i(\varphi_i(t)).$$

Subtraction of (24) from (23) yields

(25) $$[f_i(t) - f_i(s)]g_i(\varphi_i(s)) \geq [f_i(t) - f_i(s)]g_i(\varphi_i(t))$$

or, equivalently,

(26) $$[f_i(t) - f_i(s)][g_i(\varphi_i(s)) - g_i(\varphi_i(t))] \geq 0.$$

In view of (21), the second factor on the left hand side of (26) is positive. Therefore, the first factor must be nonnegative, giving us:

(27) $$f_i(s) \leq f_i(t).$$

Since $\varphi_i(s)$ and $\varphi_i(t)$ must be different, we have $s \neq t$. Moreover, f_i is monotonically increasing. Consequently, (22) holds.

Remark. It can be seen immediately that the theorem has the following consequence: $s < t$ implies $g_i(\varphi_i(s)) \geq g_i(\varphi_i(t))$. Moreover, it follows with the help of (23) that $g_i(\varphi_i(s)) > g_i(\varphi_i(t))$ implies

(28) $U_i(\varphi_i(s), \varphi_{-i}) > U_i(\varphi_i(t), \varphi_{-i})$.

It is also worth pointing out that we have

(29) $H_{i+s}(\varphi_i(s), \varphi_{-i}) > H_{i+t}(\varphi_i(t), \varphi_{-i})$

for $s < t$ and $g_i(\varphi_i(t)) > 0$, since under these conditions $i + s$ can obtain a higher payoff than $i + t$ by choosing the same strategy, $\varphi_i(t)$.

4. Monotone group strategies and aggregated games

Theorem 1 shows that group best replies have the property that the two factors of learning costs are inversely connected. Higher personal factors are connected with lower or equal linguistic factors. Loosely speaking, we may say that a group best reply must respect the order of decreasing linguistic factors. The order of decreasing linguistic factors may be looked upon as a ranking of the individual strategies for the members of a language community. In the presence of equal linguistic factors for different individual strategies, this ranking has ties. For every possible strengthened order that one can create by arbitrarily breaking all ties, we shall define a narrower "monotonicity class" of group strategies which respect this strengthened order of decreasing linguistic factors. We shall prove that every monotonicity class contains a group best reply to every *i*-incomplete strategy combination. This result will be the basis of our construction of "aggregated games".

Within a monotonicity class a group strategy is almost completely characterized by the border points of the intervals where the same individual strategy is used. These border points form a "border point vector". Within the monotonicity class the group strategies with the same border point vector form a "border equivalence class". Two members of the same border equivalence class differ at most at finitely many points, namely the border points.

In an *aggregated game*, the players are the language communities. Payoffs are "group payoffs" defined as integrals over individual payoffs. Strategies correspond to border equivalence classes within a particular monotonicity class. Formally, strategies are border point vectors.

In the presence of equal linguistic factors for different strategies for the individuals in a language community, more than one aggregated game exists. For every way of selecting a monotonicity class for each player (i.e. for each language community), we obtain a different aggregated game.

An equilibrium point of the original game is associated with every equilibrium point of every aggregated game. Therefore, the task of proving the existence of equilibrium points can be shifted from the original game to the aggregated games.

We continue to use the definitions and notations introduced in the previous sections. All statements refer to an arbitrarily selected fixed game $\Gamma(m, n, a, b, f, g)$ whose specification satisfies assumptions 1 to 5.

Monotonicity classes. A member of community i can learn up to $m-1$ foreign languages. Therefore, the number of pure strategies in the set M_i is 2^{m-1}. Let r_i be a function which assigns one of the numbers $1, \ldots, 2^{m-1}$ to every $C \in M_i$, a different one for each $C \in M_i$. We call r_i a ranking over M_i. We say that r_i is *cost compatible* if, for every pair of strategies C and D in M_i, we have:

(30) $\quad r_i(C) < r_i(D)$ for $g_i(C) > g_i(D)$.

Let r_i be a cost compatible ranking over M_i. We say that a group strategy $\varphi_i \in \Phi_i$ is *monotone* with respect to r_i if for s and t in $[0, a_i]$ the following is always true:

(31) $\quad r_i(\varphi_i(s)) \le r_i(\varphi_i(t))$ for $s < t$.

The set of all group strategies $\varphi_i \in \Phi_i$ which are monotone with respect to r_i is denoted by $\Psi_i(r_i)$. We call $\Psi_i(r_i)$ the *monotonicity class* of r_i.

Lemma 3. Let $\Psi_i(r_i)$ be the monotonicity class of a cost compatible ranking r_i over M_i. Then $\Psi_i(r_i)$ contains a group best reply φ_i to every i-incomplete strategy combination φ_{-i}.

Proof. For every t with $0 \le t \le a_i$, let $\varphi_i(t)$ be that best reply C of $i+t$ to φ_{-i} which has the lowest rank $r_i(C)$ among all best replies of $i+t$ to φ_{-i}. Obviously, φ_i is a group best reply to φ_{-i}. We have to show that φ_i is monotone with respect to r_i. In view of theorem 1, inequality (31) holds if $g_i(\varphi_i(s))$ is different from $g_i(\varphi_i(t))$. Suppose that $g_i(\varphi_i(s)) = g_i(\varphi_i(t))$, that $s < t$, and that (31) does not hold. Then $r_i(\varphi_i(s))$ is greater than $r_i(\varphi_i(t))$. Since $\varphi_i(s)$ is the lowest ranking best reply of $i+s$ to φ_{-i}, it follows that $\varphi_i(t)$ is not a best reply of $i+s$ to φ_{-i}. Therefore, we must have:

(32) $\quad U_i(\varphi_i(s), \varphi_{-i}) - f_i(s)g_i(\varphi_i(s)) > U_i(\varphi_i(t), \varphi_{-i}) - f_i(s)g_i(\varphi_i(t)).$

In view of the fact that $g_i(\varphi_i(s))$ and $g_i(\varphi_i(t))$ are equal, we can conclude:

(33) $\quad U_i(\varphi_i(s), \varphi_{-i}) > U_i(\varphi_i(t), \varphi_{-i}).$

It follows by (29) and the equality of $g_i(\varphi_i(s))$ and $g_i(\varphi_i(t))$ that $\varphi_i(t)$ cannot be a best reply of $i+t$ to φ_{-i}. This is a contradiction. Therefore, the assertion of the lemma is true.

Border point vectors. In the following we shall look at a fixed cost compatible ranking r_i. For $k = 1, \ldots, 2^{m-1}$, the strategy $C \in M_i$ with $r_i(C) = k$ is denoted by C_{ik}. For every group strategy φ_i in the monotonicity class $\Phi_i(r_i)$ of r_i, let I_{ik} be the inverse image $\varphi_i^{-1}(C_{ik})$ of C_{ik} with respect to φ_i or in other words the set of all $t \in [0, a_i]$ with $r_i(\varphi_i(t)) = k$. We use the notation z_{ik} for the upper border point of I_{ik}; moreover, we define $z_{i0} = 0$. Of course, several border points z_{ik} may coincide. Obviously, I_{ik} is a closed, open, or half open interval between the border points $z_{i(k-1)}$ and z_{ik}. For $k = 2^{m-1}$, we always have $z_{ik} = a_i$. The vector

(34) $$z_i = (z_{i0}, \ldots, z_{i2^{m-1}})$$

is called the *border point vector* of φ_i. A border point vector always satisfies the following condition:

(35) $$0 = z_{i0} \le z_{i1} \le \ldots \le z_{i2^{m-1}} = a_i.$$

The set of all vectors z_i of the form (34) with (35) is denoted by Z_i. We call Z_i the *border point vector set* of language community i. The set of all group strategies $\varphi_i \in \Psi_i(r_i)$ which have the same border point vector z_i is denoted by $\Lambda_i(z_i)$. We call $\Lambda_i(z_i)$ the *border equivalence class* of z_i. We say that two group strategies in $\Psi_i(r_i)$ are *border equivalent* if they are in the same border equivalence class. If $z_i = (z_{i0}, \ldots, z_{i2^{m-1}})$ is the border point vector of a group strategy $\varphi_i \in \Psi_i(r_i)$, then we have:

(36) $$\varphi_i(t) = C_{ik} \text{ for } z_{i(k-1)} < t < z_{ik}$$

for $k = 1, \ldots, 2^{m-1}$. Obviously, two border equivalent group strategies in $\Psi_i(r_i)$ can differ only at the border points of their common border point vector.

Combinations of rankings. A combination $r = (r_1, \ldots, r_n)$ of cost compatible rankings contains a cost compatible ranking r_i for each language community $i = 1, \ldots, n$. An *i-incomplete* combination r_{-i} of cost compatible rankings contains a cost compatible ranking for every language community j with $j \neq i$. We say than r_{-i} is *in* r if the components of r_{-i} are the corresponding components of r. The set of all cost compatible rankings of language community i is denoted by R_i; the symbol R is used for the set of all combinations r of cost compatible rankings, and R_{-i} is the set of all i-incomplete combinations r_{-i} of cost compatible rankings.

Border point systems. In the following we shall look at a fixed combination $r = (r_1, \ldots, r_n)$ of cost compatible rankings and the i-incomplete combination r_{-i} in r. The *monotonicity class* $\Psi(r)$ of r is the set of all strategy combinations $\varphi = (\varphi_1, \ldots, \varphi_n)$ with $\varphi_i \in \Psi_i(r_i)$ for $i = 1, \ldots, n$. Analogously, the monotonicity class $\Psi_{-i}(r_{-i})$ of r_{-i} is the set of all i-incomplete strategy combinations φ_{-i} whose components φ_j are in the corresponding monotonicity classes $\Psi_j(r_j)$. Let $\varphi = (\varphi_1, \ldots, \varphi_n)$ be a strategy combination in $\Psi(r)$. For $i = 1, \ldots, n$, let z_i be the border point vector of φ_i defined above. We call

(37) $$z = (z_1, \ldots, z_n)$$

the *border point system* of φ. Analogously, the *i-incomplete border point system* z_{-i} of an i-incomplete strategy combination $\varphi_{-i} \in \Psi_{-i}(r_{-i})$ contains the border point vectors of the components of φ_{-i}. The set of all systems z of the form (37) with $z_i \in Z_i$ is denoted by Z, and the set of all i-incomplete border point systems

(38) $$z_{-i} = (z_1, \ldots, z_{i-1}, z_{i+1}, \ldots, z_n)$$

with $z_j \in Z_j$ for all components z_j of z_{-i} is denoted by Z_{-i}. The *border equivalence class* $\Lambda(z)$ of $z = (z_1, \ldots, z_n)$ is the set of all strategy combinations $\varphi \in \Psi(r)$ whose border point system is z. Similarly, the border equivalence class $\Lambda_{-i}(z_{-i})$ of an i-incomplete border point system z_{-i} is the set of all i-incomplete strategy combinations φ_{-i} whose i-incomplete border point system is z_{-i}. Two border point systems or two i-incomplete border point systems are *border equivalent* if they belong to the same border equivalence class.

Group payoffs. We continue to look at a fixed combination of cost compatible rankings $r = (r_1, \ldots, r_n)$. Let $\varphi = (\varphi_1, \ldots, \varphi_n)$ be a strategy combination in the monotonicity class $\Psi(r)$, and let $z = (z_1, \ldots, z_n)$ be the border point system of φ. The *group payoff* of a language community i for φ is defined as the integral over all individual payoffs for members of the language community. As we shall see, this group payoff is a function of z. The finitely many individual strategies at border points do not matter for the integral. Therefore, we introduce the following definition of the *group payoff* $P_i(z)$ of language community i for z:

(39) $$P_i(z) = \int_0^{a_i} H_{i+t}(\varphi_i(t), \varphi_{-i})dt.$$

It follows by (12) and (13) that communicative benefits $U_i(C, \varphi_{-i})$ can be expressed as follows:

(40) $$U_i(C, \varphi_{-i}) = \sum_{j=1}^{n} \sum_{k=1}^{2^{m-1}} \int_{z_{j(k-1)}}^{z_{jk}} b_{ij}(t)\delta(C, C_{jk})dt.$$

Define:

(41) $$B_{ij}(s) = \int_0^s b_{ij}(t)dt.$$

With this notation, (40) assumes the following form:

(42) $$U_i(C, \varphi_{-i}) = \sum_{j=1}^{n} \sum_{k=1}^{2^{m-1}} [B_{ij}(z_{jk}) - B_{ij}(z_{j(k-1)})]\delta(C, C_{jk}).$$

For $j = i$, the inner sum is nothing other than $B_{ii}(a_i)$, since we always have $\delta(C, C_{ik}) = 1$. Therefore, the right hand side of (42) is a function of z_{-i} in z. We introduce the symbol $U_i(C, z_{-i})$ for the right hand side of (42):

$$(43)\qquad U_i(C, z_{-i}) = B_{ii}(a_i) + \sum_{\substack{j=1\\ j\neq i}}^{n} \sum_{k=1}^{2^{m-1}} [B_{ij}(z_{jk}) - B_{ij}(z_{j(k-1)})]\delta(C, C_{jk}).$$

For $i = 1, \ldots, n$, define

$$(44)\qquad F_i(s) = \int_0^s f_i(t)dt$$

and

$$(45)\qquad L_i(z_i) = \sum_{k=1}^{2^{m-1}} [F_i(z_{ik}) - F_i(z_{i(k-1)})]g_i(C_{ik}).$$

We refer to $L_i(z_i)$ as the *total learning cost* of language community i. Total learning cost is the integral over individual learning costs of all members of language community i. The *total communication benefit* $U_i(z)$ of language community i is defined as follows:

$$(46)\qquad U_i(z) = \sum_{k=1}^{2^{m-1}} (z_{ik} - z_{i(k-1)})U_i(C_{ik}, z_{-i})$$

for $i = 1, \ldots, n$. Obviously, $U_i(z)$ is the integral over the communication benefits of all members of language community i. The group payoff $P_i(z)$ of language community i is the difference between total communication benefit and total learning cost:

$$(47)\qquad P_i(z) = U_i(z) - L_i(z_i)$$

for $i = 1, \ldots, n$. The function P_i which assigns $P_i(z)$ to every $z \in Z$ is the *group payoff function* of language community i. The function P which assigns

$$(48)\qquad P_z = (P_1(z), \ldots, P_n(z))$$

to every $z \in Z$ is the *vector payoff function*. P_i and P are defined relative to a fixed combination $r = (r_1, \ldots, r_n)$ of cost compatible rankings. When we want to express this dependency, we shall write P_i^r and P^r instead of P_i and P, respectively.

Aggregated games. For every combination $r = (r_1, \ldots, r_n)$ of cost compatible rankings, the associated *aggregated game*

$$(49)\qquad G^r = (Z, P^r)$$

is an n-person game in normal form, where $Z_1, \ldots, Z_n$ are the strategy sets of the players 1, ..., s, respectively, and where P^r is the payoff function.

Best replies and equilibrium points. We use the notation $z_i z_{-i}$ in order to denote that border point system z which contains z_i and the components z_j of z_{-i} as components. A border point vector $\tilde{z}_i \in Z_i$ is a *best reply* to an i-incomplete border point system z_{-i} in G^r if we have:

$$P_i^r(\tilde{z}_i z_{-i}) = \max_{z_i \in Z_i} P_i^r((z_i z_{-i}). \tag{50}$$

We say that $\tilde{z}_i$ is a *best reply* to a border point system z in G^r if in G^r the border point vector $\tilde{z}_i$ is a best reply to the i-incomplete border point system z_{-i} in z. We also say that the border point system $\tilde{z}$ is a *best reply* to z in G^r if every component of $\tilde{z}$ is a best reply to z in G^r. An equilibrium point z^* of G^r is a border point system with the property that z^* is a best reply to z^* in G^r.

5. Existence of equilibrium points

In this section we shall prove the existence of equilibrium points for every game $\Gamma(m, n, a, b, f, g)$ whose specification satisfies assumptions 1 to 5. For this purpose we first show that an equilibrium point of the original game is associated with every equilibrium point of every aggregated game. We then turn our attention to the continuity and convexity properties of group payoffs, which later enable us to apply standard results from the literature to prove the existence of an equilibrium point for aggregated games.

Wherever we do not explicitly say anything to the contrary, statements refer to a fixed game $\Gamma(m, n, a, b, f, g)$ whose specification satisfies assumptions 1 to 5 and to a fixed combination of cost compatible rankings $r = (r_1, \ldots, r_n)$. The notation $G = (Z, P)$ is used for the associated game G^r. If we say that a border point vector or a border point system is a *best reply*, we mean a best reply in G.

Theorem 2 asserts the existence of equilibrium points for all aggregated games of $\Gamma(m, n, a, b, f, g)$ and therefore does not refer to a specific $r = (r_1, \ldots, r_n)$. Similarly, theorem 3, the final existence theorem, refers to all games $\Gamma(m, n, a, b, f, g)$ satisfying assumptions 1 to 5.

Lemma 4. Let $z_i \in Z_i$ be a best reply to $z_{-i} \in Z_{-i}$. Moreover, let φ_i be a group strategy in the border equivalence class $\Lambda_i(z_i)$, and let φ_{-i} be an i-incomplete strategy combination in the border equivalence class $\Lambda_{-i}(z_{-i})$. Then φ_i is an almost best reply to φ_{-i}. Moreover, for every $t \in [0, a_i]$ which does not coincide with one of the border points z_{ik} in z_i, the strategy $\varphi_i(t)$ is a best reply of $i + t$ to φ_{-i}.

Proof. Assume that φ_i is not an almost best reply to φ_{-i}. Then at least one of the open intervals of the form $(z_{i(k-1)}, z_{ik})$ must contain infinitely many points t where $\varphi_i(t) = C_{ik}$ is not a best reply of $i + t$ to φ_{-i}. Moreover, at least one strategy $D \in M_i$ must be a best reply of $i+t$ to φ_{-i} for infinitely many of these values of t. Let D be a strategy of this kind, and let S be the set of all $t \in (z_{i(k-1)}, z_{ik})$ such that D is a best reply and C_{ik} is not a best reply of $i+t$ to φ_{-i}. For every $t \in S$ we have:

(51) $$U_i(D, \varphi_{-i}) - f_i(t)g_i(D) > U_i(C_{ik}, \varphi_{-i}) - f_i(t)g_i(C_{ik}).$$

Essentially the same argument as has been applied to (17) and (18) in the proof of lemma 2 can be applied here, too. In this way it can be seen that S is a subinterval of $(z_{i(k-1)}, z_{ik})$. Moreover, since S contains infinitely many points, S has positive length. Let φ_i' be the following group strategy:

(52) $$\varphi_i'(t) = \begin{cases} D & \text{for } t \in S \\ \varphi_i(t) & \text{for } t \notin S \end{cases}$$

In view of (51) and the finite length of S, we obtain:

(53) $$\int_0^{a_i} H_{i+t}(\varphi_i', \varphi_{-i})dt) > P_i(z_i z_{-i}).$$

According to lemma 3, the monotonicity class $\Psi_i(r_i)$ contains a best reply Ψ_i to φ_{-i}. Let z_i' be the border point vector of a best reply $\Psi_i \in \Psi_i(r_i)$ to φ_{-i}. It follows by (53) that we have:

(54) $$P_i(z_i' z_{-i}) > P_i(z_i z_{-i}).$$

This is a contradiction to the assumption that z_i is a best reply to z_{-i}. Therefore, φ_i is an almost best reply to φ_{-i}.

We now prove the remaining assertion of the lemma. Suppose that one of the finitely many values of t, where $\varphi_i(t)$ is not a best reply to φ_{-i}, is a value t_0 which does not coincide with one of the border points. In some ϵ-neighborhood of t_0 all values of t with the exception of t_0 have the property that $\varphi_i(t) = C_{ik}$ is a best reply of $i+t$ to φ_{-i}, where the strategy C_{ik} is the same one for all these values of t. However, in view of lemma 2 the set of all t such that C_{ik} is a best reply of $i+t$ to φ_{-i} is an interval. It must be that t_0 belongs to the interval. This is a contradiction. Therefore, the assertion holds.

Lemma 5. Let $z = (z_1, \ldots, z_n)$ be an equilibrium point of the aggregated game G. Then the border equivalence class $\Lambda(z)$ contains an equilibrium point.

Proof. Let $\varphi = (\varphi_1, \ldots, \varphi_n)$ be an element of $\Lambda(z)$. It follows by lemma 4 that φ is an almost equilibrium point. Let Ψ be related to φ as in lemma 1. It follows by lemma 1 that Ψ is an equilibrium point. In view of lemma 4, it is clear that Ψ differs from φ only at border points in the components of z. Consequently, Ψ belongs to $\Lambda(z)$.

Lemma 6. For $i = 1, \ldots, n$, the group payoff $P_i(z)$ is a continuous and bounded function of z, and for every fixed $z_{-i} \in Z_{-i}$ the group payoff $P_i(z_i z_{-i})$ is a concave function of z_i.

Proof. Since definite integrals of bounded functions are continuous in their upper limits, the functions B_{ij} and F_i are continuous. Therefore, the continuity of $P_i(z)$ as a function of z is an immediate consequence of the definition of group payoffs. The boundedness of $P_i(z)$ follows by the compactness of Z.

In view of (46), for fixed $z_{-i} \in Z_{-i}$ the total community benefit $U_i(z)$ is a linear function of z_i. Therefore, it is sufficient to show that $L_i(z_i)$ is convex in z_i in order to prove that $P_i(z_i z_{-i})$ is concave in z_i. Equation (45) can be rewritten as follows:

$$L_i(z_i) = \sum_{k=1}^{2^{m-1}-1} F_i(z_{ik})[g_i(C_{ik}) - g_i(C_{i(k+1)})] + a_i g_i(C_{i2^{m-1}}). \qquad (55)$$

Since $F_i(z_{ik})$ is the integral over a monotonically increasing function, it is a strictly convex function of z_{ik}. In view of $g_i(C_{ik}) \leq g_i(C_{i(k+1)})$, it follows by (55) that $L_i(z_i)$ is a convex function of z_i.

Theorem 2. For every combination $r = (r_1, \ldots, r_n)$ of cost compatible rankings, the associated aggregated game G^r has at least one equilibrium point.

Proof. The strategy sets Z_i are convex and compact. Lemma 6 has shown that $P_i^r(z)$ is continuous with respect to z and concave with respect to player i's strategy z_i. Therefore, the conditions for the application of well known results on the existence of equilibrium points are satisfied (see, e.g., Friedman, 1977, p. 160, theorem 7.4). We can conclude that the assertion holds.

Theorem 3. Every game $\Gamma(m, n, a, b, f, g)$ whose specification satisfies assumptions 1 to 5 has at least one equilibrium point.

Proof. The assertion is an immediate consequence of lemma 5 and theorem 2.

6. References

Breton, Albert, & Mieskowski, Peter (1975). *The returns to investment in language: the economics of bilingualism*. Toronto: University of Toronto, Institute for Policy Analysis.

Dressler, Wolfgang U. (1982). "Acceleration, retardation, and reversal in language decay?". In *Language spread*, ed. Robert L. Cooper (Bloomington, IN: Indiana University Press), pp. 321–336.

Friedman, James W. (1977). Oligopoly and the theory of games. Amsterdam: North-Holland.

Grenier, Gilles (1985). "Bilinguisme, transferts linguistiques et revenus de travail au Québec: quelques éléments d'interaction. In *Économie et langue*, ed. François Vaillancourt (Québec: Éditeur officiel), pp. 243–287.

Hočevar, Toussaint (1975). "Equilibria in linguistic minority markets". *Kyklos*, 28, 337–357.

Lang, Kevin (1986). "A language theory of discrimination". *The quarterly journal of*

economics, 100, 363–381.
Laponce, Jean (1984). *Langue et territoire*. Québec: Les presses de l'Université Laval.
Lieberson, Stanley (1982). "Forces affecting language spread: some basic propositions". In *Language spread*, ed. Robert L. Cooper (Bloomington, IN: Indiana University Press), pp. 37–62.
Sabourin, Conrad (1985). "La théorie des environnements linguistiques". In *Économie et langue*, ed. François Vaillancourt (Québec: Éditeur officiel), pp. 59–82.
Vaillancourt, François (1985). "Les écrits en économie de la langue: brève revue et introduction au recueil". In *Économie et langue*, ed. François Vaillancourt (Québec: Éditeur officiel), pp. 11–25.
Vaillancourt, François (1985a). "Le choix de la langue de la consommation". In *Économie et langue*, ed. François Vaillancourt (Québec: Éditeur officiel), pp. 209–220.
Vaillancourt, François, & Lacroix, Robert (1985). "Revenus et langue au Québec, 1970–1980: une revue des écrits". In *Économie et langue*, ed. François Vaillancourt (Québec: Éditeur officiel), pp. 221–242.

RESISTING THE DRAFT

A PERFECT EQUILIBRIUM APPROACH[1]

by

Roy Gardner

Abstract: This paper studies conscription and possible resistance to it as an extensive form game. The draft resistance game is played in three stages. The volunteer stage consists of a call for volunteers and its possible fulfillment. In case of a shortfall of volunteers, the remaining players enter the resistance formation stage, during which a resistance movement may take shape. In the conscription stage, conscripts are called up and the resistance carries out an agreed upon threat. The paper characterizes the class of perfect equilibria for this game. Also studied are institutional variations, such as the payment of commutation fees. The theory is illustrated with data drawn from the Civil and Vietnam War.

1. Introduction

The Congress shall have power to raise and support armies.
U.S. Constitution

Hell No - We won't go
anti-Vietnam War chant

Five times in American history, beginning with the American Civil War, two World Wars, Korea and Vietnam, conscripts have been called to military service. In Europe, conscript armies have existed since the Napoleonic Wars. Conscription has often met with resistance (Chambers, 1975; Useem, 1973). This paper builds a simple model of conscription and possible resistance to it. A game model is appropriate since the actions of some players, for instance resisters, may affect the payoffs of other players by making them more likely to be drafted. Specifically, we will consider a game in extensive form, since many episodes of conscription in history have a well-defined stage structure. For instance, in the North during the American Civil War, Lincoln first called for volunteers (1861), then a draft resistance movement formed (1862), and finally conscription was introduced (1863). The solution approach

[1]The author wishes to thank M. Ali Khan, C.M. Kahn, E. van Damme, R. KcKelvey, E. Ostrom, R. Selten, R. Spector and S. Stowe for helpful comments. This research was supported in part by a stipend from the Alexander von Humboldt Foundation.

for this game will be that of subgame perfect equilibrium, due to Selten (1973). Selten has shown how fruitful a perfect equilibrium approach is to questions of industrial organization, and we hope that the results presented here will also prove as useful in answering certain questions of public economics.

Part of the success of game theory in industrial organization is due to the profit orientation of the firms competing for markets. Indeed, the use of any game theoretic model presupposes a rational choice model of the players' behavior. The phenomena of draft resistance, with its heated passion and religious overtones, may seem an entirely inappropriate subject for rational choice. Nevertheless, for such conceptually similar social phenomena as race riots (Mason, 1984), rebellion (Denardo, 1985), and tax resistance (Frohlich and Oppenheimer, 1974), such models have yielded important insights into agents' behavior. Thus, there is reason to believe that a rational choice framework for draft resistance is not completely hopeless. The ultimate test of any framework is, as always, the explanatory power of the results.

The players of the draft resistance game are the set of men deemed eligible for military service, and so classified. Since these are really large games with literally millions of players, it is useful to model them as games with a continuum of players. The existence of Nash equilibrium for such games is due to Schmeidler (1973); Schmeidler's result will be used in the sequel to show the existence of refinements of Nash equilibrium, such as subgame perfect equilibrium. Green (1984) also has results on limits of Nash equilibria as a game grows large that are relevant for the interpretation of large game models. There are two main types of players, namely those who are forced to play (whose expected payoff to playing is negative at any perfect equilibrium) and those who want to play (whose expected payoff is nonnegative at any perfect equilibrium). This later type can be further sub-divided into those who wish to serve (volunteers) and those who wish to resist, either for moral or political reasons. The former type then includes those ultimately conscripted, as well as those volunteering to avoid conscription (reluctant volunteers). The phenomenon of reluctant volunteers is extremely important to the equilibrium of the draft resistance game.

The draft resistance game is played in three stages. The first, the volunteer stage, consists of a call for volunteers. In the event that the call is fulfilled, the game ends. If not, play continues with a participation subgame, whose object is to form a resistance movement, if any, carries out an agreed upon threat. A subgame perfect equilibrium point is an equilibrium point which induces an equilibrium point on every subgame. We exhibit a unique subgame perfect equilibrium based on the distribution of agents' characteristics and the vector of government conscription parameters.

The paper is organized as follows. The next section presents the basic assumptions and notation of the model. Section III studies the equilibria of games in which resistance is not an important factor. This type of equilibrium is illustrated with data from the American Civil War. Section IV presents the more general case, where resistance does have important implications for equilibrium play. This section includes a discussion of data drawn from the Vietnam War. The last section considers institutional variations of conscription, such as commutation fees and the hiring of substitutes, both of which existed during the Civil War.

Extensive game models have recently appeared in other areas with considerable policy content such as the nuclear arms race (see Güth, 1985). Such models should receive considerable attention in the years to come.

2. Model and Notation

Let the unit interval $[0,1]$ be the player set, N. Each point t in $[0,1]$ represents a player. Each player has a utility function of the form

$$u_t(y_t, q) = y_t + z_t(q) \tag{1}$$

where y_t is an individual payoff in monetary terms depending on t's role in the draft and q is the actual size of the armed force, $0 \leq q \leq 1$. $q = 1$ denotes full mobilization; $q = 0$, abolition of the armed forces. Utility is normalized so that

$$u_t(0, \bar{q}) = 0, \tag{2}$$

where $\bar{q}$ is the size of the call for volunteers by the government. In the event of substantial draft resistance, the call-up will not be met even with conscription, so that

$$q \leq \bar{q} \,. \tag{3}$$

The size of the fighting force q plays the role of a public good (or bad). A was is *unpopular* when z_t is a decreasing function for a majority of players. As we shall see, an unpopular war leads to greater resistance. The government also controls the positive parameter b, which represents the benefit gained by volunteering for military service. This is easiest to measure in the case that b is a cash bonus to every volunteer, such as in the Civil War. The benefit can also take the form of better assignments, better treatment, lower casualty rates, and the like while in the service, in which case b represents the present value of all such benefits. For simplicity, b is taken to be the same for all players. Making b depend on players' characteristics does not drastically change the results, although it does complicate their statement.

The value of y_t depends on the player's role in the war. y_t can take on one of the following four values:

$$y_t = \begin{cases} 0, & \text{if } t \text{ remains a civilian} \\ -c_t, & \text{if } t \text{ is a draftee} \\ b-c_t, & \text{if } t \text{ is a volunteer} \\ -d_t, & \text{if } t \text{ is a resister.} \end{cases} \tag{4}$$

Thus, y_t measures the opportunity cost of non-civilian status. Each player is characterized by a pair of parameters (c_t,d_t) which measures the cost of serving in the military, or resisting the draft, respectively. Finally, $G(c,d)$ is the continuous distribution telling how agents' characteristics are distributed.

The game itself consists of two decision stages. After the government's call for volunteers, each player must decide for himself whether to volunteer or not. If the call for volunteers is fulfilled, the game is over. Otherwise the government institutes a draft, and announces how large the draft call will be. At the second decision stage, each player must decide for himself whether to resist the draft or not. Finally, the government issues individual draft notices at random. Those resisting the draft refuse induction notices, while those not resisting honor them. This decision scheme is portrayed in Figure 1, which shows the sequence of moves, but not the entire game tree.

Each player's strategic behavior can be described by a vector x_t,

$$x_t = (x_t^1, x_t^2) \tag{5}$$

where x_t^1 is the probability that t volunteers at the first decision stage, and x_t^2 is the probability that t resists at the second decision stage. Note that a player might not reach the second decision stage. If he volunteers, then he is already in the armed forces and cannot resist. He could, however, desert the armed forces, which would be tantamount to the same thing. The present model assumes there is no desertion after volunteering. In situations such as the Civil War, where desertion rates exceed 30% (Reid and White, 1985), this assumption is not palatable. In other situations, such as the Vietnam War, where desertion rates are lower than 2% the assumption is not bad. It is mainly for simplicity that we maintain it here.

The aggregate behavior at the end of the first decision is the number of volunteers v, given by

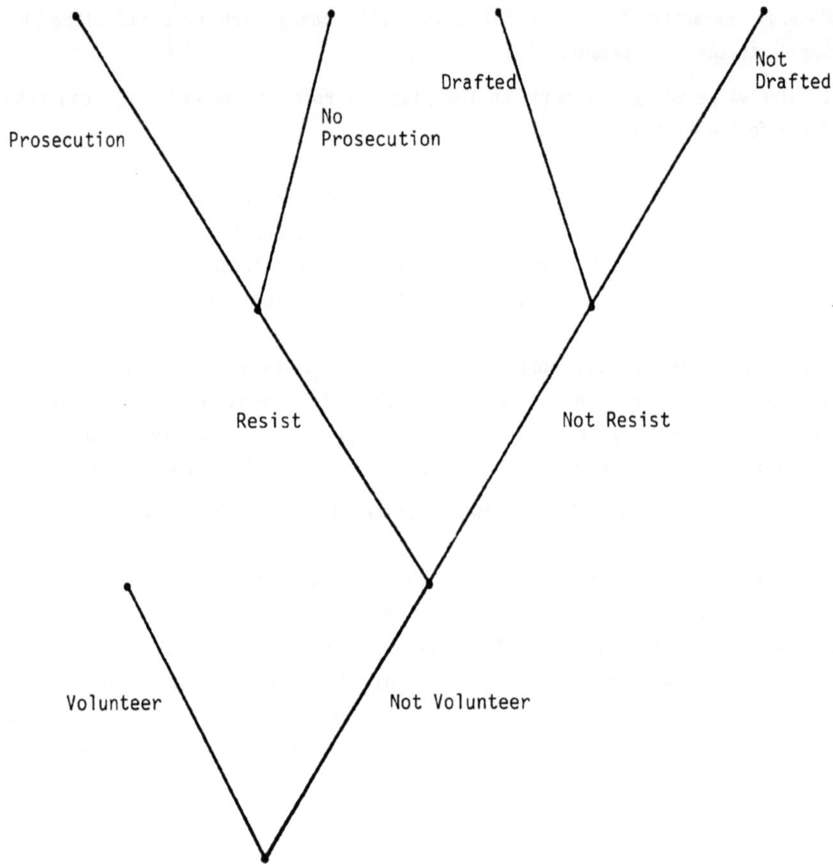

Figure 1: Schematic picture of the draft resistance game.

$$v = \int_N x_t^1 \, dt. \tag{6}$$

Clearly, $0 \leq v \leq 1$. In the event that $v \geq \bar{q}$, the call for volunteers is fulfilled and the draft resistance game ends. We shall call this behavior regime the All-Volunteer Force. One of the main tasks of the theory is to determine when All-Volunteer Force is a perfect equilibrium outcome.

In the event that $v < \bar{q}$, then the government must resort to conscription to attempt to fill its manpower needs. Here two more possible behavior regimes arise, depending on the size of the resistance. Denote by V the set of those actually volunteering during the volunteer stage. The aggregate behavior at the end of the second decision stage is the number of resisters r given by

$$r = \int_{N\backslash V} x_t^2 \, dt. \tag{7}$$

Clearly, $0 \leq r \leq 1-v$. Two cases are possible. When resistance is small, i.e., when $r \leq 1-\bar{q}$, then the government's manpower needs can be met by conscription. In this behavior regime, Small Resistance, a force of size $\bar{q}$ is recruited. When resistance is large, i.e., when $r > 1-\bar{q}$, the size of force actually recruited from volunteers and conscripts is given by

$$q = 1-r. \tag{8}$$

This behavior regime is called Large Resistance.

The draft is conducted as follows. The government calls at random $\bar{q}-v$ non-volunteers for service. Since the callup is random, a proportion $r/(1-v)$ of these will be resisters and $(1-v-r)/(1-v)$ of these will not. Resisters refuse induction and receive the payoff $-d_t$ each; for them, the game is over. If r is positive, the government does not meet its quota and samples again. The government continues to sample until it has filled its quota or it has called up everyone. One can show that under random sampling without replacement, the probability of being called up is the same for a resister as for a non-resister. This probability, and the other major characteristics of the three behavior regimes, are summarized in Table 1.

Alternative draft mechanisms are possible. The above mechanism minimizes the auditing role of the draft. A mechanism which maximizes the auditing role calls up everyone, prosecutes all

Regime constellation	Parameter	Size of force	Probability of conscription or detection, if resisting
All-Volunteer Force	$\bar{q}-v \leq 0$	$\bar{q}$	0
Small Resistance	$0<\bar{q}-v<1-v-r$	$\bar{q}$	$(\bar{q}-v)/(1-v-r)$
Large Resistance	$1-v-r < \bar{q}-v$	$1-r$	1

Table 1: Behavior Regimes of the Draft Resistance Game.

resisters, and then selects from among the non-resisters as needed. Mechanisms with intermediate auditing would fall between these two extremes. The reason for studying the minimum auditing mechanism is that it gives rise to the largest possible resistance, and so puts an upper bound on the size of all conceivable resistance movements.

3. Reluctant Volunteers and Perfect Equilibrium

Resisting the draft can be a very costly phenomenon. If prosecuted for breaking a conscription law by refusing to serve, one may face the risk of a fairly long prison term. Alternatively, if one goes underground or flees the country to avoid prosecution, one faces the costs of living in illegality or of finding a new country, with the attendant career disruption. If such costs are considerable, then rational players will choose not to incur them. This point can be made precise. At the last decision stage, if $c_t < d_t$, then resisting is dominated by waiting for the draft, and equilibrium play at this stage entails $x_t^2 = 0$. In this section, we assume that it is rational for all players not to resist the draft . Hence, from (7), $r = 0$. This allows us to focus on the phenomenon of reluctant volunteering, which is extremely important in naturally occurring conscription contexts.

A *reluctant volunteer* is a player t for whom $b-c_t < 0$, so that the player has an incentive to remain a civilian, but who nevertheless volunteers. Being forced to play the game in the first place, if the probability of conscription is sufficiently high, then getting the benefit offered by volunteering is still better than waiting to be drafted and foregoing the benefit b. As long as mobilization is less than full, only two behavior regimes arise: All-Volunteer Force or Small Resistance. This is the content of

<u>Proposition 1</u>

The perfect equilibrium of the draft resistance game is an All-Volunteer Force when the number of players for whom $b-c_t > 0$ is at least $\bar{q}$. Otherwise, the perfect equilibrium is Small Resistance with the marginal volunteer having cost c^*, defined by

$$b-c^* = \frac{\bar{q}-v}{1-v} \cdot (-c^*). \tag{9}$$

<u>Proof</u>

All the conditions of Schmeidler's result (Schmeidler, 1973; Theorem 2) are met so there exists a Nash equilibrium in pure strategies. We now refine any such Nash equilibrium using Selten's criterion of subgame perfection.

When $b-c_t > 0$, volunteering dominates waiting for the draft (indeed, it even dominates civilian status). If there are enough such players, one has an All-Volunteer Force.

Next suppose that there are not enough such players. All players for whom $b-c_t$ is positive continue to volunteer. At stage two, a player who does not volunteer faces the probability of being drafted equal to $(\bar{q}-v)/(1-v)$. Any player for whom

$$b-c_t > \frac{\bar{q}-v}{1-v} \cdot (-c^*)$$

does best by volunteering reluctantly. In particular, one has reluctant volunteers for all costs $c_t \leq c^*$. The draft quota $\bar{q}-v$ is filled entirely by conscripts with c_t higher than c^*.

It is worth noting that besides the above subgame perfect equilibrium there are also imperfect equilibria. Such equilibria are supported by the second stage strategy $x_t^2 = 1$ on the part of those not volunteering at stage 1. Everyone volunteers reluctantly, thinking that there will be a Large Resistance. Such equilibria are patently unreasonable and unlikely to be observed in practice.

To find the critical value c^* in (9), note that the number of players with cost of serving $c_t \leq c^*$ is precisely v. Substituting into (9) and rearranging, one has that

$$c^* = b \cdot \frac{1-v}{1-\bar{q}} . \tag{10}$$

Once the distribution of service costs is known, (10) can be evaluated. For example, suppose service costs are distributed uniformly on the closed interval $[0,c^{max}]$. One then has $v = c^*/c^{max}$. Substituting into (10) and rearranging, one finds that

$$c^* = \frac{bc^{max}}{(1-\bar{q})c^{max} + b} . \tag{11}$$

One sees from (11) that c^* is increasing, both in the benefit to volunteering b and in the quota $\bar{q}$. This is true in general as differentiation of (10) reveals. Thus, as mobilization becomes very nearly total, c^* approaches c^{max}; and practically everyone volunteers, however reluctantly.

The phenomenon of reluctant volunteering is empirically important. The data from the American Civil War cited in Chambers (1975) show that of the approximately two million men serving in the Union army, over 92% volunteered. The bonus for volunteering was \$100 (equal to four month's of the average wage in the North at the time), and the Draft Act of 1863 had very few exceptions -- sole support of widowed mother, for example. This data suggests many reluctant volunteers in the Union army. We will return to this data in Section 5, when we take into account some of the institutional arrangements peculiar to the American Civil War.

4. Resistance and Perfect Equilibrium

In this section, we drop the assumption that resistance costs are prohibitive. There will then exist perfect equilibria at which some players form a resistance, at least for some distributions of agents' characteristics. Moreover, all three behavior regimes can now arise as perfect equilibria, although for a given distribution of agents' characteristics, only one behavior regime will do so.

The first implication of relaxing the assumption of prohibitive resistance costs is that the regime of Large Resistance can now occur. We will study this using a backward recursion argument. At stage two of the draft resistance game, a player who has not volunteered can either wait to be drafted -- in this case a certainty -- and receive the payoff

$$-c_t + z_t(1-r).$$

Alternatively, he can join the resistance and receive the payoff

$$-d_t + z_t(1-r) - z'_t(1-r),$$

the last term on the right arising due to the marginal effect of a resister has on the size of force recruited. Thus, at stage two, the resistance consists of all players t for whom

$$c_t > d_t + z'_t(1-r). \tag{12}$$

Notice that for an unpopular war, z'_t is negative and there is an incentive to resist above and beyond purely individual considerations. At stage one, a player knows that if he does not volunteer, he will certainly be called up. Since he collects the benefit b by volunteering, however reluctantly, he will never wait to be drafted. Therefore, the choice boils down to either volunteering or resisting. Arguing as in Proposition 1, one has

Proposition 2

Under Large Resistance, the set of draft resisters are all those whose individual characteristics satisfy (12). All other players volunteer.

Instead of a single critical value c^*, one now has a critical set of players for whom

$$b-c_t = -d_t-z_t'(v)$$

at a perfect equilibrium. The outcome is highly polarized, with agents either volunteering or resisting. Indeed, no one is drafted at such an equilibrium, as all those receiving draft notices resist instead. One has the kind of situation ripe for a mutiny of the armed forces. In this event the government has lost control of the game. It might be useful to reconsider the mutinies on the Western and Eastern Fronts of World War I in 1917 from this standpoint.

We now turn to the remaining behavior regimes, All-Volunteer Force and Small Resistance. In both of these, the size of force is $\bar{q}$. At stage two of the game, a player's decision to resist or not depends solely on individual cost characteristics. The set of resister is simply those for whom

$$c_t > d_t \ .$$

At stage one, a player considering volunteering will indeed do so when

$$b-c_t > \max(-d_t,-c_t) \cdot \frac{\bar{q}-v}{1-v-r} \ . \tag{13}$$

This is summarized in the following

<u>Proposition 3</u>

In Small Resistance, the set of players volunteering consists of all t satisfying (13). The set of resisters consists of all players t with the inequality (13) reversed and $c_t > d_t$.

In Small Resistance, all four possible values of x_t are observed: volunteers, resisters, those waiting to be drafted and who are drafted, and those eligible for the draft who remain in civilian status. This is precisely the picture that emerges during the Vietnam War.

A sufficient condition for an All-Volunteer Force continues to be given by Proposition 1. One can see this by the following limiting argument. As the number of volunteers v approaches $\bar{q}$, the right-hand side of (13) approaches zero. As long as the probability of being drafted is positive, then the second stage of the game is reached. Therefore, one still needs the condition that the second stage not be reached. One should add the proviso however that for no volunteer is it the case that $b-c_t < -d_t$. Otherwise it might be that even for small probabilities of conscription, such a player would wait for the chance to resist.

It is useful at this point to give a numerical example of the solution to the draft resistance game. For simplicity assume that z_t is a constant function for all t. Let the space of agents' characteristics be the unit square, with characteristics distributed uniformly, so that $G(c,d) = 4cd$. Hence, areas in the unit square measure number of players taking a particular action. Let $b = .25$, $\bar{q} = 0.5$. The solution in the space of agents' characteristics is depicted in Figure 2. One has $v = 7/16$, $c^* = 0.5$, and $r = 7/16$. Among those waiting for the draft, the probability of being drafted is 0.5.

A routine calculation leads to the regime diagram of Figure 3. As one would expect, the All-Volunteer Force is associated with high values of b and low values of $\bar{q}$, while the opposite is true of Large Resistance. The region of Small Resistance lies intermediate between these two. The regime boundaries are linear due to the bilinearity of G. We now turn to the data generated by the conscription during the Vietnam War, all of which are drawn from Useem (1973). The behavior regime was Small Resistance. Even in the peak resistance year of 1968, there were only 6 refusing induction for every 100 inductees. The U.S. Government was able to fill its annual draft quota of some 350,000 men. In addition to these conscripts and resisters, there were also 200,000 volunteers for service. According to a Department of Defense survey, approximately one-half of these volunteers were reluctant. In the period prior to the introduction of American and North Vietnamese Combat units into South Vietnam in 1965, only about 150,000 men were being inducted annually, and about two-thirds of all volunteers classified themselves as reluctant. One sees here the effect of increasing the mobilizations note and cost of serving relative to the benefit. Not only did the cost of serving rise due to the war casualty rate, but also the cost of resisting, due to an increase in average sentence. Prior to 1965, the average sentence for draft dodging was 18 months, but this rose gradually until by 1968 it was 37 months. Unfortunately, we have no data whether those eligible for the draft were aware of these increases in the cost of resisting.

One large question surrounding the Vietnam data is whether the benefit to volunteering is positive. Volunteers hat to serve for three years in the army (four years in the other services), while draftees had to serve only two years. Thus, there was a considerable opportunity cost associated with the extra time required for a volunteer. Also, the probability of being sent to Vietnam was the same for both volunteers and draftees, about 0.5. What makes b positive is the probability of death or injury within the warzone itself. It was always possible to distinguish a soldier as volunteer or draftee on his orders, by his military schooling (typically better for a volunteer), and often by his assignment (draftees were overrepresented in infantry, underrepresented in military intelligence). The probability of being killed in Vietnam was 1.6% if one was a volunteer, but 3.9% if one was a draftee. Moreover, the probability of being seriously injured was 6.8% for volunteers, versus

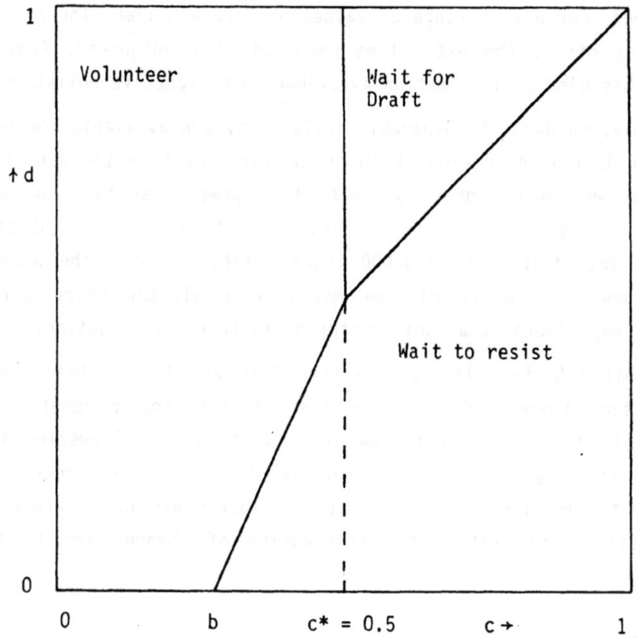

Figure 2: **Perfect Equilibrium Behavior in the space of agents' characteristics (for b = .25, $\bar{q}$ = 0.5).**

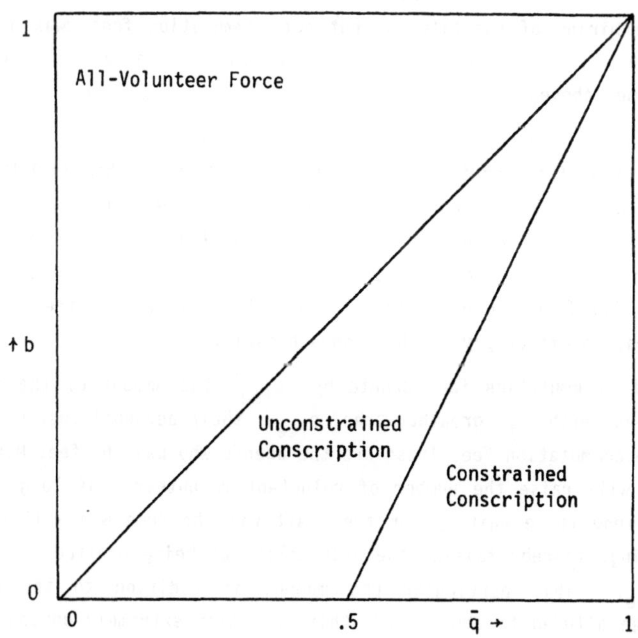

Figure 3: **Behavior Regimes in the space of draft parameters.**

11.8% for draftees. For a wide range of values of life and limb, these differentials in casualty rates offset the extra time required of a volunteer. Army recruiters typically made this kind of information available to prospective enlistees.

Unfortunately, no data of comparable reliability are available for either North or South Vietnam during the period of American involvement in the war. Clearly, the mobilization rate was much higher for both these powers, so high in fact that it could have led to regimes of Large Resistance in both. It is hard otherwise to account for the introduction of 100,000 Chinese laborers into the North in 1965, unless the manpower resources of the North were already being strained. The Vietnamese draft experience is a topic worthy of further investigation.

When the United States officially returned to an All-Volunteer Force in 1973 (for the first time since 1940), it used three factors to implement this behavior regime. First, all forces were withdrawn from Vietnam, thus lowering the cost of serving. Second, military pay levels were increased, reducing the opportunity cost of serving relative to civilian status. Finally, all forces worldwide were reduced, thus lowering the mobilization rate. This combination of changes led to the desired behavior regime.

5. Institutional Variations on the Draft

This section considers two institutional variations on conscription which wer part of the Union Conscription Act of 1863, the commutation fee and the hiring of substitutes. The hiring of substitutes, but not commutation fees, was also allowed under the Confederation Conscription Act of 1862. Each of these can be worked into the model developed above.

By reporting for service and paying the commutation fee, an inductee was thereby excused from service. The commutation fee amounted to $300 in the Civil War (roughly the annual wage of an average worker, and three times the enlistment bonus) -- a large sum not easily financed. Alternatively, by reporting for service and furnishing a substitute, an inductee was again excused from service. Since, the substitute was conscripted and therefore not eligible for the enlistment bonus, the person paying the substitute was in effect paying this bonus himself.

To model the commutation fee, denote by c_{fee} the amount of the commutation fee. For all agents with c_t greather than c_{fee}, their personal cost of serving is greater than the commutation fee. These are the agents who pay the fee. Moreover, the commutation fee will raise the number of reluctant volunteers, as long as c^* is below c_{fee}. Among those waiting for the draft will be some who will pay the fee instead of serving, thereby raising the probability of being drafted for those with cost below c_{fee}. This particular phenomenon, in addition to the problem of financing the fee alluded to above, made this provision extremely unpopular, and it was deleted from the Conscription Act in July, 1864. Approximately 50,000 men paid the commutation fee (the same number as those actually drafted) during 1863-1864.

To model the hiring of substitutes, one has to introduce an additional decision stage prior to the issuance of draft notices. In this stage, pairs of players who are not resisting the draft are allowed to bargain. A typical bargain between two players would stipulate that if one of them were drafted, the other would serve in his place at an agreed upon price. Such an agreement would be binding only in the event that exactly one of the two parties to the agreement were drafted. In a game of complete information such as the present one, one expects the substitution fee to be uniform across pairs, bounded below by the enlistment bonus and bounded above by the commutation fee. At the ensuing equilibrium, there is somewhat less incentive to volunteer reluctantly, depending on the probability that one will actually reach a substitution bargain, and that one's bargain partner will actually be called. Approximately 88,000 Northerners, and a like number of Southerners, furnished substitutes during the Civil War.

The overall picture that emerges from the Civil War data, even taking into account these two institutional variations, is one of an overwhelmingly volunteer army. The category of volunteers, two millions strong for the North (about one million for the South) is an order of magnitude or higher than any other. A word of caution is in order, however. The armies raised during the American Civil War were not professional forces, in the sense then understood in Europe or now understood in America. Discipline was extraordinarily lax, and desertion a phenomenon of massive proportions. At the end of the war, for instance, almost one-third of all Northern troops (and 40% of all Sourthern forces) were absent without leave from their units. The comparable figure from the Vietnam War is less than 2%. There are even cases of entire units, officers and men, deserting, which prompted at least one general to call his men "a mob of cowards and stragglers". Even after conscription, then, the resistance game could be played further, leading to a reduction in forces. This considerably reduced the cost of actually serving, and made volunteering even more attractive.

Another feature of rule-breaking is also worth noting. We have assumed that a player can volunteer at most once. In an unprofessional army it is possible to volunteer more than once, and many deserters did exactly that, volunteering up to six times (the known record) during the course of the war. This practice was dubbed "bounty-hunting", and represents another behavioral detail that the present model supresses.

One large difference between Civil War and Vietnam War conscription is the degree of exemption provided by the conscription system. The Union Conscription Act provided for almost no exemptions, while the Confederate Conscription Act provided some occupational exemptions. The Vietnam War draft, on the other hand, had a complex and farflung system of exemptions, deferments, and exclusions. In our analysis we have taken the eligible set of players as given. This probably does justice to a system like that of the Civil War. But in terms of the Vietnam War system, one must really model a classification stage prior to the volunteer stage. For instance, a player might have information about his eligibility, which is not available to other players or to the government. On the basis of this information, and his expected value of playing the draft resistance game once the player set has been determined, he could decide whether to invest in classification strategies that reduced his probability of eligibility. The conclusions of the present model should still hold in such an extension.

This paper has argued that mass social phenomena, in particular the military draft and possible resistance to it, can be effectively modelled as games. The paper has exhibited the perfect equilibria of one such model, which correspond at least roughly to historical experience. It will be interesting to see how well these methods apply to other mass social phenomena.

References

Chambers, John W. (1975). Draftees or Volunteers. New York: Garland.

Denardo, James (1985). Power in Numbers: The Political Strategy of Protest and Rebellion. Princeton: Princeton University Press.

Frohlich, Norman and Joe A. Oppenheimer (1974). The Carrot and the Stick: Optimal Program Mixes for Entrepreneurial Political Leaders. Public Choice, **19**: 4.

Green, Edward J. (1984). Continuum and Finite-Player Noncooperative Models of Competition. Econometrica, **52**: 975-993.

Güth, Werner (1985). An Extensive Game Approach to Modelling the Nuclear Deterrence Debate. Zeitschr. ges. Staatswiss., **141**: 525-538.

Mason, T. David (1984). Individual Participation in Collective Racial Violence: A Rational Choice Synthesis. Amer. Pol. Sci. Rev., **78**: 1040-1056.

Moore, Albert B. (1924). Conscription and Conflict in the Confederacy. New York: MacMillan.

Reid, Brian H., and John White (1985). 'A Mob of Stragglers and Cowards': Desertion from the Union and Confederate Armies, 1861-1865. J. Strat. Studies, **8**: 64-77.

Schmeidler, David (1973). Equilibrium Points of Nonatomic Games. J. Stat. Phys., **7**: 295-300.

Selten, Reinhard (1973). A Simple Model of Imperfect Competition, where 4 are Few and 6 are Many. Int. J. Game Th., **2**: 141-201.

Useem, Michael (1973). Conscription, Protest, and Social Conflict. New York: Wiley.

ALTRUISM IN SOCIAL DILEMMAS - FAILURE TO CATCH IT IN A PARAMETER*

Dirk Wendt

in cooperation with

Rainer Groggel, Isis M. Ksiensik, Margit Popp,
Bernhard Rummel, and Andreas Wettengel

Abstract

A social dilemma is a situation where everybody has a dominating strategy to maximize his payoff, but if everybody applied this strategy, everybody in the group would fare worse than if nobody applied it. A normative solution for this dilemma assumes a multiattribute utility function with a positively weighted component for the payoff for others which was termed altruism. This model was tested in a series of empirical studies with imaginary scenarios where Subjects were asked to make decisions in various social dilemma situations. Results indicate that subjects are willing to show considerable amounts of altruism but not in accordance with the prescriptions of the normative model. It was concluded that subjects actually do want to make rational decisions but that the normative model is too complicated to be applied by them successfully.

A social dilemma is a situation where everybody has a dominating strategy to maximize his payoff, but if everybody applied this strategy, everybody would fare worse than if nobody applied it. Such situations have been described and axiomatized by Dawes (1975) who gave formal criteria to decide whether a situation is a social dilemma or not; we will use his criteria here, too.

Special cases of social dilemmas, and most frequently studied in the litarature, are the prisoner's dilemma game (PDG), the union dilemma game (UDG), and the Commons dilemma game (CDG). In the PDG, there are two players each of whom can choose between a cooperative and

---

*This research was funded under Grant No. We 13/7-3 by the Deutsche Forschungsgemeinschaft. Paper given at the 3rd Conference on Social Dilemmas, Groningen, The Netherlands, July 25 - 29, 1988, and written up at ZIF, University of Bielefeld, during the Research Year on "Game Theory in the Behavioral Sciences", August 1988.

a defective strategy. Each of them would maximize her own payoff if she decides defectively whereas the other one decides cooperatively, but the sums of their payoffs would be maximized if both decided cooperatively. UDG and CDG involve, in principle, the same conflict but the decision maker is not playing against a single opponent but against a group: In the UDG, an individual could maximize her payoff by "free riding", e.g., refusing to pay a fee and benefit from others paying it, and in the CDG an individual could maximize her payoff by exploiting common resources more than others, and such at their expense.

While Hardin (1968) still complained that commons dilemma situations have "no technical solution", Messick & MacClintock (1968), and Messick (1973) offered such a solution for the union dilemma already which can easily be extended to all social dilemma situations. It assumes a multiattribute utility function with a positively weighted component for the wealth of others, i.e., it contributes to your utility if someone else in a similar situation gets something.

$$(1)\quad U(G,A) = W(G) * G + W(A) * A$$

where U(G,A) is the utility for the decision maker if he himself gets G, and the others get A,
W(G) is the weight he puts on his own gain G,
and W(A) is the weight he puts on the payoff to others, A.

The component W(A) has been termed "altruism", and several authors, like, e.g., Messick & MacClintock (1968), Griesinger & Livingston (1973), MacCrimmon & Messick (1976), Grzelak, Ivinski & Radzicki (1977), and Liebrand (1984) have built theories of social motivation around it.

These models are still normative in the sense that they do not throw over board expectation maximation completely, as do some expectation-and-risk considering model, like portfolio theory by Coombs (1975), Coombs & Avrunin (1977), or prospect theory by Kahneman & Tversky (1977), or Fishburn's (1977) target level risk theory. We still assume that some expectation has to be maximized; the only difference to classical expectation maximation is that we include in our expectation the payoff to others concerned, the amount A. We consider this expectation maximation a rational strategy, and evolutionary biologists (Wilson, 1975; Trivers, 1985)) have shown that altruism may well be part of evolutionary stable strategies.

So we tested these rational models to see if they actually guide human behavior in social dilemma situations.

But before we go to our empirical studies with various scenarios, we take a closer look at what our subjects should have done, according to the normative model.

Assume you are offered to join a union which aims at providing you (but also everybody else in the same position) with a gain G which can be obtained with probablitity $P(K)$ where K is the number of members who have joined the union already. $P(K)$ is monotonically increasing with K, and you would be the $(K+1)$st member, and thus increase the probability of success by DP from $P(K)$ to $P(K+1)$. To achieve this, you would have to pay a fee B if you join.
Your expected value if you don't join would be

$$(2) \qquad EV(G/\text{don't join}) = P(K) * G,$$

and your expected value if you join would be

$$(3) \qquad EV(G/\text{join}) = P(K+1) * G - B,$$

and this is what you should do if you were a pure expected value maximizer:

$$\begin{array}{lll} (4) & \text{join iff} & EV(G/\text{join}) > EV(G/\text{don't join}) \\ & \text{i.e., iff} & P(K+1)*G - B > P(K) * G, \\ & \text{i.e., iff} & (P(K+1) - P(K)) * G = DP * G > B, \\ & \text{i.e., iff} & DP > B/G. \end{array}$$

However, if you care for the wealth of others, and apply a multiattribute utility function like Eq. (1) to evaluate your own payoff and those to others, you should

$$\begin{array}{lll} (5) & \text{join iff} & DP * (W(G) * G + W(A) * A) > W(G) * B, \\ & \text{i.e., iff} & DP > (W(G) * B)/(W(G) * G + W(A) * A). \end{array}$$

More precisely, once you are concerned about the wealth of others and include their payoffs in your multiattribute utility function, you should also care about how many people get this payoff. Thus, if N other people get a payoff A, and you yourself get a payoff G, your utility would be

$$(6) \qquad U(G, NA) = W(G) * G + W(A) * N * A.$$

This implies a slight change in Eq.(5); if you include the numb r of other people involved, N, Eq.(5) becomes

$$(7) \qquad \text{join iff } DP > (W(G) * B)/(W(G) * G + W(A) * N * A).$$

The number of other people involved, N, may have different kinds of influences upon the amount someone is willing to invest in a union dilemma situation, depending on whether he sees them as others who just benefit from it, or who could also be expected to contribute something. In the first case, B should go up with N, in the second case it could go down.

In all these cases, your willingness to join the union will be d p ndent on how much you care for the wealth of others, i.e., on the siz of the altruism parameter W(A). Dependent on W(A), there will be a certain threshold for the fee B up to which you should be willing to join, given certain amounts of DP, G, A and N, but you should not join, according to Eqs.(5) and (7), if the fee B were higher than that threshold. Let us call this threshold Bmax, the maximal fee B up to which you would join the union, given DP, G, A, N, and the parameters of your individual utility function W(G) and W(A).

If you are a rational person, you should at least maximize expectation for your own gain, W(G), so assume that we can set

(8) $W(G) = 1.$

If we now find a threshold Bmax up to which you are willing to join a union in a given situation, we can use this to calculate your individual altruism parameter W(A) by means of

(9) $W(A) = (Bmax - DP * G)/(DP * A)$ (from Eq. 5), or

(10) $W(A) = (Bmax - DP * G)/(DP * N * A)$ (from Eq. 7).

This weighting factor W(A), calculated from subjects' behavior displayed by their respective Bmax in various situations (scenarios) is th parameter we have been hunting for in the studies to be reported in the following pages. In our first studies, we hoped to find it as a kind of personality variable like a social orientation, as presented by Kuhlman (1988), but our later studies (Wendt, Popp, Wettengel & Ksiensik, 1988, Ksiensik & Wendt, 1988) led us to expect it also to be d pendent on the situation: different social contexts may call for more or less "altruism" in our society. But now, we suspect it might as well be a random variable, at least for some of our subjects.

Our aim in the studies to be reported here was to vary contexts systematically to see how much they contribute to the amount of altruism displayed by the subjects. Thus, we tried to invent scenarios where you would expect more or less altruism considered appropriate in our society given the social context. Unfortunately, we could not test our models in real life but only in imaginary situations, in "scenarios", as we called them.

In our scenarios, we tried to catch some rather diverse situations of social dilemmas, because we got the impression, from former studies (e.g., Wendt, Popp, Wettengel & Ksiensik, 1988), that the amount of altruism displayed by subjects depends on the social context of the situation: In a plain gambling or business situation where it is th declared goal to maximize expected payoff there would be less altruism than in a peer group situation where people are more personally involved.

Study 1

Our first study was a situation where you would expect a lot of altruism, in a real life situation: Students like to study for their exams from so called "scripts", i.e. syllabi which are put together by students for students from lecture notes, excerpts from books and articles, and collections of exam questions from previous exam p riods (together with the answers to the questions they consider right). Learning from such scripts saves them a lot of time in preparation for the exam since these scripts contain exactly what has to be learned to pass the exam, and so are more concentrated than are books and articles. Thus, students love such scripts, and try to prepare them, or rather get them for as many subjects as possible. Preparing such a script, however, requires manpower; some students have to work more to collect the material, and write the script. Once it is ready, it is available in the library for everybody to be read or xeroxed, no matter whether he contributed to it or not. Thus, we have a union dilemma situation here: Some students have to contribute working hours, and all students will profit from it, if they invested or not.

At the University of Kiel, we were in the position that, for a certain subject, there existed no such script but a considerable demand for it. So we took the opportunity to use this situation for the following experiment: With questionnaires, we found out from th previous generation of students (i.e., from those who had passed this exam already) how much working time (in hours) students expected to save in their exam preparation, and considered this the "gain" G if the enterprise to put together a script was successful. So we had som realistic figures to present as possible gain G in the union dilemma situation to follow.

Another group of 50 students (who had not yet passed the respective exam) were asked in another questionnaire, under a variation of conditions, how much time in hours they would be willing to invest into the preparation of such a script. Conditions were varied in a 2x2x2x3 ANOVA design, with two levels of gain for the decision maker himself (G

= 30 hours or G = 60 hours), two levels of gain for others (A = 30 hours or A = 60 hours), two levels of number of students who would benefit from the script (N = 60 ,i.e., one year of students in psychology at the Univerity of Kiel, or N = 180, three years of students), and three levels of a-priori probability of success (K = 3, 5 or 7 students who allegedly had signed up already to work on the script - participants were told that it would take at least 10 students to work on the script to complete it so that 7, 5 or 3 more students were needed to accomplish the task.).The levels of G (30 vs. 60 hours saved) were realistic figures, taken out of the range of expected savings in preparation time indicated by the group of older students. It was also true that some students expected to save more than double as many hours by using a script than others did.

In this situation, we expected students to display a lot of altruism. They were peers, in the same situation, and some solidarity would be not only highly social desirable but actually advantageous for surviving the common "danger" of the exam. Indeed, some of them stated that they would invest as much time as was needed to complete the script no matter how long it took. We also expected the amount of time to be invested to go up with the number of (other) students who would benefit from the investment, since these others were students who were not yet in the position to contribute, but would enter the University in future years.

The results showed that subjects were willing to contribute, on the average, 20.8 hours if 60 students would benefit from their work, and 23.1 hours if it was for 180 students, so this difference was in the right direction as predicted by the normative model. With 30 hours to be saved - no matter if for themselves or for others - they were willing to contribute, on the average, 17.8 hours, and 25.8 hours with 60 hours to be saved - which is also in the right direction but by far not as strong as required by the model: with twice the gain, they should be willing also to double their efforts. A completely selfish person would have invested only DP * G = 3 h with G = 30 h to be saved, and 6 h with G = 60 h to be saved. So subjects were willing to invest much more time than requested for pure payoff maximation, but were not sensitive enough to the parameters of the situation. If we calculate the altruism parameter W(A) from these data, we find W(A) = 0.05 for N = 60, and W(A) = 0.02 for N = 180 from Eq. (7), i. e. assuming larger investment with larger N, and W(A) = 3.62 for N = 60 and W(A) = 4.13 for N = 180 if we calculate it from Eq. (5), i.e., disregarding N in the calculation so it looks like the truth lies somewhere between these two models -- or outside both of them. The independent variable K (number of students contributing already) had mixed influences which are hard to generalize: Some of the students felt it would not be worthwhile to invest much if there were so many others contributing already whereas others felt obliged to contribute just in that case, since their contribution could be needed to reach the critical number.

Study 2

We expected less altruism in a pure business situation which was our first scenario in Study 2: A group of condominium or home owners see their property endangered by the city planning a thoroughfare, a large road which would attract lots of traffic close to their apartments and which, if built, would considerably reduce the value of the apartments because for the noise, dirt, and air pollution involved. To avoid this, some of them form an initiative to hire a lawyer to protest, an hopefully force the city to change their planning such that the apartment are no longer affected. If you were in the position of such a home owner, how much would you invest in such an enterprise? This is a union dilemma, again: all home owners would profit if the road were not built, no matter if they contributed or not, but if nobody contributes, the road will certainly be built, and hurt all of them. But, we did not expect them to develop any kind of peer solidarity in this situation, but just as much altruism as needed to avoid the financial loss.

In this situation, the appropriate amount to be invested (B), according to the rational model, should be the higher the larger the loss in the value of the apartments to be avoided (G), and higher with a larger increase (DP) in the probability of success (P), and also higher if more home owner were concerned (N) - for, if you care at all for the income of others, you should care more if more others are affected. But, the amount to be invested (B) should be independent of the a-priori probability of success (P), as can easily be figured, and only move up with the increase of that probability (DP) achieved by the contribution.

These last two features of the normative model, i.e., the increase of the appropriate fee with group size, and its independence of a-priori probability, seem to be rather counterintuitive, as we will see.

The second scenario in Study 2 was the situation of a group of hobby gardeners cultivating allotments in a "Kleingartenverein". They plan to drill a new well to improve the irrigation possibilities for their gardens and - if they find water - save them working hours (G) to fetch water as it has to be done now. Working hours (B) have to be invested to build the new well, and the probability of success (P) increases the more work is invested. But once the well is built, nobody can keep those who did not contribute from using it too, so that we hav a union dilemma situation again. How much working time (B) would you contribute in such a situation? Here, we expected more altruism than in the home owner scenario since these "Kleingartenvereins" are a kind of clubs which develop a lot of social life, and their members oftentimes share some common interests and beliefs, if not ideologi s.

The third scenario in Study 2 was a joint advertising campaign of a group of house owners who live in a vacation or recreational area and let rooms for rent to tourists and who hope to attract more tourists to th region by the campaign. The campaign, if successful, would benefit (G) all who let rooms in the region, no matter what they contributed (B) to the campaign. How much would you contribute in this case ? We expected less altruism in this situation than in the previous one because these landlords and landladies have little or nothing in common except that they all rent rooms to tourists, and, moreover, th y compete for them. So "altruism" in this situation should go only as far as needed to fill their beds and tills.

The three scenarios in Study 2 were displayed to participants by means of the BTX medium of the Deutsche Bundespost: People connected to this network by their telephone lines could dial a certain number, and get the pages of our questionnaire onto their tv screen or home comput r if they had one. They could answer our questions by keyboard, or simply by dialing the numbers associated with the possible answers on the dial of their telephone if they had no keyboard. The answers are transfered to the other party's electronic "mailbox", and can be call d from there to the addressee's tv screen or home computer whenever he wants to see them. Our project was advertised in some magazines specially addressed to BTX participants, and also transferred by lectronic mail into the "mailboxes" of all non-commercial BTX participants in the BRD. We got over 400 more or less useful replies; most respondents quit after the first questions, and finally we were l ft with 125 complete answeres which could be used for data analysis. W knew nothing about these respondents and their motivation to call us; some may just have been curious and peeked into our game. We had promised 200 DM for the best solution, so some may have hoped for an easy win, and then quit when they noticed that our tasks looked like work.

On the average, these subjects reacted correctly, i.e. in the direction predicted by the normative model, to the variation of gain G, and of increase in probability DP, but they reacted also to th a-priori probability P, and in the context of adv rtising for rooms for rent, they reacted in the wrong direction with respect to the number of people involved: They agreed, on the average, to pay less when more people were concerned. This is, however, a phenomenon not compl tely new and surprising: It is well known from other studies (without normative model) that cooperation goes down with increasing group size. Behind this strategy, there may be another normative assumption, nam ly that everybody has to contribute less if more people share the burden - which was but not implied by our model.

Study 3

In Study 3 (Groggel, 1988), we repeated the scenarios of Study 2, but this time with 86 students as subject at the terminal of a lab comput r. Their respective altruism parameters W(A) were immediately calculated from their reaction so that we were able to test the mod l. We inferrred their altruism parameters W(A) from the first set of their decisions in the experimental session, and then presented them a second set of decision situations in the same context where we could predict from the first set and their respective altruism parameters how they should decide if they behaved according to the normative model, i.e., if they maximized expected utility as given by the multiattribut utility function with their individual altruism parameter W(A) as inferred from their first set of decisions.

The scenario of hobby gardeners jointly digging for a well was replaced in Study 3 by a scenario of hobby fisherman who jointly owned a lake where they wanted to breed a new species of fish - again a union dilemma: the more of these new fish could be bought to be put into the lake, the more likely they would reproduce and settle in the lake, but after that, everybody could catch them. This replacement was done because we tried, in Study 3, also a commons dilemma situation which was also in the context of hobby fishing: In this additional scenario, subjects had to decide how much fish they would be willing to take out each season, give the reproduction function of the fish population in addition to the other independent variables such as group size N, and gains G and A for the decision maker and other participants.

In the scenarios of Study 3, subjects again reacted correctly to the variation of G and DP, but not to N, and in the home owner scenario incorrectly to P, i.e., investing more with larger a-priori probabilities P.

Individual strategies

However, the average trends reported above are by far not representative for the whole group, neither in Study 2, nor in Study 3. In both groups, many subjects displayed individual strategies to which they stuck consistently over situations, in spite of the fact that these strategies were all "wrong" from a normative point of view. We collected these individual strategies, or "fallacies", as we called them, to see what subjects actually do in such situations. These individual strategies, and their respective frequencies of occurrence in our subject groups, are displayed in Tables 1 through 7. These Tables are to be read as follows: The outer margins show the variation of the independent variables. The inner margins show how the dependent variable, Bmax as indicated by the subjects, varied with respect to the independent variable on the outer margin next to it. Figures in the body of the tables indicate the number of subjects who reacted that way. To save space, Bmax has been abreviated to B. "B+" indicates that Bmax increased, "B-" indicates that Bmax decreased, and "B=" means that Bmax was independent of the independent variable next to it at the outer margin. As an example, let us take the "12" in the upper left corner of Table 1. This tells us that 12 people increased their Bmax with increasing P (left margin), that they increased it also with G = A (right margin), and also with N (top margin).

In Tables 1 through 7, we underlined those numbers of subjects who reacted as predicted by the normative model, with respect to all independent variables involved.

Table 1 shows that only 4 out of 49 subjects did so in Study 1: those 4 who did not react to P (left margin, B=), increased their Bmax with G = A (right margin, B+) and with N (top margin, B+).

In Table 2 we find only one subject (out of 126) behaving according to the model in the home owner scenario of Study 2, and in Table 3 we find two such persons (out of 86) in Study 3.

We are glad to report that these two subjects in Study 2 fulfilled also the consistency test mentionned above,in all three scenarios, so that we can assume that they really behaved consistently according to the normative model.

In the scenario of advertising for rooms for rent, the situation looks slightly better: 8 subjects consistent with the model in Study 2, and 17 in Study 3. But in the well digging scenario of Study 2 there is, again, only one subject left (Table 6), and two of them in the corresponding fishery scenario in Study 3 (Table 7).

'p'		'n' increases				'G = A'
		B +	B =	B -		
i	B +	12	3	2	B +	i
n		1	1	-	B =	n
c		-	-	-	B -	c
r	B =	<u>4</u>	3	3	B +	r
e		-	3	-	B =	e
a		-	3	-	B -	a
s	B -	5	6	4	B +	s
e		-	1	-	B =	e
s		1	-	-	B -	s

Table 1: Syllabus (absolute frequencies of strategies and fallacies) Study One

'p'		(A ≤ G) 'A' increases				'A'
		B +	B =	B -		
i	B +	2	-	3	B +	i
n		1	23	8	B =	n
c		6	6	26	B -	c
r	B =	<u>1</u>	2	-	B +	r
e		-	12	-	B =	e
a		-	-	4	B -	a
s	B -	2	-	3	B +	s
e		-	4	2	B =	e
s		1	3	16	B -	s (A ≥ G)

Table 2: UDG 1 - Homeowners against planning a thoroughfare (absolute frequencies of strategies and fallacies (Study Two)

'p' increases		(A ≤ G) 'A' increases				'A' increases (A ≥ G)
		B +	B =	B -		
	B +	5	4	-	B +	
		2	19	1	B =	
		2	10	7	B -	
	B =	2	-	-	B +	
		-	9	-	B =	
		-	3	-	B -	
	B -	1	-	-	B +	
		-	5	-	B =	
		2	4	9	B -	

Table 3: UDG 1 - Homeowners against planning a thoroughfare (absolute frequencies of strategies and fallacies (Study Three)

'n' increases		`G = A` increases		
		B +	B =	B -
	B +	8	1	1
	B =	40	2	1
	B -	49	3	9

Table 4: UDG 2 - Advertisement for letting rooms for rent (absolute frequencies of strategies and fallacies (Study Two)

'n' increases		'G = A' increases		
		B +	B =	B -
	B +	17	4	1
	B =	33	2	-
	B -	21	1	-

Table 5: UDG 2 - Advertisement for letting rooms for rent (absolute frequencies of strategies and fallacies (Study Three)

'n' increases		'Dp' increases				'p' increases
		B +	B =	B -		
	B +	13	3	1	B +	
		1	-	-	B =	
		2	1	1	B -	
	B =	6	8	-	B +	
		7	8	-	B =	
		3	2	1	B -	
	B -	36	3	2	B +	
		3	3	-	B =	
		1	5	3	B -	

Table 6: UDG 3 - Initiative for drilling a well for a community of allotment owners (absolute frequencies of strategies and fallacies) (Study Two)

'n'		'Dp' increases				'p'
		B +	B =	B -		
i	B +	6	5	-	B +	i
n		2	-	-	B =	n
c		1	-	-	B -	c
r	B =	14	18	-	B +	r
e		5	4	-	B =	e
a		4	2	-	B -	a
s	B -	9	6	-	B +	s
e		-	1	-	B =	e
s		2	1	1	B -	s

Table 7: UDG 3 - Fishery project (absolute frequencies of strategies and fallacies) (Study Three)

Study 4

Some subjects in Studies 2 und 3 obviously used models different from our normative one, and rather consistently so. We tried to find out what the rationale behind these might be. In a follow-up study Rummel (1988) presented the same scenarios as in Study 3 to a new set of subjects, again on the terminal of the lab computer, with the same program. But this time subjects were asked individually what influence each independent variable had upon their decisions, and why. Only a few of them named the relevant variables implied by the normative model. Most of them thought (correctly) that their own gain G would be relevant, (erroneously) that the a-priori probability P, and (correctly again) that the increase in probability DP would count, but only half of them considered the number of others concerned, N, and their possible gains, A. Most interesting (but also disappointing) in Study 4 was the discrepancy between what people said, and what actually did influence their behavior. So, although we do have many extensive protocols of what subjects claimed to do, and what the actually did in this Study, it seems not worthwhile reproducing them here. This result is in accordance with findings by Nisbett & Wilson (1977) who reviewed many studies from various fields where no correspondence between subjects' reports on their mental processes and their respective behavior has been found, also in situations of problem solving and decision making.

Study 5

After our experiences from Studies 2, 3 and 4, we had the impression that the normative model might be too complex to be understood and applied by our subjects - if you have to prove it to your statistically sophisticated research assistents with pencil on paper that the a-priori probability P should not count in your decision on Bmax, how could you expect your unaided subjects to know?

So we tried, in Study 5, to create a very simple gambling situation where all the relevant variables were visible on the table, so to speak. We invented a die rolling game: A regular die was rolled on the table in front of the subjects, and the face which came up decided if the subjects would get a prize G ,or not. If certain faces came up, everybody in the group would get the prize G, but the winning faces of the die had to be paid for first. In the beginning, no die face was winning unless somebody in the group bought it. Subjects could buy die faces in a kind of an auction, in a Marschak bidding procedure: They could bid a price up to which they were willing to buy a die face, and the computer came up with a random number. If the random number was above their bid, they got nothing at all, but if the random number was below their bid, they got it at the price of the random number. Thus, it was always best for them to bid exactly the maximum price they were willing to pay for a die face. In each round, as many die faces were "sold", i.e., made winning die faces as there were offers above the computer's random number. Then the die was rolled, and everybody got a prize G if the face that came up was a winning one.

This situation is a union dilemma, again: you can sit there and pay nothing and get a prize G if a die face comes up which someone else has paid for, but you can also pay and lose your money if a face comes up which nobody has paid for.

In Study 5, we varied group size N in three levels: N = 3, 6, or 12, gain G in two levels: G = 100 or 200 points which were actually paid to them in Pfennige, and P in three levels: P = 0, 1/3, or 2/3. The results, i.e., average bids Bmax, and the respective altruism parameter W(A) resulting from them according to Eq.(10), are displayed in Table 8.

<table>
<tr><td>Means in subsample</td><td colspan="6">Ss 1 - 12</td><td colspan="6">Ss 13 - 24</td></tr>
<tr><td>Group size N</td><td colspan="3">N = 12</td><td colspan="3">N = 6</td><td colspan="3">N = 6</td><td colspan="3">N = 3</td></tr>
<tr><td>played first/last</td><td colspan="3">first</td><td colspan="3">last</td><td colspan="3">last</td><td colspan="3">first</td></tr>
<tr><td>Bmax
W(A)</td><td colspan="3">106.04
.29</td><td colspan="3">60.28
.28</td><td colspan="3">99.59
.60</td><td colspan="3">59.26
.69</td></tr>
<tr><td>a-priori P</td><td colspan="2">P = 0</td><td colspan="2">P = 1/3</td><td colspan="2">P = 2/3</td><td colspan="2">P = 0</td><td colspan="2">P = 1/3</td><td colspan="2">P = 2/3</td></tr>
<tr><td>Bmax
W(A)</td><td colspan="2">103.99
.40</td><td colspan="2">78.98
.27</td><td colspan="2">56.09
.16</td><td colspan="2">95.47
.80</td><td colspan="2">83.54
.67</td><td colspan="2">51.24
.30</td></tr>
<tr><td>gain size G</td><td colspan="3">G = 100</td><td colspan="3">G = 200</td><td colspan="3">G = 100</td><td colspan="3">G = 200</td></tr>
<tr><td>Bmax
W(A)</td><td colspan="3">52.40
.27</td><td colspan="3">106.45
.27</td><td colspan="3">54.82
.65</td><td colspan="3">111.80
.67</td></tr>
<tr><td colspan="13">Interactions :
There was a significant interaction of gain G and a-priori probablitity P, in both subsamples, in the sense that subjects invested more than expected under the condition P = 2/3 and G = 100, and less than expected under the condition P = 2/3, G = 200. Another interaction existed between group size N and a-priori probability P: Both subsamples tended to invest more with P = 2/3 and the smaller group size, and less with P = 2/3 and the larger group size.</td></tr>
</table>

Table 8: Main effects and interactions in Study 5, Die Rolling Game 1

These results, finally, do show some consistency with the normative model, but we suspect that some subjects may have been doing the right thing for the wrong reason: In the experimental procedure described above, we gave them the range of the random number against which to bid th ir maximum buying price for a die face, to make sure they would not name numbers outside the range of the random numbers from the computer. But some subjects took these ranges of possible random numbers as guidance to decide on their maximum buying price Bmax, by bidding certain proportions of the maximal random number indicated. Since the maximum random number of the computer was related to the parameters of the respective situation, the variation of bids tied to the maximum random number would be in some accordance with the normative mod 1, too.

So we avoided this source of errors in a repetition of the die rolling game. We left group size N = 12 out because such large groups were hard to get together, and we did not expect too much new information from them. We left the variation of P out, and varied DP instead, in two levels: DP = 1/6 and DP = 1/3 (Subjects could bid a maximum buying price for two faces of the die).

The results of this repetition are shown in Table 9. They are still in the direction predicted by the normative model, although less drastically so than in the first die rolling game. For comparison, Table 10 shows the extreme possible bids Bmax for the situations of the second die rolling game, with the extremes of W(A), W(A) = 0 and W(A) = 1. Thus, subjects' actual bids Bmax are within the right range but do not vary as much as predicted by the normative model with the independent variables in the situation.

So, we finally conclude that there actually is some altruism involved in subjects' decisions in these social dilemma situation, and even more than recommended by the normative model, but that the model is probably too complicated to be used to describe subjects' behavior.

Means in subsample	Ss 1 - 12		Ss 13 - 24	
Group size N	N = 3	N = 6	N = 3	N = 6
played first/last	first	last	last	first
increase in probability of winning: DP = 1/6 Bmax W(A)	43.38 .53	37.05 .20	28.4 .23	43.75 .27
DP = 1/3 Bmax W(A)	52.48 .19	-	37.9 .04	-

Table 9: Mean bids in Study 5, Die Rolling Game 2

Bmax with G = A = 100	DP = 1/6	DP = 1/3
with W(A) = 0	16.7	33.3
with W(A) = 1 N = 3 N = 6	 50 100	 100 200

Table 10: Normative optimal bids in Die Rolling Game

Alternative strategies and models

We spent some thoughts on what else might have influenced our subjects' behavior if not our normative model, particularly in those cases where they seemed to apply, consistently, some different rules.

As we have seen from Tables 2 and 3, in the home owner scenario where G and A varied independently, many subjects decreased their maximal investment (Bmax) with increasing gains (A) for others: In Study 2 about half of the subjects (62 out of 125 BTX participants) did so if their own gain G was below that of others, $G < A$, and the same number did so if the reverse condition was given, $G > A$, and 46 out of the 125 did so in both cases. The respective figures from Study 3 were less drastic: 37 out of 85 students at the lab computer terminal decreased Bmax with increasing A if $G < A$, and only 17 out of 85 if $G > A$, and only 16 in both cases. In the case of $G < A$ this kind of behavior could be explained in terms of "envy": People don't like to give too much if others benefit from it more than they themselves although they are still altruistic in their choice of Bmax. Güth (1986) suggested to include this "envy" component into the multiattribute utility function, with a weight of its own: $W(G-A) * (G - A)$.

Another peculiarity of our subjects' behavior that might be worthwhile modelling is the inverse reaction of many subjects to the increase of group size N, particularily in the rooms-for-rent scenario in Studies 2 and 3, but also in one subgroup of the Die Rolling Game 2 in Study 5. This could be explained by the assumption that there is a certain fixed cost (e.g., of advertising in the rooms-for-rent scenario, or to buy all the faces of the die in the Die Rolling Game) to be shared -- and if the cost is fixed, then it seems just and rational that every participant's share should be the smaller the more people are concerned. Some comments in this direction were made by subjects in the Die Rolling Game, and there it seemed to make sense: Once all die faces have been bought, nobody could be interested in buying more. However, subjects in this situation overlooked the fact that, given the auctionning character of the Marschak bidding procedure, nobody would get any die face if they all bid too low. And in the room-for-rent scenario, some subjects may have misunderstood the implication that every additional contribution would attract more tourists to the region, no matter how much had been invested already by others -- so there were no fixed costs to share, but subjects may have thought so.

Arjaan Wit (1988) suggested we should teach our subjects the normative model, and how to use it in their subsequent decisions, in order to test our hypothesis that it might be too complicated for them to follow it at all. Although this was not the initial problem of the studies reported here, we can try that in later experiments.

As a more general criticism, we might question our assumption of altruism as a unidimensional concept at all, as we did when trying to catch it in a single parameter, W(A). Instead, it might be a whole set, or vector of parameters, with components for different situations and social contexts -- as we suspected already when concocting the various scenarios. We may find altruism from pure ethical reasons, regardless what it costs, as indicated by some of our subjects in Study 1, or altruism displayed only because of its social desirability, or altruism for peers in order to help the peer group survive (which may well be a selfish desire in the long run), or altruism in order to secure other long term resources also for oneself, or reciprocal altruism (where it pays to be altruistic expecting others to be so too -- for a game theoretic analysis of this kind of altruism, see Güth, 1984, and Messick, 1988), or non-reciprocal altruism, or altruism because of some religious belief that one will be paid back later (in which case we may doubt it if there is altruism involved in it at all.

For our initial hypothesis of altruism or social orientation as some kind of a relatively stable personality trait, it was supportive already that in all our ANOVAs the largest source of variance was between subjects. The other independent variables, like G, DP, P, and N, showed significant effects, if at all, only after removal of the between-subjects variance from the error variance. If we follow this line of thought, we may come up with a general linear model to describe an individual's amount of altruism in a specific situation: The larg st component would be the individual (as our ANOVAs show), and second might be some situational, or context variable. But, for the pres nt data, error variances and inconsistencies are too large to do such an analysis.

References

Coombs, C.H. (1975): Portfolio theory and the measurement of risk. In: M. Kaplan & S. Schwartz (eds.): Human Judgment and Decision Processes. New York: Academic Press, 64-85.

Coombs, C.H., & Avrunin, G.S. (1977): Single-peaked functions and the theory of preference. Psychological Review, 84, 216-230.

Dawes, R. M. (1975): Formal models of dilemmas in social decision making. In: M.Kaplin & S. Schwartz (eds.): Human Judgment and Decision Processes. New York: Academic Press, 87-107.

Fishburn, P.C. (1977): Mean-risk analysis with risk associated with below-target returns. American Economic Review, 67, 116-126.

Gri singer, D.W., & Livingston, J.W. (1973): Toward a model of interpersonal motivation in experimental games. Behavioral Science, 18, 173-188.

Groggel, R. (1988): Kontextabhängigkeit von 'Egoismus' und 'Altruismus' in experimentellen Spielen. Diplomarbeit, Kiel: Institut für Psychologie der Universität.

Grz lak, J.L., Ivinski,T.B., & Radzicki, J.J. (1977): "Motivational" components of utility. In: H.Jungermann & G. DeZeeuw (eds.): Decision Making and Change in Human Affairs. Dordrecht, NL: Reidel, pp. 215-230.

Güth, W. (1984): Egoismus und Altruismus. Eine spieltheoretische und experimentelle Analyse. In: Todt, H. (Hrsg.): Normengeleitetes Verhalten in den Sozialwissenschaften. Berlin: Duncker & Humblot, pp. 35-58.

Güth, W. (1986) private communication at Bielefeld Conference on Bounded Rational Behavior in Experimental Games and Markets, Sept. 21-25,1986.

Hardin, G.R. (1968): The tragedy of the commons. Science, 162, 1243-1248.

Kahneman, D., & Tversky, A. (1979): Prospect theory: An analysis of decisions under risk. Econometrica, 47, 262-291.

Ksiensik, M.I., & Wendt, D. (1988): Normative and individual strategies in social dilemmata. In: R.Tietz, W.Albers, & R. Selten (eds.): Bounded Rational Behavior in Experimental Games and Markets. B rlin: Springer, pp.21-36.

Kuhlman, (1988): Paper given at the 3rd Conference on Social Dilemmas, Groningen, The Netherlands, July 25-29, 1988.

Liebrand, W. (1984): The effect of social motives, communication and group size on behavior in an n-person multistage mixed-motiv game. European Journal of Social Psychology, 14, 239-264.

MacCrimmon, K.R., & Messick, D. (1976): A framework for social motives. B havioral Science, 21, 86-100.

M ssick, D.M. (1973): To join or not to join : An approach to the unionization decision. Organisational Behavior and Human Performance, 10, 145-156.

Messick, D.M. (1988): On the evolution of group-based altruism. Paper given at Symposium on Evolutionary Game Theory, July 11-15, 1988, ZIF, University of Bielefeld.

M ssick, D.M., & MacClintock, C.G. (1968): Motivational bases of choice in experimental games. Journal of Experimental Social Psychology, 4, 1-25.

Nisb tt, R.E., & Wilson, T.D. (1977): Telling more than we can know: Verbal reports on mental processes. Psychological Review, 84, 231-259.

Rumm l, B. (1988): Rationales und reales Verhalten in sozialen Dil mmata: Konsistenz zwischen Situationen und zwisch n Einstellung und V rhalten. Manuscript, Kiel: Institut für Psychologie der Universität.

Triv rs, R. (1985): Social Evolution. Menlo Park, Calif.: Th Benjamin/Cunnings Publishing Company.

Wendt, D., Popp, M., Wettengel, A., & Ksiensik, I.M. (1988): Egotism and altruism in social dilemmata. In: A.Upmeyer (ed.): Attitude and Behavioral Decision Making. New York: Springer, pp. 149-161.

Wilson, E.O. (1975): Sociobiology: The New Synthesis. Cambridge, Mass.: The Belknap Press of Harvard University Press.

Wit, A. (1988) private communication at ZUMA Workshop with Anatol Rapoport,Mannheim, June 13-15, 1988.

THE STRATEGY OF CHALLENGES:
TWO BEHEADING GAMES IN MEDIAEVAL LITERATURE

by

Barry O'Neill

Abstract. I use game models to analyse two mediaeval tales about remarkable challenges. The goal is to understand their plots and to clarify in general why challenges are made and accepted. Children's contests of daring provide a simple context to study challenging, and I argue that children seek a certain reputation; they want to be known for placing a high payoff weight on others' estimate of that very weight. This definition might seem circular, but it leads to well-specified, solvable games. The model for the children's dares is modified in steps to fit the mediaeval stories. Game theory has treated language as a way to transmit information, but here verbal challenges are speech acts, or "performatives" in Austin's sense, that trigger the concern for reputation. The analysis of challenging is relevant to international conflicts where large powers' make commitments and struggle in the Third World over "credibility."

The Middle English poem *Sir Gawain and the Green Knight* and its Old Irish ancestor *The Feast of Bricriu* recount a remarkable stranger's challenge to the hero, in effect, "You can chop off my head if you'll let me return the blow." Old romances and sagas teem with challenges, but this one puts the hero in a genuine dilemma. This paper analyses the situation using the theory of games, to clarify some obscure plot elements and to show that the best action is strategically interesting, that it exacts real risks, costs and tradeoffs. A game approach reveals some of the reasons why their storylines are so successful.

SIR GAWAIN AND THE GREEN KNIGHT

Sir Gawain and the Green Knight survives in one small manuscript (British Library, Cotton Nero A.x). Its script and language indicate that it was written in the late 1300s. Authorship then conferred less personal celebrity and the poem bears no signature. It would likely have been read aloud to a gathering, and its length of 2500 lines would have taken up several nights, like a television mini-series. The poem was unknown to the literary world for four and a half centuries, until the palaeographer Sir Frederic Madden published it in 1839. Since then it has prompted a great scholarly outpouring: the number of books and papers is nearing one thousand, and only the writings on *Beowulf* and Chaucer's poetry surpass this amount for an English work of such antiquity. It has been popular with general audiences, and can be read in Breton, Turkish, Japanese and several other languages, or viewed as a motion picture starring Sean Connery as the fierce green man.

Laura Hibbard Loomis (1959, p.528) borrows the Green Knight's epithet for Gawain to praise the poem itself, "The hero of *Gawain and the Green Knight* is likened to a pearl beside a pea (vs. 2354), and so might the poem be reckoned among its contemporaries. It moves over an almost flawless structure as smoothly as supple skin moves over the bones of the hand. With the exception of Chaucer's *Troilus and*

Criseyde, no other Middle English romance approaches its artistic and spiritual maturity, its brilliant realism, its dramatic vigour, its poetic sensitivity to nuances of word and mood, its humour, its nobility of spirit." While most mediaeval works allow us a glimpse into a vanished past, this one engages modern readers by addressing modern problems. Gawain tries to maintain his ideals when the world brings them into opposition with each other, forcing him to sacrifice one or another.

The poem recounts the green stranger's intrusion at Camelot's Christmas feast, his offer to allow anyone to behead him in exchange for a return blow, Gawain's acceptance of the dare, and the latter's eventful search for the Green Knight to give him his turn. The challenger is of immense size and entirely green. As Gawain deliberates whether to accept, he and the reader will ask the same questions: Why would the Green Knight propose this bizarre game? Is he supernatural or deranged, or neither? What are the stakes for me, and what are my prospects? Both the reader and Gawain strive to construct the game.

This paper offers two game analyses of the poem. The first game, of which I describe three versions, portrays Gawain as taking up the dare for the sake of his reputation. This thesis seems reasonable, almost obvious, but it is elliptical: his reputation for what? A reputation is a general belief about a person and must have content. At first thought, the reputation is simply for bravery, but this hypothesis gives no role to the Green Knight. If Gawain had needed to boost his name for risktaking he could have performed a display without waiting for someone to challenge him. And he could have devised better proofs of bravery than this beheading match where he runs the risk for no higher end. The Green Knight questioned the Round Table's bravery but offered no evidence to put it in doubt, beyond the fact that the Court was not taking up his bizarre challenge.

The Green Knight himself identifies the issue addressed in this paper. When his first offer produces silence, he furrows his green brow, draws himself up to his full height, and speaks to the Round Table:

> What, is þis Arþureȝ hous, quoþ þe haþel þenne
> Þat al þe rouse rennes þurȝ ryalmes so mony?
> Where is now your sourquydrye and your conquestes
> Your gryndellayk and your greme and your grete wordes?
> Now is þe reuel and þe renoun of þe Rounde Table
> Ouerwalt wyth a worde of on wyȝes speche
> For al dares for drede withoute dynt schewed. (309-315)

(Excerpts are from the edition of Tolkien and Gordon, 1925.) This is surely not Arthur's house renowned through many kingdoms, he is declaring. The Round Table's pride and fierceness and bragging talk have vanished, its fame has been upset by one man's speech, *his words alone*, for all are cowering without a blow being dealt. How could the Green Knight's mere words make them play his game or suffer disgrace? The paper will try to answer this question.

Gawain's goal, I will argue, is a reputation for placing a high value on his reputation. This objective seems circular, but as the first model shows, it is well-specified in the sense that it allows Gawain to decide on an action. Defining payoffs recursively in this way leads to the consequence that sensible actors sometimes choose to fight over an issue of no direct value to them whatsoever. Because of this recursiveness, the goal is confusing when stated in English, but it is easy to formulate in mathematics, and it is also within the intuitive reach of ordinary people, even though they may be unable to articulate it. It seems to underly children's contests of daring. The power of the Green Knight's verbal challenge in triggering Gawain's display of bravery is explained here through a theory of daring contests, combined with John

Austin's theory of performative utterances, which describes how certain sentences function to change the social relationships of the speaker and listener. A dare is one such performative utterance, and the Green Knight's words infuse Gawain with a new motive to establish a particular type of reputation. The Green Knight's attempt to set up a compelling dare explains many text elements.

The paper then examines a much earlier beheading story, the final episode of *The Feast of Bricriu.* Events are superficially parallel to the Gawain story, but the underlying game is different, and the difference explains changes in the dialogue and the procedure of the beheading. An informal strategic analysis clarifies some obscure events in the plot. The next section of the paper presents a second game, one played wholly within Gawain, a conflict between his dutiful nature and his self-preserving nature. Finally I add some comments on how game theory can supplement conventional literary analysis, and how the model of daring clarifies the motives behind superpower proxy wars in the Third World.

SUMMARY OF THE BEHEADING SCENE

Arthur was the noblest king of England, and here is an extraordinary episode, a marvel that befell him, that the storyteller heard told "in town": On New Year's the Round Table is celebrating the fifteen-day Christmas festival. True to his holiday custom and his youthful spirit, the king will not eat until he has heard of some wonder or seen some feat of arms. As his courtiers sit down to their feast, an awesome figure rides into the hall, a large man, perhaps a giant, handsome, sturdy, finely dressed, and all over bright green. He wears no armour, but carries a cruel battle-axe in one hand and a holly bough in the other. The metal of his axe, even his horse, are as green as he is. Offering no one greeting, he brusquely asks who is the head of the group. Arthur welcomes him and invites him to dismount. The stranger says he will not spend time with them, but has come to propose a Christmas "gomen," a game or sport. The fame of the Round Table is universal and he hopes his request will be granted here. Arthur promises that he will have single combat if he wants it. The Green Knight replies that his is a different quest, that in any case none here could match him; next to him they are like beardless boys. He challenges any person to deal him one blow in exchange for another. Anyone who is rash-headed, hot-blooded, enough to accept will use his green axe to strike the first blow, and keep the axe as a gift. The Green Knight will not resist, and he will not demand his turn before a year and a day.

When no one steps forward, the Green Knight taunts the company, declaring that he has now exposed its reputation as counterfeit. Pricked by shame and anger, Arthur leaps forward to seize the axe. The Green Knight dismounts and bares his neck, but before the king can strike his blow, Gawain interrupts from the banquet table. In an elaborate and courteous speech he asks that he stand in for Arthur, and the counsellors consent. The Green Knight says he is happy to be facing Gawain, and has him recite the terms of the bargain. Gawain asks the Green Knight who he is and where he lives? The Knight promises to answer if he survives the blow, and Gawain must then seek him out. Of course, if afterwards he can reveal nothing, Gawain will be free of obligation.

The Green Knight again lifts his long hair to expose his neck. The axe shears through flesh and through bone. The head rolls across the floor and into the crowd, blood streams down the green chest, but the body neither staggers nor falls. It springs after the head and grabs it. The body mounts the horse, holds the head up by the hair, the eyelids open, and the head commands Gawain to appear one year hence at the Green Chapel or be branded a coward. The body rides out of the hall. When the hoof-beats have receded into the distance, Arthur and Gawain laugh nervously, stifling their awe at what they have witnessed.

This is the poem's opening episode. The narrative moves on to Gawain's search for the Green Knight, his encounter with the lady of the castle and her tests of his chivalry. He passes them creditably but not perfectly. In the final scene, the Green Knight feigns preparations to behead Gawain, but only wounds him with a nick on the neck, which will thereafter signify his imperfection.

If Loomis is right that the poem fits its structure like skin over the hand, there ought to be some logically consistent account of what has happened here. How are we to understand Gawain's view of the Green Knight's challenge? The poet's narrative style is unusual in that he reports all of the dialogue but nothing about Gawain's thoughts. The reader can speculate: Does Gawain think the Green Knight is trying to disgrace the Round Table, or perhaps to kill him by shaming him into a bargain tantamount to suicide? Why does Gawain accept?

CHILDREN'S DARING CONTESTS

Gawain is competing for reputation, I argue, in an adult version of children's contests of daring. (The fullest accounts of children's dares are given by Opie and Opie, 1969, and Fowke, 1988.) Under the simplest rules, one child challenges another to perform some bold or risky feat: knock on a door and run away, hail a cab and walk through it, make a prank phone call, enter an empty old house, lie down on a railroad track and let a train pass over you, tell a schoolmate of the opposite sex that you like them or otherwise. In Bielefeld, children dare each other to swallow one of the large slugs that crawl out during damp times.

Clearly children accept dares, sometimes dangerous ones, to promote their reputation, but it makes little sense to speak of reputation attached to no specific character trait. We have to ask, as we did for Gawain, what personal attribute do they want their friends to reassess? Toughness, perhaps? But "reputation for toughness" has no clear meaning. Taking the case of a dare to eat a slug, the reputation could not be one of indifference to eating slugs. Having that name would make you a celebrity, but it would not raise your degree of respect or dominance in the group. If it became known that you savoured slugs, the dare would not be regarded as a real test, and the group would simply choose another task. For the daree's purposes, the audience must know that he or she finds the task aversive but is strong-willed enough to do it anyway for the sake of the reputational goal. Eating the slug functions like a measuring marker, to show how highly the daree values reputation.

REPUTATION FOR VALUING REPUTATION

My thesis is that the content of the reputation at stake in dares can be defined recursively. The reputation is for valuing reputation. The child is in effect saying, "I don't like to eat this, but I'll do it. I attach great importance to your estimate of this very importance. When you see me eat this slug, you raise your estimate of that importance." In its simplest form I am comparing the payoffs
(my reputational weight) x (Other's estimate of my reputational weight given I accept) - c versus
(my reputational weight) x (Other's estimate of my reputational weight given I decline), where c is the cost of performing the challenge. (Other papers where game payoffs depend directly on others' beliefs independent of their actions, are those of Gilboa and Schmeidler, 19°°, Geanakoplos, Pearce and Stacchetti, 1989, and Nalebuff, 1990.) It is surprising that, while the goal of "valuing reputation" is incomplete, that

of "valuing reputation for valuing reputation" leads to a well-defined game.

A reputation for valuing reputation might be desirable exactly because it is free of links to specific traits. Onlookers can generalize it to other contexts more readily, in that they infer that the daree would face other risks and discomforts for the sake of reputation. Compare a *dare* with a *bribe* like eating a slug in exchange for a dollar bill. If you take the bribe, should I conclude you are particularly tough, or particularly greedy for money? One interpretation is as valid as the other. The same argument can be made against an institution of performing unpleasant chores in hopes of a *specific* reputation: "Eat this slug, and I'll tell everyone you can speak French." But when you accept a dare, the audience can assess you as willing to suffer for your reputation in many contexts, someone to be reckoned with in general.

Children play many types of the daring contests. The simplest involves a dare leading to a simple performance or a refusal. Another version is Truth, Promise, or Dare, where the target chooses either to give an honest answer to an embarrassing question, which the group will select, for to forfeit some object, or to perform a dare. A risqué version of this game was popular with adults in centuries past (Opie and Opie, 1969). Follow the Leader requires the darer to act first, after which the daree may accept or decline. In Darers Go First the daree can either do the task, or simply refuse, or answer back, "Darers go first!" Then the darer must perform the deed or the contest is off. Another custom is Double-daring. This concept seems to have different meanings, but from my memory someone who declines can be double-dared, and the second challenge carries greater coercive power. "I double-dog-dare you!" is stronger still. Not all dares fit the pattern I have described, that of building a reputation for valuing reputation; some are more to prove one's ease in doing the task. A child who can smoke a cigar without throwing up is proving that he or she is almost an adult, but this type of dare falls outside the present model.

DARING AS A FORMAL GAME: VERSION I

The following game models one type of daring contest. Following it are Versions II and III, each meant to be successively closer to the poem. That the three games have non-trivial solutions shows that my interpretation of daring, whether right or wrong, is at least strategically coherent.

To simplify the analysis I will take the person who dares as a given, so there is no issue of me trying to jump in and dare you before you can dare me. In reality, social rules often limit who can make the dare: groups are more able to do the challenging, and a younger child or newcomer is more often the target. The task will also be taken as a given. It would be interesting to examine the darer's strategic choice of a task, but the paper does not address this question. The contest requires both the darer and the daree to perform the task if the challenge is accepted. This assumption is closer to the Green Knight's rules.

The sequence of play is shown in Figure 1. Player K (the Green Knight) chooses to dare or not to dare, and Player G (Gawain) either accepts or declines. If K does not dare, reputation remains non-salient and the two players receive 0. (I could have assumed that K suffers a reputational loss for failing to dare, but the present approach fits the poem better since the Green Knight could have simply stayed hom and never issued a challenge.)

If G accepts, both do the deed, and the each receives a difference-of-reputations payoff minus the cost of the deed: $Fw_K(E[w_K|dare]-E[w_G|accept]) - 1$ and $Fw_G(E[w_G|accept]-E[w_K|dare]) - 1$, respectively. Each component of these payoffs will now be explained. (The corresponding two payoffs in the case that G declines are shown in Figure 1, and have corresponding meanings.) A cost of 1 is subtracted from thei payoffs as the burden of performing the task. Without loss of generality this cost can be taken as 1 for both, and it then defines each player's unit of utility. The coefficients w_K and w_G, both assumed to be greater than zero, are the weights that the two players attach to their relative reputations. The expression $E[w_K|dare]$ is the onlookers' and Player G's estimate of K's weight w_K after they observe that K has made the dare. Likewise, other expectations in the two payoffs involve estimates of Player G's weight, given an acceptance or a refusal of the dare. To derive their values the game must be solved. The expression $E[w_K|dare] - E[w_G|accept]$ in K's payoff is the difference in reputations. These payoffs portray the challenge in the poem as competitive, a contest of relative reputations. Also, in the case of the poem, the reputation must be understood as not just Gawain's own but that of Arthur's court, and likewise the "onlookers" are really the Round Table's whole world, everyone who matters. For simplicity I will talk about personal reputation in the eyes of those present.

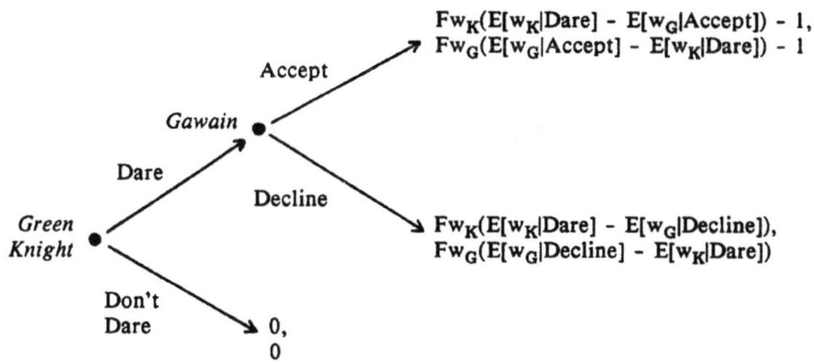

FIGURE 1: Sequence of play for Versions I, II and III.

The remaining element in the players' payoffs is the strictly positive constant F, which stands for the "felicity," a term introduced by Austin to analyse performative utterances like daring. It refers to how effectively the darer sets up a compelling dare. A dare is felicitous if it has real force. Thus, as F approaches zero, a dare loses its power to change the daree's motivation. Felicity is determined by both the content and context of the dare, as will be discussed later after the game's solution has been derived.

SOLUTION OF VERSION I

This game involves incomplete information, since each player is uncertain about the other's value for reputation. They know all the other factors. A (subgame perfect) equilibrium solution will comprise a pair of sets: the values of w_K for which Player K will dare, and the values of w_G for which Player G will accept. In the Appendix it is shown that at any equilibrium a player who dares or accepts a dare for a value of w_K or w_G, will dare or accept for all higher values. An equilibrium therefore is specified by a pair of thresholds, k and g: Player K makes a dare for $w_K > k$, and Player G accepts for $w_G > g$, and they do not act for lower values. When the weight is just at the threshold either behaviour is part of an equilibrium.

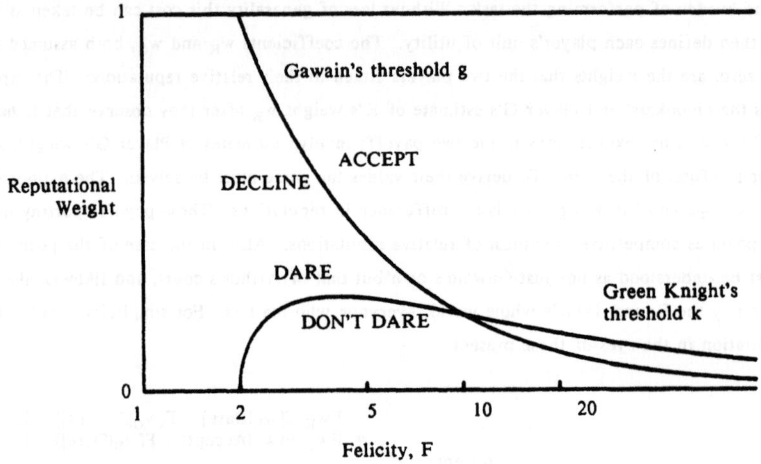

FIGURE 2: Thresholds for Version I.

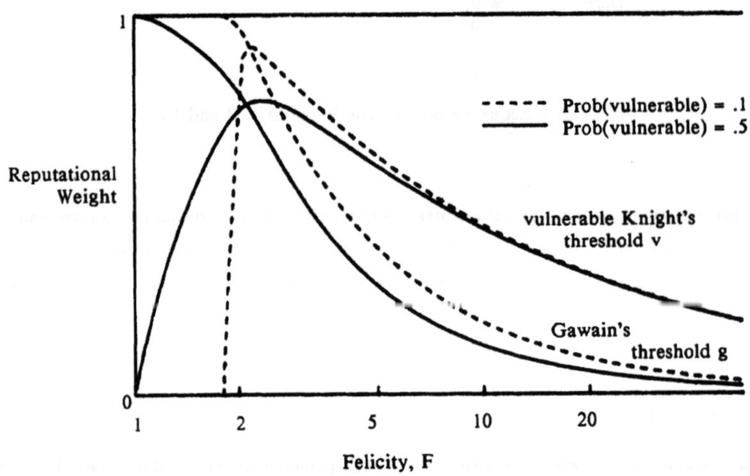

FIGURE 3: Thresholds for Version II. A strong Knight always dares.

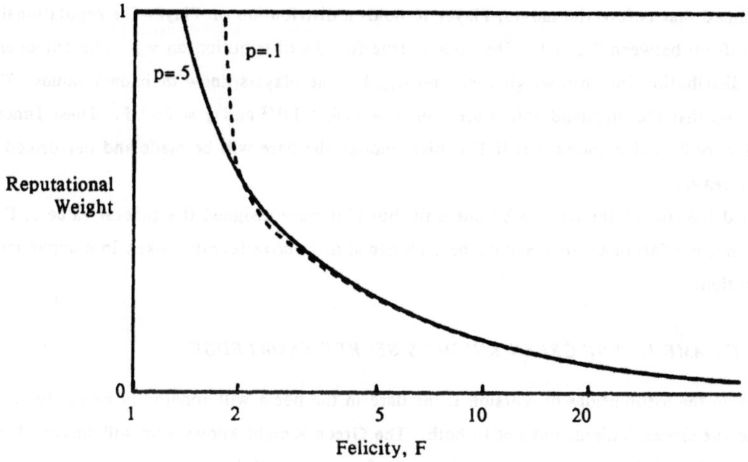

FIGURE 4: Thresholds for Version III.

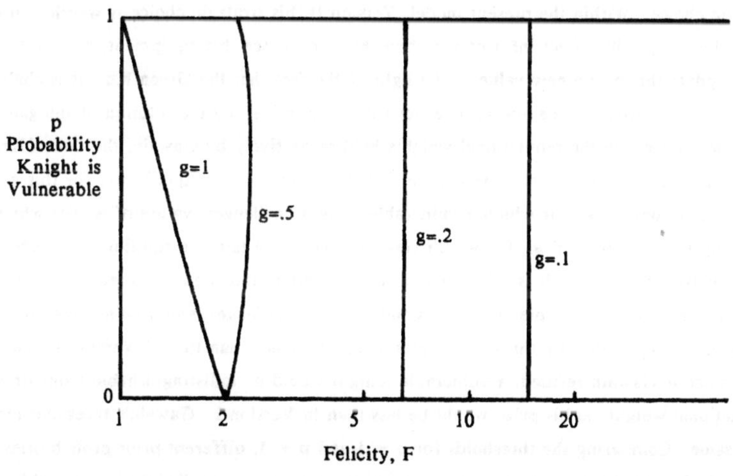

FIGURE 5: Version III, values of p and F for which Gawain with a given threshold accepts.

It is assumed that before the moves, Player K holds a distribution on Player G's reputational weight w_G that is uniform between 0 and 1. The same is true for G's distribution on w_K. The onlookers hold the uniform distribution for both weights w_K and w_G, but the players know their own values. The Appendix shows that the threshold values are then $k = F(F/2-1)^{1/2}$ and $g = 2^{1/2}/F$. These functions are graphed in Figure 2, which shows that if F is high enough the dare will be made and performed, a result that stands to reason.

I specified the cost of the dare to be one unit, but if it were assigned the general value c, F would become F/c in these formulae, so F should be understood as *relative* felicity, taken in comparison to the cost of the action.

THE DARING GAME II: THE GREEN KNIGHT'S SECRET KNOWLEDGE

Contrary to the assumptions of Version I, the dare in the poem will apparently be costly either to Gawain or to the Green Knight, but not to both. The Green Knight knows who will suffer. Either he is invulnerable, a shapeshifter, or he is vulnerable, a bluffer, who will die by his own axe, and is challenging in hopes that the Round Table will decline and lose face. In Version II, the costs of performing the dare are 0 to Gawain and 1 to the Green Knight if the Green Knight is vulnerable, or 1 and 0 if the Green Knight is invulnerable. The sequence of play and payoffs are the same as Version I, shown in Figure 1 -- only the information conditions are different.

Let p be Gawain's and the onlookers' prior probability that the Green Knight is vulnerable. This probability is "prior" in the sense that it is based on everything they know about the Green Knight short of his strategic choice. Within the present model, Version II, his strategic choice is whether to challenge the Round Table, so p will reflect information about his appearance, his equipment, his words, etc. Gawain will update this p to a new value in the light of the fact that the Green Knight is challenging him. This new probability will be lower, its exact value determined by the solution of the game.

Let w_G, w_V and w_S be the reputational weights held respectively by Gawain, the Green Knight if he is vulnerable, and the Green Knight if he is invulnerable ("strong"). An equilibrium for this game involves three sets: values of w_V for which a vulnerable Knight challenges, values of w_S for which a strong Knight challenges, and values of w_G for which Gawain accepts. Again the equilibrium is essentially unique and involves thresholds. It is calculated in the Appendix, and shown in Figure 3 for various values of F and p. For $p = .5$, for example, a strong Knight always challenges, and a vulnerable one challenges only if his value of w_V is above a threshold, which is much higher than that of Version I. This rise makes sense since if Gawain refused, a vulnerable Knight would be indistinguishable from strong Knights of low reputational weight, so his prize would be less than in Version I. Gawain's threshold remains roughly the same. Comparing the thresholds for $p = .1$ and $p = .5$, different prior probabilities of vulnerability, Gawain is more likely to accept, the more he believes the Green Knight is vulnerable.

THE DARING GAME III: PLAYING BY PRETENDING TO PLAY

This final version of the game most accurately describes the events in the poem. It posits that the Green Knight and Gawain know the truth about each other. (This is supported by textual evidence which I will summarize later.) Gawain knows that the Green Knight is invulnerable, and knows that for some unexplained motive he wants a contest, that he has no motive to disgrace him by intimidating him into a refusal. Gawain regards the Green Knight as having chosen his words, appearance, and the terms of his offer, in order to recruit someone into playing. On his part, the Green Knight knows that Gawain's value for his reputation is high. The audience, however, is as uninformed as it was in Version II, and in fact thinks that Version II is the game being played; it is uncertain about the value for reputation held by the Green Knight and by Gawain, and also about the Green Knight's vulnerability. The view of the onlookers is important since they will give Gawain his reputational payoff. Like a poker player who has spied on the other's hand but must hedge the use of this information or be accused of cheating, Gawain must consider the onlookers' view of the game.

Gawain's optimal choice in Version III is found by comparing his reputational payoff from accepting minus his performance cost of 1, with his payoff from declining, both reputational payoffs being calculated for Version II played with a Green Knight who would be revealed as invulnerable. These payoffs, and therefore Gawain's action, will be functions of F and w_G for a given p, as Figure 4 shows. For the purpose of later discussion, Figure 5 graphs his strategy on different axes than Figure 4, using p and F, the variables the Green Knight is manipulating.

Several sections of the text support the postulates of Version III that the Green Knight wants his challenge accepted, unlike a normal daring contest, and that Gawain knows that the Knight is invulnerable. That the Green Knight wants a contest is consistent with his goading of Arthur. When the gathering was silent to his first challenge, he could have ridden out to spread the news. Instead he chides the Round Table and especially Arthur, working on the king's pride in his name. The Green Knight seems determined to have an exchange of blows, and his tactic succeeds on the volatile king. Further, the Green Knight's demeanour while Arthur is preparing to swing the axe reveals that he wants an exchange. He is as calm as if "any burne vpon bench hade broȝt hym to drynk of wyne" (337-338), as if someone at the table had approached him with a drink of wine. Previously, when his challenge was unanswered, he had been restlessly shifting in the saddle (304-306), but now events are proceeding as he intended. Finally, note the Green Knight's remark, made and repeated, that he is happy it is Gawain who will be striking the blow (387-391). He leaves us ignorant about why he prefers Gawain in particular, but he does want a contest.

Gawain seems sure that the Green Knight will live through the blow. In his request to substitute for Arthur he speaks of his life as being the least valuable to lose (355). Arthur declares that a deft blow will end the match (372-374), but Gawain gives no word of agreement. When he promises to seek the challenger out (402), his words are unconditional; he does not qualify them with "if you survive my blow." Also revealing is the poet's focus when the decapitated body stays standing. We learn about the onlookers' shock as the head rolls into their midst, their fear as the body remounts the horse and the head gives orders to Gawain, but not until the visitor has galloped off are we told of Gawain's reaction. Although his fate is uniquely tied to the news that the Green Knight has survived, the poet does not report Gawain's response, and the suggestion is that he expected the Green Knight to live. Version III incorporates these ideas, that the Green Knight and Gawain know the truth about each other.

The game analyses of Versions I and II clarify not the actual contest being played, but the social conventions through which the Green Knight traps the hero into the bargain. The real planning and strategizing is close to Version III. The Green Knight's problem is to select Gawain from the company for his adventure. He presents himself to the Round Table in a way that invokes the social convention of a challenge, that is not so effective that everyone volunteers, but strong enough to recruit the opponent he wants. In making his offers, the Green Knight manipulates the prior probability p of his vulnerability and the felicity F of his challenge. To describe how he does this, I first look at the determinants of the felicity of a dare in general.

AUSTIN'S THEORY OF PERFORMATIVES

A dare is both a statement and an act; felicity is its effectiveness, regarding it as an act. When John Austin delivered the 1955 William James Lectures, the philosophical majority held that sentences assert possible events. He expounded a theory, to become widely influential in the years following his death, that some sentences do not say things, but rather do things (Austin, 1962). *Performatives* do not primarily make claims about the facts, but are more like actions, and they typically cannot be labelled true or false. One example is a promise. If I promise to loan you my car and never do, perhaps never intended to, you cannot say my promise was "incorrect," that I claimed to be promising but I was not. I may have promised insincerely and broken my word, but I did indeed make a promise.

A promise is one type of performative and there are many others. Austin estimated one or several thousands. Sentences that thank, blame, apologize, welcome, find a defendant guilty, pronounce a couple married, all do things beyond their assertive function. Just what do they do? Many change the social relationship between actors by inducing new utilities, duties or rights in that relationship. A jury finding a defendant guilty gives the legal system the right to punish. Forgiving constitutes a promise to accord the other respect in your future relationship, in spite of a transgression.

A performative is judged not by its truth, but by its effectiveness, and Austin labelled this concept *felicity*. Felicity depends on both the immediate context around the speaker and listener, and on the societal customs supporting the performative. For example, if I make a promise but at the same time show you my crossed fingers, I am not really promising. My own action undermines my words. If I declare, "I apologize for X," but I am reciting a line as an actor in a play, then I am not really apologizing. Taking an example of Austin's, I come upon a ship ready to be launched, and seizing the occasion, declare, "I name you the *S.S. Joseph Stalin*." For good measure I break a bottle of champagne over the bow and kick out the chocks. However if I am not the person appointed to christen the ship, my action is void.

Felicity is not all or none: it can be a matter of degree (promising versus solemnly promising), or it can be debatable (promising my cat a treat). A performative thus is not a magical incantation whose mere enunciation does the job. Performatives "come off" to a greater or lesser extent, depending on their appropriateness.

THE GREEN KNIGHT'S MANIPULATION OF FELICITY

The requirements of felicity depend on the particular performative. I will suggest some requirements for a successful dare. (Neither Austin nor later philosophers seem to have investigated daring. Greenberg, 1990, gives an interesting account of the language of challenges to duel in the antebellum American South, but he does not use the framework of performatives.) First, the two actors must be part of a *shared culture* that supports the practice of daring. What goes over with children may not with adults: if one grownup challenges another to eat a slug, the dare will fail. It also seems unlikely that an adult could dare a child; daring is not a practice parents can use to manage their children. The folklorist Roger Abrams (1963) described black youths in Philadelphia playing the "dozens," a back-and-forth game of sexual insults, "Somewhere between the ages of sixteen and twenty-six, playing begins to lose its effect . . . When someone indicates that he wants to start, the one who is supposed to be insulted may reply, 'Oh man, don't play with me.' If he needs a more clever retort, he may rely on the proverb, 'I laugh, joke and smoke, but I don't play.'" (p.51). A particular mode of daring must have cultural support.

Second, the dare must have *apparent purity of motive*. If one child were to dare another to hand over $5, the trick would surely fail. Dares deal in the commodity of reputation, and a blatant scheme to make money lacks felicity.

Third, for those dares where both parties are to perform the deed, there must be *apparent symmetry* of the action. "Let's run through the boys' locker room, I dare you!" This might work between little girls, but not from a boy to a girl. Fairness of this type is necessary to compare the players' weights on reputation.

Much of the Green Knight's self-presentation is aimed at satisfying the three requirements of felicity of his dare. Regarding the need for a shared culture that supports daring, the rules of courtly behaviour do provide such support for challenging, and the Green Knight himself says he is challenging "bi ryȝt" (274). His problem is to establish himself as member of the culture, a bone fide knight whose challenges compel a response. His fine clothing, his articulate speech, his horse obedient to the rider's command ("To þe gome he waþ ful gayn," 179) serve this end. When he is offered a fight, he replies that he did not come for that, although at home he has a hauberk, a helmet, a shield and sharp shining spear (268-269). Cataloging his equipment can be taken partly for its face meaning, as an elaboration of how he has not come to do battle, but he is also letting it be known that he is a fellow knight as he possesses the accoutrements of one (Burrow, 1965).

In line with the need for purity of motive, the Green Knight stresses the reason for his challenge: it is a game, for sport (263, 274). He mentions no other goal. His manner is almost jocular. Only after Gawain accepts does he become businesslike (Davenport, 1973). His size ensures that physical prowess is not the issue, since no one could stand before him. The Round Table has drawn him here by its reputation, he says (258), and it will lose face if it does not take up his challenge (313).

The third requirement is apparent symmetry of the action. Here the Green Knight has a difficult problem. Getting his bizarre terms accepted will not be easy. He is either mortal and vulnerable, or magical and invulnerable, and in either case his beheading proposal is unfair. He must obscure this truth, so he loads the details of the agreement with symmetries that give a facade of fairness. You strike at me, I strike at you. I offer no resistance, neither will you. I sought you out at your abode, you seek me out at mine, and come at this same time next year.

THE GREEN KNIGHT'S MANIPULATION OF THE APPEARANCE OF MAGIC

Having described the ways through which the Knight ensures felicity F, I turn to how he sets the perceived likelihood p that he is vulnerable. Several authors have noted the "richness and variety of suggestion" in the Green Knight's description (Benson, 1965; Burrow, 1973). He is "bristling with ambiguity." Within the framework of Version III, this dual nature stems from the Green Knight's manipulation of p, the audience's probability that he is invulnerable. The poet is careful not to place him squarely in this world or the other. He promises to tell us of a "selly in siȝt summe men hit holden . . . an outtrage awenture . . ." (28-29), a marvel, *so some men say*, an extraordinary adventure, but he refrains from unequivocally promising a tale of magic. The Green Knight hovers between the human and the supernatural by the way he enters, his physical form and his words. He does not appear in a puff of smoke, but rides in as a mortal could. He is large, the poet says, perhaps a half-giant, surely the biggest of men, but in any case handsome. The poet describes his shape and size, anchoring us in the first impression that he may be human, and after a dozen lines of detail tells us that punchline, that he is bright green. In the Green Knight's own words, deliberately chosen, he is a "wyȝe" (315), a term that usually denotes a person but could mean a living being of any kind (Borroff, 1962, p.112). In battle, he claims, he would defeat his adversary through armour and weapons, not by dint of magic. He addresses Arthur with the familiar second person pronoun. Is this the manner of a hostile but human challenger or that of a being who has no obligation to human kings? Within Version III's approach, uncertainty over the Green Knight's nature, measured by p, functions to recruit Gawain but to make his decision hard.

The effect of the Green Knight's manipulation of F and p can be understood through Figure 5. Depending on the joint values of F and p, a challenge will select a knight of the Round Table whose weight on reputation is above a certain value. (A further refinement would portray the situation as an n-player game including all the members of the court.) By setting his parameters correctly the visitor induces Gawain, the pearl among peas, to step forward for the adventure that awaits him.

THE FEAST OF BRICRIU

The earliest surviving version of *The Feast of Bricriu* is found with other important Irish sagas in <u>The Book of the Dun Cow</u>, transcribed about 1100 A.D. The book's title derives from the tradition that the ancient champion Fergus, dead for centuries, returned to life to narrate the deeds of the heroes of Ireland to St. Ciaran who copied them onto the hide of his cow. *The Feast of Bricriu* was first written down in the Eighth Century, as indicated by its language, so it was almost as old when the Gawain poet lived as his work is now. The tale describes a more warlike, less refined society than Arthur's court, and probably originated far earlier than the Eighth Century. Although it is highly regarded, the literature on it is sparse (e.g., Thurneysen, 1924), and spends little attention on the plot structure. This summary of the final episode in the story follows the translations of Henderson (1899) and Meyer (1893).

SUMMARY OF THE FINAL BEHEADING SCENE IN THE FEAST OF BRICRIU

Bricriu of the Poisonous Tongue is a gossip and a troublemaker. (His motto: "Clearer to me is a whisper, than to anyone else a cry.") To foment strife, he invites three of Ulster's eminent warriors, Loigaire the Triumphant, Conall the Victorious, and Cuchulainn ("Kuh-<u>hoo</u>-lin"), to a feast, maliciously advertising that only one shall have the champion's portion of the meal, and his wife preeminence over the women of Ireland. The three warriors refuse to come for fear of the trouble Bricriu might stir up, but when he threatens to use his poetic skill to incite slaughter across all Ireland, they relent. Knowing that the heroes will not permit him to attend his own banquet, he builds a sun-room of coloured glass from which he can look down on the great hall. The three heroes and their wives perform various feats, some grotesque and bawdy, to gain the champion's share. Cuchulainn wins consistently, but the other two find excuses for denying his victory. In the final section of the tale Cuchulainn is publicly proven to be the deserving one, through the following beheading contest.

The Ulstermen return to the Royal Court after a day of athletics; the three champions are absent. A *bachlach* (a rough herdsman) of great size and ugliness, dressed in an old hide, enters the hall. From his head protrude ravenous yellow eyes, each the size of a cauldron to hold an ox. (His description is so fearsome that the Eleventh Century monk recopying the manuscript here inserts into the margin the Sign of the Cross.) In the bachlach's left hand is a block, in his right hand an axe; "its handle would require a ploughteam to move it; its sharpness would lop off hairs, the wind blowing them against its edge." He strides to the high arched forkbeam by the fire, and Duach of the Chaffer Tongue immediately speaks up to ridicule his height: "Is the hall lacking in room for you, that you find no other place than by the forkbeam, unless you wish to be a domestic lightholder? Only sooner will a blaze be to the house than brightness to the household." The giant stays calm: "Whatever property may be mine, you will agree, no matter how big I am, that the whole household will be brightened, while the hall will not be burnt."

Bearing a light is not his only skill, he says. He has searched the world, yet found no one to give him fair play. As the Ulstermen excel all, he expects someone among them will grant his request. The company agrees to grant him fair play. He exempts the king and his relation from his challenge, but to the others, "Come whoever of you that wishes to do it, and I may cut off his head tonight, and he will cut off mine tomorrow night." (This, of course, is the reverse of the Green Knight's game.) Duach doubts that anyone will volunteer, and Munremar mac Gerreind springs to the floor to propose that the order of the blows be switched. "Were that my quest, I could have got that anywhere," the bachlach objects, but Duach goads him and he accepts, adding the comment, "strange as that may seem to you." Pledging Munremar to come back for a return blow, the giant lays his head on the block. Munremar swings the axe, and the head flies across the hall. Blood streams from the giant's neck and fills the house. He gathers the head, axe and block to his chest and strides out.

The giant returns the following night, but Munremar is not to be found. Loigaire, one of the three contending heroes, is there, and the giant summons him by name. He beheads the giant, but, like Munremar, shirks his part of the bargain. Conall is challenged and behaves identically. Finally the bachlach calls Cuchulainn to play, and despite the latter's innovation of smashing the head after severing it, bending the rules as Gawain would later do, the bachlach returns intact next night to demand his turn. Cuchulainn has fallen into deep gloom, but, unlike the other two, keeps his word and lays his head on the block. The giant mocks the reputation of the court and teases Cuchulainn, saying that he cannot kill him yet because his neck does not reach across the block. Cuchulainn complains that he is being tormented and asks to be dispatched forthwith. He extends his neck so that a man might put his foot between each of the vertebrae, and it stretches across the whole block. As the giant raises the axe, the hides that cover him creak like a forest in a winter storm. He brings it down on Cuchulainn's neck, but blunt side downwards. He bids him rise and accept the champion's portion. The giant disappears. He was the magician Curoi mac Daire who came to prove to all that Cuchulainn was the deserving one.

THE BACHLACH'S STRATEGY

The narrative seems less polished than *Sir Gawain and the Green Knight*'s, more like a disconnected dream, but an informal game viewpoint reveals the functions of some obscure plot elements. Some of the puzzles are these:

- Arthur's court falls silent when the Green Knight enters the hall, out of fear and courtesy according to the poet, but Duach speaks up to insult the bachlach. What is the function of Duach's remark?
- Why does the bachlach propose the unfair rule that he strike the first blow, but quickly agree to reverse the order?

- Why does the giant include Munremar in list of beheaders when he is not one of the three contending champions, and in fact he has no other role in the poem?
- Why does the bachlach summon each champion by name?

The key to these riddles is the different strategic structure of the contest compared to Gawain's dilemma, and the different trick the challenger uses to induce the heroes to accept. The Green Knight strove to make the contest look fair between himself and Gawain, but the bachlach has a different problem. His aim is to establish fairness among the three champions who he is comparing publicly, so he must first ensnare them into the game and then design their tests to be equal.

In inducing the heroes to take up his challenge, the giant is clever. He first extracts a pledge of fair play, without revealing the game. He then proposes that he strike the first blow. When Munremar objects that the order should be switched, the giant readily accepts. Munremar is trapped; he cannot refuse the bargain since he himself proposed it. The situation is something like Darers-Go-First, except that we suspect that Munremar made his retort without realizing that he was falling in with the bachlach's plan. To his chagrin, Munremar learns that the giant did not issue his initial proposal naively; the giant's remark that he is ready to go first "strange as that may seem to you," reveals that he knew his earlier plan had no chance of being accepted.

Regarding the giant's goal of making the contest fair among the three, the difficulty is that they will play the game sequentially, and this ordering introduces unfairness in two ways. One is that the person offered the first turn will be more ready to accept, since he is not sure that the giant can replace his head. Later players will know about the magic and will be less likely to play. Consequently, the audience cannot gauge their courage by noting who accepts and who declines. This first difficulty is solved by introducing Munremar, whose trial displays the bachlach's wizardry to the later players, the actual rivals for the champion's portion.

The bachlach's second problem involves the order in which the champions play. He wishes to show the world who will accept the game and keep his pledge to return. The giant knows that Cuchulainn alone is worthy among the three, and must ensure that he is the final player. If Cuchulainn plays too early, the others will learn that the giant will not behead them. The giant accomplishes his goal by asking each of them by name to step forward and play.

There is a strategic equivalence between Munremar in *The Feast of Bricriu* and Arthur in *Sir Gawain and the Green Knight*. The Green Knight incites a first-taker by angering Arthur. Once his king has accepted the challenge, Gawain can scarcely maintain that the terms are unfair and confer no obligation on anyone to accept. The bachlach accomplishes the same end with a more subtle trick, by maneuvering Munremar into proposing and accepting the terms of the bargain.

Another strategic correspondence between the poems is the Green Knight's bough of holly and Duach's initial insult to the bachlach. Both establish immediately that, despite his threatening appearance, the stranger has not come to fight. The bachlach and the Green Knight must shift the issue to one of sport and fair play. The latter cites the holly as a symbol of his peaceful intentions -- he carries it in his hand where one would expect a weapon or shield. The bachlach proves his peacefulness when he stays unruffled by Duach's suggestion that he could not hold a light for the household. Each device fits its setting: Arthur's assembly could not have insulted the Green Knight without violating its code of courtly behaviour, and the bachlach would have looked ludicrous holding a holly twig or a flower.

A ONE-PERSON TWO-PLAYER GAME

Another portrayal of Gawain's dilemma has the hero playing against himself. Here the focus is on the Green Knight's enigmatic character and on Gawain's choice of what he should believe about it. The knight offers plenty of evidence that he is malicious, but plenty that he is chivalrous and will give Gawain fair play. Gawain has two natures, a dutiful one and a self-preserving one. Each can regard the Green Knight as malicious or chivalrous. If the Green Knight were a clearly malicious being, his challenge need not be accepted; submitting to him would be suicide. However if he is a chivalrous fellow knight, Gawain owes him fair play and the Green Knight will somehow reciprocate it. Neither case is entirely convincing from the evidence available before decision. Either choice presents a risk, one of abandoning duty, the other of dying, but the two natures weight these risks differently. His dutiful half worries more about shirking an obligation, and his self-preserving half more about being killed.

If Gawain's two natures disagree on the Green Knight's character, Gawain will experience tension, an aversive mental state. They want to agree, but one nature will have to compromise by overlooking the evidence supporting the other half's worries. The question of what to believe is here a strategic choice; each nature can focus on part of the evidence and try to rationalize away the rest. However each nature can ignore its worries with only partial success, and each would prefer that the other did the compromising.

	Gawain's self-preserving nature	
Gawain's dutiful nature	**Regard Knight as Malevolent**	**Regard Knight as Chivalrous**
Regard Knight as Malevolent	GAWAIN REFUSES risk of shame, 1 — ideal 2	PARALYSIS tension 0 — tension 0
Regard Knight as Chivalrous	PARALYSIS tension 0 — tension 0	GAWAIN ACCEPTS ideal 2 — risk of death, 1

FIGURE 6: A game between Gawain's two natures. The dutiful nature's payoffs are on the left.

This approach is somewhat Freudian, with its duality of the individual's psyche and its idea of minimizing mental tension. Freud would call the two natures the superego and the id, and expect that the superego would be logical and verbal while the id would be intuitive and inarticulate. This dichotomy fits the poem well. In his speech, the Green Knight talks like a chivalrous knight in the role of a challenger (Burrow, 1973), but the unearthly aspect of him, his green hue, stays entirely off the spoken record. Neither he nor the court make any reference to it. Freud's framework suggests that the dutiful nature of Gawain will listen to the speech, while his basic nature will focus on the huge green figure. The notion

of choosing beliefs to minimize tension is prominent also in cognitive psychology, as in the theory of dissonance reduction. Seeking out evidence to confirm what you have already chosen to believe appears in decision-making psychology -- Janis and Mann (1977) call it "bolstering."

These assumptions describe a Battle-of-the-Sexes game, with payoffs shown in Figure 6. It has two strong equilibria and at each one or the other nature compromises. Since there are two equilibria the game has no obvious solution. If one were to substitute cardinal payoffs in the diagonal cells, one selection solution proposal, justified by axioms (Harsanyi and Selten, 1988; O'Neill, 1988), would multiply the payoffs in the equilibrium cells to calculate the Nash products, and choose the higher of the two. However this rule is rather esoteric, and the essential situation has an outcome that is equivocal. This indeterminacy makes for a more suspenseful plot.

GAME THEORY AND LITERARY ANALYSIS

Given the memorable impact of the beheading scene in *Sir Gawain and the Green Knight*, it is surprising how few papers analyse the characters' motives and choices. The conventional literary repertoire seems to lack a good method to analyse plot per se. The field of narratology, for example, investigates how the story is told but not the story itself.

Several formal techniques have been developed to analyse plot. Vladimir Propp (1928) initiated the "linguistic method," categorizing the elements in Russian folktales by their abstract functions, and describing the patterns that appear in the corpus of tales. Kissam (1977) applied the technique to mediaeval romances. More recent work has paralleled developments in linguistic theory and has investigated "story grammars" (e.g., Alker, 1987, and the works he cites). Prince (1973) tackled the interesting question of what makes a story a story rather than a mere string of facts. More than Propp's work, this study deals with the causal and relationships among the elements, and in general these researchers are moving from a syntactic view of a story toward a semantic one. However they themselves express frustration in their strivings to get at the "meaning" of the plot (Ryan, 1981). "Meaning of the plot" is a linguist's way of putting it; a game theorist would say that they do not deal with the strategic structure of the game, they do not treat characters' acts as directed towards goals.

Further formal methods to study plots have sprung from the "Romanian school of poetics," led by the mathematical linguist Solomon Marcus (1970; in English, see for example, Marcus, 1977, 1984, and Brainerd and Neufeldt, 1974). An example is the use of signed graphs along with the hypothesis that events move towards a state of "structural balance" (Martindale, 1976; Lalu, 1977). Teodorescu-Brinzeu (1984) represents the sequence of thought in soliloquies using symbolic logic. The statistical analysis of joint presence of pairs of characters on stage is another theme. Although these techniques are interesting, they work at the periphery of the plot while game theory aims at the center.

More than a dozen papers have applied game theory to stories, as Table 1 shows. Oskar Morgenstern presented the Holmes/Moriarty dilemma non-mathematically in 1928 and 1935, and later in his book with von Neumann defined it as a formal game. Rapoport's analysis of *Othello* was detailed and advanced for its time. Other papers range from informal game analogies to Mehlmann's use of differential games. He addresses the puzzle of why Faust was saved after he had lost his wager with the devil. Mystery and detective stories, heavy on plot, are frequent subjects, as are stories with explicit bargains detailing each possible outcome, e.g., *The Merchant of Venice*, *Faust*, and the present study. One topic not in the table is informal game models of an author trying to maintain suspense versus the reader trying to discover the story, e.g. Bruss (1977) and Davey (1984). Brams (1991) discusses the prospects for game theory in the humanities.

Conventional scholars have ignored these studies, and some of the papers give them good grounds for doing so. Some models do not take the literary work seriously; they are inconsistent with important elements in the story, and do not compare their game interpretation with details from the text. I will give the two examples that are the easiest to relate, admitting that my criticism is somewhat unfair since the authors devised them more to illustrate game theory rather than to study a plot.

PAPER	LITERARY WORK	GAME MODEL
von Neumann/Morgenstern 1944	*The Final Problem*, Conan-Doyle	2-person zerosum
Williams 1954	*Merchant of Venice*, Shakespeare	2x2 zerosum
Rapoport 1960	*Othello*, Shakespeare	various
" 1962	*La Tosca*, Puccini	Prisoner's Dilemma
Vorobev 1968	*Eugene Onegin*, Pushkin	informal, duels
"	*The Gift of the Magi*, O.Henry	Battle-of-the-Sexes
Howard 1971	*The Caretaker*, Pinter	metagames
" 1980	*Dr. Zhivago*, Pasternak	metagames
Teodorescu-Brinzeu 1977	*Othello*, Shakespeare	3x4 non-zerosum
Steriadi-Bogdan 1977	*The Mousetrap*, Christie	ext've, perfect info
Lalu 1977	*Richard III*, Shakespeare	ext've, perfect info
Brams 1980	*Old Testament*	various finite games
Hintikkas 1982	*Silver Blaze*, Conan-Doyle	informal ext've form
Carlson 1984	Numerous mystery novels	informal
Mehlmann 1989, 1990	*Faust*, Marlowe, Goethe	differential games

TABLE 1. Game theory studies of literary plot.

The first is the opera *La Tosca* treated as a Prisoner's Dilemma. The Chief of Police Scarpia has condemned Tosca's lover to death but offers to fake the execution in exchange for Tosca's favours. She consents, but after he has sent off the order, she seizes a knife and stabs him. He, however, had already doublecrossed her, and had written an order that her lover be really shot. The pair come to an inefficient outcome like the players of a Prisoner's Dilemma, but if we scrutinize the model, flaws appear. If Scarpia knew that Tosca might stab him, why did he leave a knife lying out? And if she knew his set of strategies, could she not have demanded to see her lover safe out of the country first, and so on? Prisoner's Dilemma is a valid model only if the information conditions are right; ending up at an inefficiency is not enough. The model cannot explain why they limited their moves in this way, or why they made the bargain in the first place. The analysis of O. Henry's short story *The Gift of the Magi* also overlooks the data. A fond husband and wife plan to surprise each other with Christmas gifts. She cuts her hair to buy him a watchchain, but, ironically, he has sold his watch to buy a clasp for her hair. Vorobev cast it as a Battle-of-the-Sexes game, symmetrical with two strong equilibria as in Figure 6. (The interpretation of the payoffs in this "battle" has each worrying mainly about the other's material welfare.) The model accurately reproduces what happens in the plot; since the players in the abstract game have no way to coordinate their choices, it is reasonable that they might end up at an inefficient non-equilibrium. However O. Henry's trademark was the surprise ending, and an analysis that predicts that his story will turn out as mi^ht be expected, clearly has not captured the story's essence. When they reveal their gifts to each other, the characters are stunned. This fact is inconsistent with the simple matrix analysis. We need methods that go deeper, perhaps using recent work on information and knowledge in games.

Lacking a method for examining characters' overall strategies, scholarship on the plot of *Sir Gawain and the Green Knight* has taken several directions. A few critics have tried to construct new motives for the actions of Gawain, the Green Knight or Morgan Le Fay. However most conventional analyses of plot (Bloomfield, 1961) have dealt with the symbolism of the characters (Does the Green Knight's hue symbolize nature or Christmas or death or the fairy world?), or with the social meaning of various elements (Is his manner to Arthur rude, or appropriate to a challenger? Is Arthur's demeanour seemly for a king?), or with motifs shared with other stories (Where else do we see a challenge by a stranger, or an exchange of blows?) This third subject of recurrent themes is prominent in the critical literature on the poem (Catalini, 1979; Benson, 1965; Kittredge, 1916; Brewer, 1973). The findings are fascinating; for some deep reason, challenges by strangers to test one's virtue, exchanges of blows and magical survivals of deadly wounds abound in folk and mediaeval literature. However the comparative study of plot elements is limited in what it can reveal. One could not learn much about checkers and backgammon by comparing the physical shapes of the pieces. The element can have different functions in different stories, or conversely, similar functions can be fulfilled by different elements. The Green Knight's bough of holly and Duach's insult to the giant correspond strategically, as I argued, but are different physically. Munremar's challenge to the bachlach's offer has the same role as Arthur's impulsive seizing of the axe, even though the events themselves are dissimilar. The element of Munremar confronting a stranger appears in *Scela Mucce meac Da Tho*, The Tidings of Mac Datho's Pig, another Old Irish claims problem about the champion's portion of a meal and the ownership of a marvellous hound (Chadwick, 1927), but there its strategic role is quite different. Mediaeval writers borrowed and combined familiar ideas without compunction, and the meaning of an element was not inherent, but arose from the context. Strategy-oriented methods can bring out similarities of function across different settings.

SUPERPOWER RIVALRY IN THE THIRD WORLD

Mediaeval challenges possibly had a defensive purpose. A knight ready to fight in single combat would be seen as more reliable in a battle, so courts looked for ways to display their willingness to fight. In the language of modern strategy studies, chivalric challenging may have served to enhance deterrence, and the same could be said of children accepting dares. Nuclear weapons have reemphasized the problem of reputation in deterrence, although it is now called the problem of credibility (Jervis, 1979; Morgan, 1985; Powell, 1990). Because a nuclear crisis could lead to a catastrophe beyond precedent, each government has sought to convince the other beforehand that it will resist if its interests are crossed.

A model along the lines of the daring game can be devised for the international case. It could take this form: Country 1 commits itself to a given issue or does not commit itself, then Country 2 fights with it on that issue or does not. If the commitment is made and challenged, both lose pay a conflict cost, and also receive the reputational payoff. They have a strong desire for this reputation, a desire generated through its usefulness in future situations. The added self-referential element in a side's payoff, the value for reputation for valuing reputation, outweighs the cost of resisting in some conflicts and make threats credible.

To check the validity of the model we can identify its key features and look for them in international interactions. The important elements are the same three as in the daring model: first, a language act that triggers the reputational goal; second, the goal of reputation for valuing reputation, as incorporated in payoffs like reputational weight x [others' estimate of reputational weight given your action] - conflict cost, or the more competitive form of Figure 1; and third, the model's consequence that the parties will fight over issues where winning is of no benefit to them. To the extent that Cold War contests show these three elements, the model deserves more credence.

The first feature of the model, the performative utterance, is prominent in the international context. One side often declares or reasserts its "commitment" to defend some interest. At first it is puzzling why words should matter in the *realpolitik* of world struggle, but this issue was the very one raised by the Green Knight's remarks to Arthur, and the answer again involves Austin's theory of performatives. A difference between the mediaeval and the modern case is a *dare* comes from the disturber of the status quo, whereas a *commitment* is issued by the status quo supporter, but both declarations function as performatives in Austin's sense. Verbal threats by governments also fall under his theory.

In regard to the second feature of the model, seeking reputation for valuing reputation, some explanations for U.S. foreign policy seem to treat reputation, will or credibility as autonomous entities. James Payne, a supporter of U.S. policy, writes, "Hostile states tend to oppose moves by each other and each other's allies to alter the status quo because they desire to avoid the loss in threat coerciveness (or reputation for determination) which would result from the successful alteration" (1970, p.117). Robert Tucker, also a supporter, argues, "When engaged in a contest for global stakes, what may appear as a marginal interest will be invested with a significance it would not otherwise have, for almost any challenge is likely to be seen by the challenger and by third parties as a test of one's will" (1981, pp.144-145, quoted by Schoultz, 1987). Robert Johnson, a former U.S. State Department analyst and a critic of these policies, asserts, "Concern with credibility itself defines U.S. interests and plays the major role in determining the particular commitments the United States undertakes and the resolve with which it carries out those commitments; the specific situation is relatively less important" (1985, p.43). Referring to U.S. policy in Vietnam and Nicaragua, James Chace writes, "the Soviet challenge has ignited an obsession with credibility so that concerns about it determine commitments, which in turn expand U.S. interests" (1988, p.10).

These rationales are puzzling, as they construe determination, will and credibility as standing by themselves. Payne stresses reputation for determination, but "determination" to do what? If the situation involves one's interests only marginally, as Tucker allows, why is it a "test of one's will"? And what could Johnson's phrase "credibility itself" mean? These were the types of questions that led to a recursive definition of reputation in dares. The international model would state that governments seek reputation for valuing reputation, that they want to make it credible that they place a high value on that credibility. The idea is confusing in English, but innately simple and consistent, as the model shows.

The third feature of the model was the possibility of struggles over issues of no innate importance. Finding an international question of absolutely no importance is hard, of course, but based on their involvement in past policy debates, Johnson and Chace hold that commitments often lead to interests, rather than the other way around as common sense would expect. The model explains how this phenomenon could happen. Others have doubted that international struggles over pure non-issues could be sensible. Desch (1989) lists some past writings on the subject and concludes that the idea deprives credibility of any objective or rational basis (p.95). Jervis (1979, p.315) states, "Commitment can only be built on a foundation of intrinsic and strategic interest." As theory, their claims seem to be incorrect; the present model suggests that a logical government might commit itself and fight for credibility per se. According to the model, in fact, it does not matter who wins the struggle, as long as each exerts the effort of struggling.

More conflict over non-issues arise in arms building decisions and in arms control negotiations. When each superpower has over 10,000 strategic nuclear weapons, exact counts are militarily meaningless, but the US and USSR have increased their arsenals over the years, first in numbers, then in quality. Negotiations have stalled over small differences in armaments to be given up, even over the banning of weapons that neither side possessed or wanted to build, like multiple warhead cruise missiles (Talbott, 1979). Other instances of conflict over nothing were the "chicken of the sea" games played by Soviet and American warships, heading at each other and swerving at the last moment (Zumwalt, 1976, p.322), or shouldering each other off during refuelling (Lynn-Jones, 1985). Before the 1972 Prevention of Incidents at Sea Treaty, this perilous sport went on one or two dozen times a year, apparently tolerated by naval and civilian authorities.

The model predicts that players will sometimes welcome a contest. During the Middle courts went beyond responding to individual challenges, and held jousting tournaments (Vale, 1981) and *pas des armes*, in which a group of knights seized a location such as a hill or bridge and invited others to dislodge them. Modern governments often seem glad when commitments are challenged, since they have an opportunity to increase their credibility.

Does the model mean that the superpowers have acted sensibly in waging proxy wars in the Third World? Not at all, since a set of beliefs and actions may be self-consistent yet invalid. Indeed some empirically-based studies have suggested that credibility is overemphasized (e.g., George and Smoke, 1974; Lebow, 1981; Huth and Russett, 1984, p.519; see also the references given by Jervis, 1989, p.193). Those holding power worry about credibility, however, and this paper's attempt to state the argument precisely is a step in scrutinizing it. A further reason that the model does not rationalize proxy wars, is its narrow scope. It treats what to do in a single situation, not whether Third World rivalry is sensible as a continuing practice. Looking at a larger game, the superpowers have moves that could end this system of "security." Its demands seem ill-defined, open-ended and immensely harmful. Rather than imposing its costs on the privileged, it inflicts suffering on peoples who have no power to end it.

Acknowledgements: I would like to thank Steve van Evera, Gabrielle MacDonald, David Pearce, Mancur Olson, Janice Newton, Steve Brams, Robert Chibka, Elaine Bennett, Neta Crawford and Catharine and Dorothy O'Neill for their help and good suggestions. Also appreciated are the ideas of David Rothman, Donna Gregory, Jeff Smith, Art Stein and other participants in UCLA's Jacob Marschak Colloquium on Mathematics in the Behavioral Sciences. This work was done with support from an SSRC/MacArthur Fellowship in International Security and a grant from the United States Institute of Peace.

APPENDIX. CALCULATING THE EQUILIBRIA.

Version I. To show that all equilibria must be of the threshold type: Let the onlookers' estimates of the difference between Player G's (Gawain's) and Player K's (the Green Knight's) reputational weights, be designated A and D if G declines or accepts respectively. Assume, contrary to the threshold claim, that there are weights w and w' such that a G holding w would accept, a G holding w' would decline, and $w' > w$. The payoffs with the lower weight w from accepting and declining are FwA-1 and FwD, and the existence of an equilibrium involving acceptance implies $A > D$ and $w \geq 1/F(A-D)$. However declining at higher weight w' yields $w' \leq 1/F(A-D)$, a contradiction. The existence of a threshold in Player K's equilibrium strategy can be shown similarly.

Define w_K and w_G as K's and G's respective weights and k and g as their thresholds. To derive the thresholds: By the uniform prior on the weights, if Player K chooses to dare, the others' expectation for w_K is $(1+k)/2$. Also, if Player G accepts, others' expectation for weight w_G will be $(1+g)/2$, but a refusal produces estimate $g/2$. Substituting these in the payoff expressions in Figure 1, K and G get $Fw_K(k-g)/2 - 1$ and $Fw_G(g-k)/2 - 1$, respectively, when a dare is accepted, and $Fw_K(1+k-g)/2$, $Fw_G(g-k-1)/2$, when one is declined. If w_G equals the threshold g, G is indifferent between accepting and declining, so $Fg(g-k)/2-1 = Fg(g-k-1)/2$, yielding $g = 2^{1/2}F^{-1}$.

Also, at an equilibrium Player K has likelihoods 1-g and g of a dare being accepted or declined, respectively. Using these probabilities to calculate the expectation from daring, it follows that when K's weight is at the threshold and K is thus indifferent between the actions, then $(1-g)[Fk(k-g)/2 - 1] + gFk(1+k-g)/2 = 0$. Combining this formula with the expression derived for g, gives $k = F(F/2-1)^{1/2}$.

Version II. To simplify the description of the equilibria, we will assume that the weights are uniform on the *open* interval (0,1). Derivation of the threshold property is similar to Version I. Gawain and the onlookers hold probability p that the Green Knight is vulnerable and the costs will be 1 for the Green Knight and 0 for Gawain. With probability 1-p the costs will be 0 and 1. Abbreviate Gawain's threshold by g, a vulnerable Knight's threshold by v, and a strong Knight's by s. Everyone else's probability that the Knight is vulnerable given that he challenges will be $p' = (1-v)p/[1-s(1-p)-vp]$. This expression is used to calculate the following payoffs: If a strong Knight challenges and Gawain accepts, the Green Knight and Gawain get $Fw_S(s/2-g/2)$ and $Fw_G(g/2-s/2)-1$ respectively.
If a vulnerable Knight challenges and Gawain accepts: $Fw_V(v/2-g/2)-1$ and $Fw_G(g/2-v/2)$.
If a strong Knight challenges and Gawain declines:

$Fw_S[(1-p')(1/2+s/2)+p'(1/2+v/2)-g/2]$ and $Fw_G[g/2-(1-p')(1/2+s/2)-p'(1/2+v/2)]$.

If a vulnerable Knight challenges and Gawain declines:

$Fw_V[(1-p')(1/2+s/2)+p'(1/2+v/2)-g/2]$ and $Fw_G[g/2-(1-p')(1/2+s/2)-p'(1/2+v/2)]$.

If there is no challenge: 0 and 0.

To calculate a strong Knight's expectations from challenging, weight the appropriate payoffs above by likelihoods of accepting or refusing (1-g) and g. Simplifying the expression yields that a strong Knight expects $(Fw_S/2)[s(1-p'g)+p'gv]$ from challenging, compared to 0 from not challenging. If the expression in brackets is strictly positive any type of strong Knight will challenge (that is, $s = 0$). For a possible equilibrium with $s > 0$, the expression in brackets must be zero (it cannot be negative), implying $p' = g = 1$ and $v = 0$. Equilibria of the latter type cannot exist, however, since it can be shown that any strong Knight would be motivated to challenge given the knowledge that Gawain will surely decline. Therefore we turn back to investigate the remaining possibility of equilibria with $s = 0$. The conditions equating the payoffs for each action for a vulnerable Knight and Gawain when their weights are just at threshold, are respectively $(1-g)[Fv(v/2-g/2)-1] + gFv[(1-p')(1/2)+p'(1/2+v/2)-g/2] = 0$ and $p'Fg(g/2-v/2) + (1-p')(Fg^2/2-1) = Fg[g/2-(1-p')/2-p'(1/2+v/2)]$. The first yields $g = (2-Fv^2)/[2-(1-p')Fv^2]$, and the second $g = 2(1-p')/F$. Equating these and including the further condition given $s = 0$, that $p' = (1-v)p/(1-vp)$, gives a quartic equation in v which can be solved numerically and interpolated for a diagram. Figure 3 uses the technique of parabolic blending.

Version III. Here Gawain is paid as if he were playing Game II, but in fact he knows that the Green Knight is invulnerable. Let A(F,p) and D(F,p) be the onlookers' assessment of his reputational weight minus a strong Knight's, given he acfcepts or declines respectively. Given the above analysis of Version II, $A(F,p) = g/2$ and $D(F,p) = g/2-(1-p')/2 -p'(1/2+v/2)$, where $p' = (1-v)p/(1-vp)$ and v and g are the equilibrium values calculated in Version II (thus functions of F and p). Gawain then expects $Fw_GA(F,p) - 1$ from accepting and $Fw_GD(F,p)$ from declining. He will accept if the former is greater, i.e., if $w_G > 1/F[A(F,p)-D(F,p)] = 2/F(1+v)$. Figures 4 and 5 plot the thresholds.

References

Abrams, Roger. Deep Down in the Jungle. Hatboro, Pa.: Folklore Associates. 1964.

Alker, Hayward. Fairy tales, tragedies and world histories. Behaviormetrika. 21, 1-28, 1987.

Austin, John. How to Do Things with Words. Cambridge: Harvard University Press. 1962.

Benson, Larry. Art and Tradition in *Sir Gawain and the Green Knight*. New Brunswick, N. J.: Rutgers University Press. 1965.

Bloomfield, Morton. *Sir Gawain and the Green Knight*: an appraisal. PMLA. 56, 7-19, 1961.

Borroff, Marie. *Sir Gawain and the Green Knight*, A Stylistic and Metrical Study. New Haven: Yale University Press. 1962.

Brainerd, Barron, and Victoria Neufeldt. On Marcus' method for the analysis of the structure of a play. Poetics. 6, 31-74, 1974.

Brams, Steve. Biblical Games: A Strategic Analysis of Stories in the Old Testament. MIT Press: Cambridge. 1980.

Brams, Steve. Game theory and the humanities. History of Political Economy. Forthcoming, 1991.

Brewer, Elizabeth. From Cuchulainn to Gawain: Sources and Analogues of *Sir Gawain and the Green Knight*. Totowa, N.J.: Rowman and Littlefield. 1973.

Bruss, Elizabeth. The game of literature and some literary games. New Literary History. 9, 153-172, 1977.

Buchanan, Alice. The Irish framework of *Gawain and the Green Knight*. PMLA. 48, 315-339, 1932.

Burrow, John Anthony. A Reading of *Sir Gawain and the Green Knight*. London: Routledge and Kegan Paul. 1965.

Carlson, Lauri. "Well" in Dialogue Games: A Discourse Analysis of the Interlocutive *well* in Idealized Conversation. Amsterdam: John Benjamins. 1984.

Catalini, C.V. Gawain-narrative. Plot Components in Five Mediaeval Romances. Bologna: C.L.E.U.B. 1979.

Chace, James. A new grand strategy. Foreign Policy. 70, 3-25, 1988.

Chadwick, Nora. An Early Irish Reader. New York: Cambridge University Press. 1927.

Davenport, W. A. The Art of the Gawain-poet. London: University of London Press. 1978.

Davey, Lynda. Communication and other games of theatre. Poetics. 13, 5-15, 1984.

Desch, Michael. The keys that lock up the world: identifying American interests in the periphery. International Security. 14, 86-121, 1989.

Fowke, Edith. Daring games. Ch. 9 in Red Rover, Red Rover: Children's Games Played in Canada. Toronto: Doubleday. 1988.

Geanakoplos, John, David Pearce, and Ennio Stacchetti. Psychological games and sequential rationality. Game Theory and Economic Behavior. 1, 60-79, 1989.

George, Alexander, and Richard Smoke. Commitment theory. Ch. 19 in Deterrence in American Foreign Policy. New York: Columbia University Press. 1974.

Gilboa, Itzhak, and David Schmeidler. Information-dependent games: can common sense be common knowledge? Economics Letters. 27, 215-221, 1988.

Greenberg, Kenneth. The nose, the lie, and the duel in the Antebellum South. American Historical Review. 95, 57-74, 1990.

Harsanyi, John, and Reinhard Selten. A General Theory of Equilibrium Selection in Games. Cambridge: MIT Press. 1988.

Henderson, George, ed. and trans. *Fled Bricrend, The Feast of Bricriu*. London: Early Irish Texts Society. 1899.

Hintikka, Jaakko and Merrill. Sherlock Holmes confronts modern linguistics: toward a theory of information-seeking questions. 55-76 in E.M. Barth and J.L. Martens, eds. Argumentation: Approaches to Theory Formation. Amsterdam: John Benjamins. 1982.

Howard, Nigel. Paradoxes of Rationality. Cambridge: MIT Press. 1971.

Howard, Nigel. The plot of *Dr. Zhivago*. Conan Newsletter. 2, 2-4, 1980.

Huth, Paul, and Bruce Russett. What makes deterrence work? World Politics. 36, 496-526, 1984.

Janis, Irving, and Leon Mann. Decision-Making: A Psychological Analysis of Conflict and Commitment. New York: Basic. 1977.

Jervis, Robert. Deterrence theory revisited. World Politics. 31, 298-324, 1979.

Jervis, Robert. The symbolic nature of nuclear strategy. Ch. 6 in The Meaning of the Nuclear Revolution. Ithaca: Cornell University Press. 1989.

Johnson, Robert. Exaggerating America's stakes in Third World conflicts. International Security. 10, 32-69, 1986.

Kissam, Margaret. The Narrative Structure of Middle English Romances. Ph.D. thesis, City University of New York. 1977.

Kittredge, George Lyman. A Study of *Gawain and the Green Knight*. Gloucester, Mass.: P. Smith. 1916.

Lalu, Iolanda. Richard III: Balance and game in the study of theatre. Poetics. 6, 339-350, 1977.

Lebow, Richard Ned. Between Peace and War: The Nature of International Crisis. Baltimore: Johns Hopkins University Press. 1981.

Loomis, Laura Hibbard. *Gawain and the Green Knight*. 528-540 in R.S. Loomis, ed. Arthurian Literature in the Middle Ages. New York: Oxford University Press. 1959.

Marcus, Solomon. Mathematische Poetik. Frankfurt: Athenaum Verlag. 1973. Translation of Poetica Mathematica. Bucharest: Editura Academiei Republicii Socialiste Romania. 1970.

Marcus, Solomon, ed. The Formal Study of Drama, special edition of Poetics. 9, 1977.

Marcus, Solomon, ed. The Formal Study of Drama, II, special edition of Poetics. 16, 1984.

Martindale, Colin. Structural balance and the rules of narrative. Poetics. 8, 57-67, 1976.

Mehlmann, Alexander. Applied Differential Games. New York: Plenum Press. 1988.

Mehlmann, Alexander. De Salvatione Fausti. Vienna: Faude. 1989.

Meyer, Kuno. The Edinburgh version of the *Cennach ind Ruanado* (Bargain of the Strong Men). Revue Celtique. 10, 454-491, 1893.

Morgan, Patrick. Saving face for the sake of deterrence. 125-152 in Robert Jervis, Richard New Lebow and Janice Stein, eds. Psychology and Deterrence. Baltimore: Johns Hopkins. 1985.

Morgenstern, Oskar. *Wirtschaftprognose*. (Economic Prediction.) 1928.

Morgenstern, Oskar. Vollkommene, Voraussicht und wirtschaftliches Gleichgewicht. Zeitschrift fur Nationalokonomie. 6, 337-357, 1935. English translation: Perfect foresight and economic equilibrium. In Andrew Schotter, ed. Selected Economic Writings of Oskar Morgenstern. New York: New York University Press. 1976.

Nalebuff, Barry. Rational deterrence in an imperfect world. Forthcoming in Michael Intriligator and Urs Luterbacher, eds. Cooperative Game Models in International Relations. 1990.

O'Neill, Barry. Rational probabilities for the outcomes of games with two strong equilibria. Mimeo, School of Public Affairs, University of Maryland. 1988.

Opie, Iona and Peter. Daring games. Ch. 9 in Iona and Peter Opie. Children's Games in Street and Playground. Oxford: Clarendon. 1969.

Payne, James L. The demonstration of will. Ch. 5 in The American Threat: The Fear of War as an Instrument of Foreign Policy. Chicago: Markham. 1970.

Powell, Robert. Nuclear Deterrence: The Search for Credibility. New York: Oxford University Press. 1990.

Prince, Gerald. A Grammar of Stories. The Hague: Mouton. 1973.

Propp, Vladimir. Morfologii Skazki. 1928. Tranlation, Morphology of the Folktale. Bloomington: Indiana University Press. 1958.

Rapoport, Anatol. Fights, Games and Debates. Ann Arbor: University of Michigan, 1960.

Rapoport, Anatol. The use and misuse of game theory. Scientific American. 108-118, 1962.

Ryan, Marie-Laure. Linguistic models in narratology; from structuralism to generative semantics. Semiotica. 28, 127-155, 1979.

Schoultz, Lars. National Security and U.S. Policy Towar Latin America. Princeton: Princeton University Press. 1987.

Steriadi-Bogdan, Mariana. The evolution of plot and problems of strategy in a detective play. Poetics. 6, 375-382, 1977.

Talbott, Strobe. Endgame. New York: Harper and Row. 1979.

Teodorescu-Brinzeu, Pia. A systems approach to theatre. Poetics. 6, 351-374, 1977.

Teodorescu-Brinzeu, Pia. The monologue as dramatic sign. Poetics. 13, 135-148, 1984.

Thompson, Raymond. "Muse on þi mirrour . . .", the challenge of the outlandish stranger in the Engish Arthurian verse romances. Folklore. 87, 201-208, 1976.

Thurneysen, Rudolf. Die Irische Helden- und Konigsage bis zum Siebzehnten Jahrhundert. Halle: Niemeyer. 1921.

Tolkien, J.R.R., and E.D. Gordon, eds. *Sir Gawain and the Green Knight*. Oxford: Clarendon Press. 1925.

Tucker, Robert. The Purposes of American Power: An Essay on National Security. New York: Praeger. 1981.

Vale, Malcolm. Chivalric display. Ch. 3 in War and Chivalry. London: Duckworth. 1981.

von Neumann, John, and Oskar Morgenstern. The Theory of Games and Economic Behavior. Princeton: Princeton University Press. 1944.

Vorobev, Nikolai. Khudozhestvennoe modelirovanie konflickty i teoria igr. (Literary conflict modelling and the theory of games). 348-372 in B.S. Meilakh, ed. Sodruzhestvo Nauk i Tainy Tvorchestva. (The Close Relationship of the Sciences and the Secrets of Artistic Creation.) Moscow: Izkustvo. 1968.

Williams, John. The Compleat Strategyst. Santa Monica: RAND Corporation. 1954.

Zumwalt, Elmo. On Watch: A Memoir. New York: Quadrangle. 1976.

BALANCE OF POWER IN A PARLOR GAME

R. Selten
University of Bonn, FRG[1]

1. Introduction

The idea of balance of power in a system of sovereign states has a long history. Already 300 B.C., Artha-Sastra Kautilya has written on the subject (Kautilya, 1957). A recent book by Bernholz (1985) looks at the historical facts in the light of the theory. A more detailed account of the literature can be found here.

The term *balance of power* is not always used in the same sense, but it seems to apply to a state of affairs, where potential aggressors are deterred from military actions whose immediate or remote probable consequences involve the elimination of one of the actors. This does not necessarily exclude wars which do not endanger the survival of any actor. Wars may happen, but the system of actors remains the same. It is also possible that a situation of long lasting stable peace is secured by a balance of power

Game theory is a general theory of rational behavior in interactive decision situations of conflict and cooperation. Balance of power theories (Kaplan 1957, Zinnes 1967. Bernholz 1985) try to explain the typical behavior of nation states as the rational pursuit of their interests. However, traditional verbal theories do not offer a logically stringent explanation. For this purpose, the use of game theory seems to be unavoidable.

An interesting game theoretic approach to the balance of power problem has been developed by Niou and Ordeshook (1986, 1987). Their model is a cooperative game, to which they apply a version of the bargaining set (Aumann and Maschler 1964). Strategies of individual players cannot be discussed in the framework of a cooperative game. Therefore, we think that a non-cooperative game model is preferable. The use of non-cooperative game theory does not mean that cooperation is excluded. On the contrary, even if this may seem to be paradoxical, the explanation of cooperation requires non-cooperative game theory.

---

[1] I am grateful to Peter Bernholz for stimulating discussions and valuable suggestions.

A non-cooperative model of a balance of power system has been proposed by Harrison Wagner (1986). His analysis is heuristic rather than precise, but nevertheless suggestive. In his model, the players have to decide on the deployment of continuous resources and on the military use in continuous time. The strategy space remains unclear. Regardless of how this gap is filled, a precise analysis is bound to be difficult.

In this paper, a non-cooperative game model will be presented which takes the form of a parlor game. Theoretical results need to be confronted with the behavior of subjects in experiments. Therefore, playability is an important modelling consideration.

The rules of the parlor game will be given in ordinary language. The playing board and other paraphernalia are described in detail and can easily be prepared by anybody who wants to play. The game is called *Changing Alliances*. Playing the game does not require any knowledge of game theory.

Practical experience shows that *Changing Alliances* is not without entertainment value, even if this aspect has not been a major concern in the construction of the game. The rules have been kept simple in order to enhance the usefulness of the game as a research tool.

The game is not meant to be a realistic description of a system of nation states, but rather a radically simplified picture which concentrates on the essence of the subject matter. Features of historical systems which are not essential for the reasoning underlying balance of power theories are intentionally neglected.

The players in the parlor game represent sovereign states in a world consisting of 30 *provinces* but without any geographical structure. Provinces are resources which can be won or lost in war. The game is played over 30 *rounds*. In each round an *alliance* and a *counteralliance* can be built up by formal moves on the board. Alliances may change from round to round.

After the description of the parlor game in Section 2, reasons for details of the rules will be given in Section 3. In particular, it will be discussed why a fixed number of rounds is specified. An infinite number of rounds is theoretically more convenient but practically infeasible.

In Section 4, a modified version of the game with only 6 provinces and an infinite number of rounds will be introduced; a long run average definition of payoffs is used. Unlike the much more complex parlor game with its 30 provinces, this simpler model permits a game theoretical analysis without serious difficulties. A "solution" determined by symmetry, stationarity and local non-cooperative equilibrium conditions is derived in Section 5. In order to clarify technical problems concerning global equilibrium conditions, discounted payoffs are considered in Section 6.

The solution of the 6-province model is discussed in Section 7. An interpretation of the game theoretical results will try to exhibit resons for stability and instability.

2. Changing Alliances - A Parlor Game

2.1 Inventory

1 playing board (see Figure 1)
12 player tokens, marked 1,...,12
12 country boards (see Figure 2)
12 player cards, marked 1,...,12 (see Figure 3)
30 province cards, marked 1,...;30 (see Figure 2)
30 province stones, marked 1,...,30 (see Figure 4)
1 province stone box (see Figure 3)
360 resource chips (see Figure 2)

2.2 The Object

Each player represents a country. Countries can form changing alliances and counteralliances and fight wars. Provinces are won and lost as a result of war. A player should try to obtain as many provinces as possible, but it is even more important to secure continued existence. A small advantage is attached to staying out of wars which do not permanently increase one's territorial holdings.

Scores are computed for each player at the end of the game. It is not important to have a higher score than other players. Everybody should aim at a high score for himself without being concerned about the scores of other players.

2.3 Preparation for Play

The game is played by at least 3 and up to 12 players. The players sit around a table. The playing board is put on this table (see Figure 1). Each player receives a country board placed in front of his or her seat (see Figure 2). The number on the country board becomes the number of the player.

The provinces are distributed among the players. At first, each player receives one province card which is put on his or her country board. Then, each of the remaining province cards is allotted as follows. The player cards (see Figure 3) of those players who do not yet have 12 province cards are mixed and one is drawn at random. This player receives the card. The province card is put on his or her country board. For each province, the player cards of those players who do not yet have 12 province cards, are mixed again and the same procedure is repeated until all province cards have been distributed.

Finally, the number of province cards owned by every player must be written down on a slip of paper, since it will be important for the computation of scores at the end of the game.

Each player receives a token showing the player's number. The tokens are placed on the field "wait for decision", each on the square bearing its number.

The game begins with the selection of the first decision maker for the first round.

The game is played over 30 rounds. Every round is played in the same way. It will now be explained how a round is played.

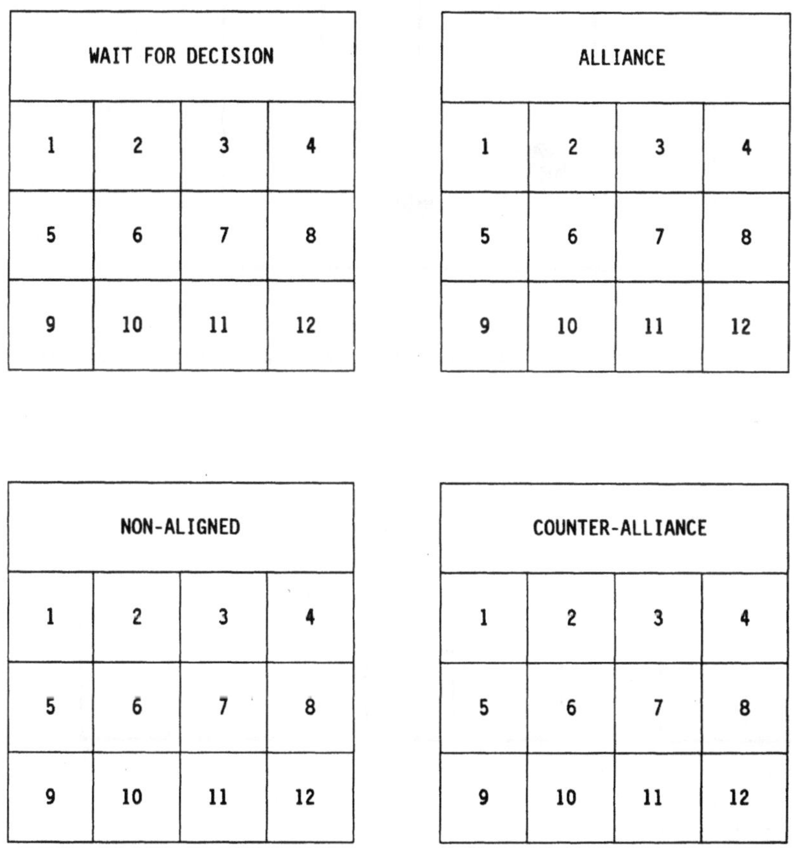

Figure 1: The playing board

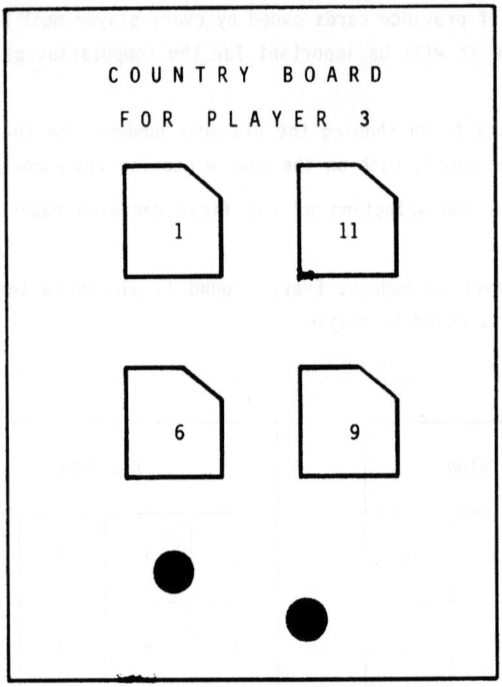

Figure 2: Player 3's country board with four province cards and 2 resource chips. (The upper right corner of province cards is cut off to distinguish 6 and 9.

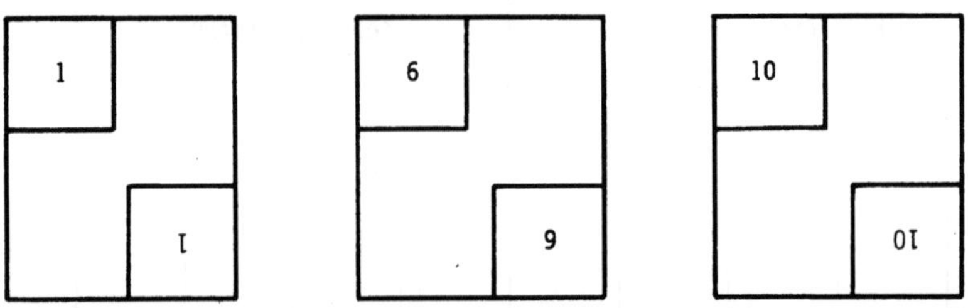

Figure 3: Player cards

2.4 Selection of a Decision Maker

As long as there are players with tokens on the field "wait for decision", the next decision maker is chosen among these players. For this purpose, their player cards are mixed and the upmost card in the deck is turned up. The player whose card has been turned up becomes the next decision maker.

2.5 Pre-Attack and Post-Attack Decision Maker

The choices which are open to a decision maker depend on previous decisions. As long as no player has attacked another one, the decision maker is in the position of a pre-attack decision maker. After an attack, the decision makers are post-attack decision makers, as long as there are tokens on the field "wait for decision".

Of course, the first decision maker is a pre-attack decision maker, randomly chosen among all players in the game.

2.6 Choices of a Pre-Attack Decision Maker

A pre-attack decision maker has the following options:

a) Stay non-aligned: In this case, the token of the decision maker is moved to its square on the field "non-aligned". Nothing else is changed.

b) Attack another player: The decision maker names the player he wants to attack. The attacked player can be any other player who owns at least one province, no matter whether the token of the attacked player is on the field "wait for decision" or on the field "non-aligned". The attack has the following consequences:

- The token of the decision maker is moved to its square on the field "alliance".
- The token of the attacked player is moved to its square on the field "counteralliance".
- After the tokens of the decision maker and the attacked player have been moved, any tokens left on the field "non-aligned" are moved back to their squares on the field "wait for decision".

2.7 Choices of a Post-Attack Decision Maker

A post-attack decision maker has the following options:

a) Stay non-aligned: In this case, the token of the decision maker is moved to its square on the field "non-aligned". Nothing else is changed.

b) Join the alliance: The token of the decision maker is moved to its square on the field "alliance". - All tokens on the field "non-aligned" are moved back to their squares on the field "wait for decision".

c) Join the counteralliance: The token of the decision maker is moved to its square on the field "counteralliance". - All tokens on the field "non-aligned" are moved back to their squares on the field "wait for decision".

2.8 Stages of a Round

A round has two stages. The explanations given up to now concern the alignment stage, in which pre-attack decision makers and post-attack decision makers make their alignment decisions. Sooner or later, a situation is reached where no tokens are left on the field "wait for decisions". At this point, the alignment stage comes to an end and the round enters its allotment stage.

2.9 End in Peace

If at the end of the alignment stage all tokens on the board are on the field "non-aligned", the round ends in peace. In this case, the allotment stage is very short: Every player with a token on the field "non-aligned" receives one resource chip.

2.10 Allotment Procedure

If at the end of the alignment stage some tokens are on the fiels "alliance" and "counteralliance", the territorial consequences of the military confrontation have to be determined. One of both groups always emerges as the "winning party" and the other becomes the "losing party". A province is transferred from a member of the losing party to a member of the winning party. Which province is lost by whom and to whom is determined by an allotment procedure explained in the following.

- First, the strength of the alliance and the counteralliance is determined. The strength of each of both groups is measured by the number of provinces owned by its members. If one group is stronger, it becomes the winning party and the opposing group becomes the losing party.
- All province stones are put into the province stone box (see Figure 4). Then, one stone after the other is drawn at random without replacement until it becomes clear which province is lost by whom to whom.
- If one group is stronger, then the first stone with a number owned by a member of the winning party determines the "winner". The player who owns this province is the winner.
- If both groups are equally strong, then the first stone with a number of a province owned by a member of one of both groups determines the winner. The player who owns this province is the winner. His group becomes the winning party and the other group becomes the losing party.
- The first stone with a number of a province owned by a member of the losing party determines the province lost. The owner is the loser. The loser has to transfer the province card with this number to the winner.

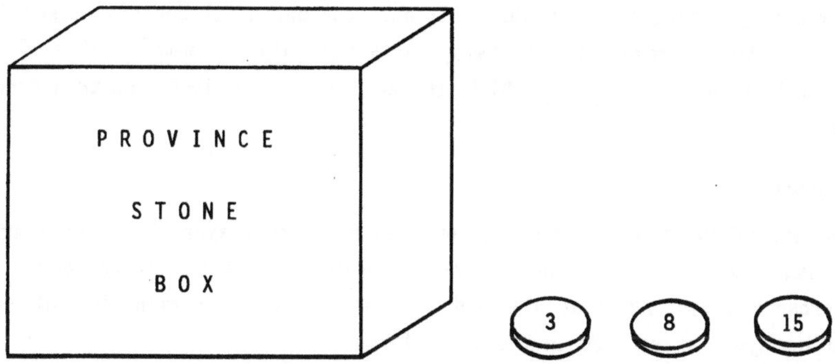

Figure 4: Province stone box and province stones.

2.11 Giving Away a Province Won

A player who has won a province can offer this province to another player if he wants to do this. The other player may be any other player, including the loser or a player who does not own a province. The player to whom the offer is made can accept or reject it. If the offer is accepted, the winner transfers the province card won to the concerning player. If the offer is rejected, the winner keeps the province. A second offer to give away the province is not permissible.

2.12 Resource Chips

At the end of the alignment stage of a round, each player with a token on the field "non-aligned" receives one resource chip.

A player who loses his last province does not lose his resource chips. Resource chips can be owned by players without provinces.

2.13 End of the Game

The game ends after 30 rounds. (Another fixed number of rounds can be agreed upon in advance.)

2.14 Beginning of a New Round

At the beginning of a new round, all tokens of players who own at least one province are put on their squares in the field "wait for decision". The tokens of players without a province are removed from the playing board. Then, the round is played according to the rules explained above.

2.15 Simplification of Random Procedures

In some cases it is not necessary to perform the random procedures exactly as prescribed by the rules given above. If only one potential decision maker is left, no card needs to be drawn. If the losing party has only one member, this player is the loser and it does not matter which of the loser's provinces is transferred to the winner.

2.16 Scores

At the end of the game, scores are computed for each player. The scores are based on the numbers of provinces owned at the beginning and the end of the game. The number of resource chips owned at the end also enters the determination of scores. The following symbols are used in the score formula:

a_i number of provinces owned by player i at the beginning of the game

m_i number of provinces owned by player i at the end of the game (after the last allotment)

r_i number of resource chips owned by player i at the end of the game

e_i this number is 1 if player i owns at least one province at the end of the game and zero otherwise

S_i score of player i.

The score formula is as follows:

$$S_i = 100e_i + 30(m_i - a_i) + r_i. \qquad (1)$$

2.17 Game Director

The game runs more smoothly if one person does not actively participate, but, instead of this, functions as a *game director*, who performs the random choices which determine the order of moves in the alignment stage, moves the tokens on the board, runs the allotment stage, and pays out resource chips. The game director can concentrate on strict obeyance of the rules. Thus, he should not forget to move all tokens back from "non-aligned" to "wait for decisions" after a player has joined the alliance or the counteralliance. the greater the number of players, the more it is advisable to appoint a game director.

2.18 Playing for Money

Groups of players who want to play for money can agree to compute deviations from the average score and to determine wins and losses accordingly. However, if this is done, the payoff of a player does not only depend on his own score but also on the scores of the others. For theoretical reasons, it is preferable to employ a payoff scheme which avoids the dependence on other players' scores.

If one participant functions as a game director, one can proceed as follows: an agreed upon "base score" S_0 is subtracted from each players' score S_i in order to obtain a "net score" S_i-S_0. According to some fixed value of a score point, a player wins the corresponding amount if his net score is positive and loses the corresponding amount if his net score is negative. If there is a surplus of losses over gains, the game director receives it; otherwise he has to pay the surplus of gains over losses.

The net score S_0 must be determined before the beginning of the game. One way of doing this is an auction for the position of the game director. Every participant secretly writes his name and a proposed base score S_0 on a piece of paper. The position of the game director goes to the lowest bidder (if there are several lowest bidders, a random choice has to be made). The lowest bid becomes the base score S_0.

One could also agree to fix S_0 at 130 and to donate the surplus of losses over gains to a humanitarian cause. For S_0 = 130 there cannot be a surplus of gains over losses.

3. Comments on the Rules of *Changing Alliances*

3.1 The Number of Rounds

Practical experiences were gained with an earlier version of the game which involved a fixed stopping probability. After each round, two dice were thrown and the game stopped in the case of two sixes. In view of experiences with actual plays of the game, this rule was later abandoned in favor of a fixed length of 30 rounds.

A theoretical objection against a fixed stopping probability is based on the impossibility of playing such games in the laboratory. The subjects cannot be kept in the laboratory for an unlimited time. This imposes an upper bound on the number of rounds. It may not be clear how many rounds can be played at most, but a finite upper bound can be found below which the length of the game must remain. Eventually, the game must be stopped against the rules, if the random event required for stopping does not occur before the upper bound is reached.

In some games, e.g. in supergames of the prisoners' dilemma, it is of crucial theoretical importance whether the number of rounds has a finite upper bound or not. A finite upper bound supplies an anchor for the familiar induction argument, which shows that all equilibrium plays involve non-cooperative choices only (Luce and Raiffa, 1957). Contrary to this, the infinite supergame has subgame perfect equilibrium points whose equilibrium plays involve cooperative choices only.

It is quite common in parlor games that infinite plays are not excluded (e.g. in Parchesi). However, even if one ignores the theoretical objection raised above, one still has to face a completely different behavioral difficulty. In one of the actual plays with the stopping rule of a throw of two sixes, the game did not end in the first 36 rounds. After 36 rounds, several players expressed their belief that now the end must come soon. One of the players had a thorough training in statistics and knew that his confessed belief did not conform to probability theory. This kind of behavior is an example of the ***gambler's fallacy***. The gambler's fallacy has a strong influence on decision making which should not be overlooked in the design of experiments.

The crucial theoretical difference between finite and infinite prisoner's dilemma supergames does not seem to be behaviorally relevant. This is shown by experimental studies (e.g. Selten and Stoecker, 1986). Apart from a few rounds near the end, observed behavior of experienced subjects achieves cooperation in sufficiently long finite prisoners' dilemma supergames. In view of these findings, it seems to be justifiable from the point of view of descriptive game theory to apply the analysis of infinite games to long finite games. We shall take this approach in the analysis of the model presented here.

3.2 Parameters

Numerical parameters like the number of provinces or the coefficients in the function for the determination of scores can be varied in the abstract model. It is of particular interest to look at cases with very few provinces, since there a complete theoretical analysis is easier to obtain.

3.3 Formation of Alliances and Counteralliances

It is important that all non-aligned players move back to "wait for decision" after a player has joined the alliance or the counteralliance. This rule has the purpose to make the random order of alignment decisions less important. A player can decide to enter an alignment only after some other player, even if he is the first one to make a decision.

3.4 Interpretation of a Round

Only one province can change hands during a round. This rule is based on the idea that alignments must be permitted to change after each change of the situation. Within a round, only a war episode takes place. The war can be continued in the next round with changed or unchanged alignments.

3.5 Allotment

If resources were infinitely divisible, the allotment procedure could be modelled deterministically as in Wagner's model (Wagner, 1986). In view of the indivisibility of the provinces, a random allotment is needed in order to make expected games and losses proportional to resources held. In his book, written 1832-34, Carl von Clausewitz emphasized the random element in military conflict (von Clausewitz, 1963).

The random element of the allotment procedure of Changing Alliances is rather weak, since the stronger bataillons always win.

3.6 Resource Chips

Resource chips provide an incentive not to try to gain a province whose ownership cannot be maintained in later rounds. The value of a resource chip is just small enough to exclude a disincentive to enter an alignment which is sure to win if in the case of winning a province, the concerning player can count on having one province more at the end. The minimum probability of winning a province for a member of the winning side is 1/29. This is the winning probability of a player with 1 province in an alliance with 29 provinces against a counteralliance of 1 province. The probability of 1/29 of having 1 province more at the end corresponds to an expected score gain of 30/29 which is slightly greater than 1, the value of 1 resource chip.

A player who loses all his provinces still keeps his resource chips earned in the past. One may think of resource chip as money accumulated on a Swiss bank account.

It can be argued that the first part of rule 2.12 should be changed as follows: At the end of a round, each player with a token on the field "non-aligned" receives one resource chip for every province earned. - Clearly, this rule would be more natural and it would have similar incentive effects. However, the simpler rule of 2.12 is easier to follow in the actual play of the parlor game.

3.7 Scores

As has been explained above, incentive considerations suggest a value of 30 resource chips for 1 province. It is reasonable to attach a positive value to existence, but there seem to be no theoretical reasons for a particular choice of this "existence value".

The score function is meant to be a reasonable representation of preferences of nation states in the simplified hypothetical environment of the game. Obviously, Changing Alliances is not a constant sum game. The score parts $30(m_i-a_i)$ due to the numbers m_i of provinces owned at the end sum up to zero, but each player who ends up without provinces decreases the score sum by 100 and each resource chip earned increases the payoff sum by 1. The greatest score sum is obtained if all 30 periods end in peace.

3.8 Playing for Money

A money payoff scheme based on deviations from average scores has the undesirable effect of transforming the game into a zero-sum game. The transformation creates an additional incentive for the elimination of players. Moreover, in the transformed game it is advantageous to prevent others from gaining resource chips. Therefore, it is preferable to use one of the alternative money payoff schemes proposed in 2.18. These proposals result in a money payoff which is a positive linear transformation of the score. Of course, one may prefer to determine the game director by another type of auction. The first price scaled bid auction proposed in 2.18 is only one of many possibilities.

4. A Theoretical Model

4.1 The Model

A modified version of *Changing Alliances* will now be introduced as a basis for a game theoretical analysis. In 3.1, it has been argued that there are good reasons for a fixed number of rounds in the parlor game. However, in a purely theoretical model, it is justifiable and maybe even preferable to take a different approach. The deviations of the modified model from the parlor game are as follows:

1) The number of rounds is infinite
2) There are only 6 provinces
3) There are no resource chips
4) There is no possibility to give away a province
5) Payoffs are based on the long run average of temporary payoffs for the number of provinces held in each round (see 4.3).

Apart from these modifications and simplifications, the rules for the alignment stage and the allotment stage are the same as in the parlor game.

4.2 Strategies

Technically, the model is an infinite extensive game with perfect information. A precise mathematical description of the game will be omitted here in order to avoid cumbersome formalism. The structure of the extensive game is sufficiently clear by the rules of the parlor game modified by 1) to 4). However, it is necessary to add some remarks to the formal definition of strategies and payoffs.

Our analysis will be focused on stationary behavior strategies. Stationarity means that decisions do not depend on anything else than the current province distribution and the board position. This means that a player's behavior is the same at all nodes of the game tree which agree with respect to the current province distribution and the board position.

Since there are only six provinces, it is not necessary to consider more than six players. Therefore, we assume that there are six *players* 1,...,6. A *province distribution*

$$m = (m_1,\dots,m_6) \tag{2}$$

assigns a number of provinces m_i to every player $i = 1,\dots,6$. The m_i are non-negative integers which sum up to 6. The set of all province distributions is denoted by M. In order to describe the board position, we introduce the following notation:

W the set of players with tokens on the field "wait for decision"

A the set of players with tokens on the field "alliance"

C the set of players with tokens on the field "counteralliance"

N the set of players with tokens on the field "non-aligned".

These four sets form a *board position*:

$$B = (W,A,C,N). \tag{3}$$

A *decision situation* is a triple

$$D = (m,B,j) \tag{4}$$

composed of a province distribution, a board position, and a decision maker j. A player with at least one province is called *active*, the other players are called *passive*. Let I be the set of active players in m. A board position B partitions the set I into four sets W,A,C,N with the interpretation given above. W must be non-empty and j is one of the players in W. The set of all decision situations $D = (m,B,j)$ is denoted by Δ.

As in 2.5, we distinguish between *pre-attack* decision situations with $A = C = \phi$ and *post-attack* decision situations with $A \neq \phi$ and $C \neq \phi$. In a pre-attack decision situation $D = (m,B,j)$, the choice set has as many elements as there are active players. Player j can choose "non-aligned" or he can attack one of the other active players. In a post-attack decision situation, the decision maker has three choices, namely "non-aligned", "join the alliance", and "join the counteralliance". It is not necessary to add a formal description of the choice sets.

We shall always look at the case of an initial province distribution (1,1,1,1,1,1), where each of the six players has one province. This does not involve any loss of generality since all other cases are subgames of this game.

A *local strategy* s_D at $D = (m,B,j)$ is a probability distribution over the choice set at D. We shall not be interested in stationary behavior strategies in general, but only in symmetric behavior strategies played by all players. We use the term *global strategies* for the sake of shortness. Formally, a *global strategy* s is a function which assigns a local strategy s_D at D to every $D \in \Delta$, such that two symmetry requirements (i) and (ii) are satisfied:

(i) Symmetry with respect to players: s is invariant with respect to a renumbering of players.

(ii) Symmetry with respect to alignments: s is invariant with respect to an exchange of the sets A and C, i.e., two board positions $B = (W,A,C,N)$ and $B' = (W,A',C',N)$ with $A' = C$ and $C' = A$ are treated symmetrically.

Of course, (i) and (ii) are not really mathematically precise statements, but since the meaning is sufficiently clear, we can avoid the formalism which would be necessary for a precise definition. The set of all global strategies is denoted by S.

A province distribution $m = (m_1,\ldots,m_6)$ is called *basic* if we have:

$$m_1 \geq m_2 \geq m_3 \geq m_4 \geq m_5 \geq m_6. \tag{5}$$

The set of all basic province distributions is denoted by M_0. A decision situation $D = (m,B,j)$ is called *basic* if $m \in M_0$ holds. The set of all basic decision situations is denoted by Δ_0. In view of (i), a global strategy s is completely determined by its restriction s_0 to Δ_0. In fact, s_0 is still redundant as a description of s, since (ii) permits a further reduction of the set of decision situations which need to be considered. A *basic strategy* is restricted to the set Δ_0 but otherwise defined in the same way as a global strategy. We say that a global strategy s and a basic strategy s_0 *correspond* to each other if s_0 is the restriction of s to Δ_0.

4.3 Payoffs

We first assign a *temporary payoff vector*

$$h(m) = (h_1(m),\ldots,h_6(m)) \tag{6}$$

to every province distribution $m = (m_1,\ldots,m_6)$. The *temporary payoffs* are given by

$$h_i(m) = \begin{cases} m_i & \text{for } m_i > 0 \\ -F & \text{for } m_i = 0 \end{cases} \tag{7}$$

Here, F is a positive constant called the *existence value*. Payoffs will be defined as long-run averages of temporary payoffs. It will not be sufficient to look at payoffs in the whole game. It is also necessary to define local payoffs for every node which can be reached by a play. Consider first the situation at the beginning of a round t. Assume that the current province distribution is m and that in rounds t,t+1,... all players always play the same global strategy s. This stochastic process yields a sequence m(t), m(t+1),... of province distributions. The sequence can be looked upon as the realization of a finite Markoff chain (Feller, 1968). Depending on $m(t) = m$, we obtain uniquely determined stationary distributions over M. Expected temporary payoffs $H_i(m,s)$ can be computed with the help of these distributions. The vector

$$H(m,s) = (H_1(m,s),\ldots,H_6(m,s)) \tag{8}$$

is the *payoff vector* for m if s is played. We now define local payoffs $L(D,r_D,s)$ for every triple D,r_D,s, where $D = (m,B,j) \in \Delta$ is a decision situation, r_D is a local strategy at D and s is a global strategy:

$$L(D,r_D,s) = E(H_j(m(t+1),s)) \tag{9}$$

with $m(t) = m$. Here, E is the expectation operator with respect to r_D. It is not necessary to express the identity of the player by an index at L, since the local payoff is always a payoff for the decision maker in D.

4.4. Equilibrium Strategies

The set of all local strategies at a decision situation $D = (m,B,j)$ is denoted by S_D. Local strategies which cannot be used by global strategies in view of the symmetry conditions (i) and (ii) in the definition of a global strategy are not excluded from S_D. A local strategy q_D at $D \in \Delta$ is a *local best reply* to a global strategy s if the following is true:

$$L(D,q_D,s) = \max_{r_D \in S_D} L(D,r_D,s). \tag{10}$$

A global strategy s in an *equilibrium strategy* if for every $D \in \Delta$ the local strategy s_D assigned to D by s is a local best reply to s:

$$L(D,s_D,s) = \max_{r_D \in S_D} L(D,r_D,s). \tag{11}$$

The equilibrium condition (11) is purely local. It is necessary but not sufficient for subgame perfect equilibrium in the usual sense. This is due to the definition of payoffs as a long run average of temporary payoffs. If the same decision situation is reached infinitely often, it is possible that nothing is gained by deviating just once, but payoffs can be improved by infinitely many deviations.

In finite games with perfect information, local equilibrium conditions like those expressed by (11) are sufficient for subgame perfect equilibrium. The same is true for the infinite game if payoffs are defined as discounted sums of temporary payoffs. This payoff definition seems to be the more adequate one, if one looks at the infinite game as an approximation of long finite games. In Section 6, we shall investigate payoffs defined as discounted sums of temporary payoffs in order to justify the neglect of global equilibrium conditions in the definition of an equilibrium strategy.

4.5 Classification of Province Distributions

Let m be a basic province distribution. We say that a province distribution m' is of *type* m if m can be obtained from m' by a renumbering of the players. In order to determine the payoff vector $H(m',s)$ for m', it is sufficient to determine $H(m,s)$ for the basic province distribution m of the same type.

A province distribution $m \in M$ is called *stable with respect to the global strategy* s if for all pre-attack decision situations $D = (m,B,j)$ the local strategies at D choose "non-aligned" with probability 1. Obviously, in this case we have:

$$H_i(m,s) = m_i \quad \text{for} \quad m_i > 0. \tag{12}$$

If m is stable, then every round ends in peace after m is reached as long as s is played.

We say that a province distribution m is *viable with respect to the global strategy* s if with probability 1 the set of active players remains the same for all remaining rounds after m is reached. In view of $F > 0$, this is the case if and only if we have

$$\sum_{i \in I} H_i(m,s) = 6 \tag{13}$$

where I is the set of active players.

The total number of provinces is always 6. If m is not viable, then the sum of the left hand side is diminished by the expected number of players which will lose their existence in the future multiplied by F.

A province distribution which is not viable with respect to s is called *transitory* with respect to s. After a transitory province distribution m has been reached, the set of active players must be eventually diminished with probability 1 if s is played. This is a consequence of the stationarity of s.

Obviously, stability implies viability, but not vice versa. A province distribution m which is viable but not stable with respect to s is called ***fluid*** with respect to s. The classification into stable, fluid and transitory province distributions with respect to s partitions the set M into three classes. Every province distribution belongs to exactly one of these classes. We say that s assigns the ***stability status*** "stable", "fluid" or "transitory" to a province distribution m, if m is stable, fluid or transitory resp. with respect to s. In view of the symmetry properties of global strategies, the stability status of m with respect to s depends only on s and the type of m.

4.6 Conservative Equilibrium Strategies

The game does not have a uniquely determined equilibrium strategy. However, some equilibrium strategies seem to be unreasonable and therefore should be eliminated by some equilibrium selection criterion. Recently, a general equilibrium selection theory has been developed (Harsanyi and Selten, 1988). However, this theory is restricted to finite games and therefore cannot be applied here. Instead of this, a criterion will be applied which is suggested by the context of our analysis. We shall make use of the following notation:

M_k the set of province distributions with k active players (players with at least one province);

$M_{<k}$ the set of province distributions with less than k active players;

$P_k(s)$ the set of all province distributions with k active players which are viable with respect to s.

The subgames beginning with province distributions in M_k cannot move to province distributions with more than k active players. Therefore, it is natural to search for an equilibrium selection by first looking at M_1, then at M_2, then at M_3, etc., and finally at M_6. The analysis of equilibrium conditions for province distributions in M_k can rely on results obtained for $M_{<k}$. However, usually other province distributions in M_k can be reached from a province distribution in M_k. Therefore, the analysis of the game is not completely recursive as in the search for subgame perfect equilibrium points in finite games with perfect information. All province distributions in M_k have to be considered simultaneously.

We say that an equilibrium strategy s is *conservative* if for k = 2,...,6 there is no other global strategy r_k which satisfies the following three conditions (a), (b), and (c):

(a) The global strategy r_k agrees with s for decision situations $D = (m,B,j)$ with $m \in M_{<k}$.

(b) The global strategy r_k assigns local best replies to r_k to all decision situations $D = (m,B,j)$ with $m \in M_k$.

(c) $P_k(s)$ is a proper subset of $P_k(r_k)$.

The concept of a conservative equilibrium strategy is based on the idea that the players form expectations recursively in the order of increasing numbers of active players and that at each inductive step as much priority as possible is given to the assumption of viability.

Viability means that no active player is in danger to be eliminated. In this sense, viability is a weak form of balance of power. Even if the province distribution keeps changing, the actors in the system always remain the same. Conservativeness postulates a process of expectation formation which favors balance of power in the weak sense of viability. To some extent, balance of power needs the expectation of balance of power. This can be shown by a simple example which arises in the analysis of our game.

Consider the province distribution (3,3,0,0,0,0). If one of the active players attacks the other, each of them has half a chance to conquer the world and to obtain a payoff of 6 and half a chance to be eliminated and to receive $-F$. The attacker has a local payoff of $3-F/2$ for the decision to attack. Obviously, it is better for both active players not to attack each other. Nevertheless, the local equilibrium conditions applied to decision situations with province distributions with two active players do not exclude a global strategy where both of them attack the other one whenever there is an opportunity. If the other one is going to attack anyhow, one can just as well attack oneself.

As we shall see, equilibrium strategies exist, with respect to which (3,3,0,0,0,0) is stable. Conservative equilibrium strategies must have this property. Conservativeness excludes the possibility of unreasonable distrust in the balance of power between the two active players in (3,3,0,0,0,0).

5. The Solution of the Theoretical Model

5.1 Character of the Solution

In view of the symmetry properties of global strategies, the equation $H_i(m,s) = H_j(m,s)$ holds for every province distribution m with $m_i = m_j$. The payoff $H_i(m,s)$ of player i does not depend explicitly on i, but only on m_i, the number of provinces held by i. It will be useful to look at $H_i(m,s)$ as a function of m_i. Therefore, the following notation is introduced:

$$V_k(m,s) = H_i(m,s) \quad \text{with} \quad m_i = k > 0. \tag{14}$$

Of course, $V_k(m,s)$ remains undefined for $k = 0$ and in the case that $m_i \neq k$ holds for $i = 1,\ldots,6$. Let S^* be the set of all conservative equilibrium strategies. We write $V_k(m)$ instead of $V_k(m,s)$ with $s \in S^*$, if $V_k(m,s)$ does not depend on s within S^*. In this case, we call $V_k(m)$ the *value for* k *provinces in* m. We say that values for conservative equilibrium strategies are *uniquely determined*, if for every $m \in M$ and every k, for which $V_k(m,s)$ is defined, $V_k(m,s)$ does not depend on s within S^*. Similarly, we say that conservative equilibrium strategies *uniquely determine* the stability status, if every $s \in S^*$ assigns the same stability status to each $m \in M$.

We say that the game is *solvable* for a given existence value F if for this F conservative equilibrium strategies exist and uniquely determine values and stability status. If this is the case, the values $V_k(m)$ and the classification of province distributions according to their stability status form *the solution* of the game. In order to avoid unnecessary notation, a more formal definition of the solution is not given here.

5.2 The Range of Existence Values

In this paper, attention will be concentrated to a plausible middle range of existence values:

$$1/3 < F < 3. \tag{15}$$

An extension to existence values outside this interval is possible but will not be attempted here, since additional lengthy arguments would become necessary.

It is our intention to show that the game is solvable for every F in the interval (15). Within this range of F, the stability status does not depend on F and the values are linear functions of F. We shall refer to the interval (15) as the *middle range*.

5.3 The Solution in the Middle Range

Table 1 gives an overview over the solution in the middle range. Only basic province distributions are listed in Table 1, since values and stability status depend only on the type of the province distribution.

In view of the symmetry of global strategies with respect to players, it is convenient to introduce a manner of speaking which permits us to focus on numbers of provinces held instead of identification numbers. Therefore, the number of provinces held will be called the *size* of a player and a player of size k will be referred to as a *size-k player*. The *size compositions* of alignments are described by the juxtaposition of the sizes of their numbers; thus, 221 stands for an alignment of two players who own 2 provinces and one player who owns 1 province.

We have to show that conservative equilibrium strategies exist and that values and stability status are uniquely determined. Existence is proved by the construction of a particular conservative equilibrium strategy. This strategy is described by Tables 2, 3, and 4. We shall look at province distributions essentially in the order of increasing numbers of active players. At each step, we shall show for the province distributions under consideration that the decisions listed in Tables 2, 3, and 4 satisfy conditions required by the definition of a conservative equilibrium strategy and that values and stability status are uniquely determined.

The decisions specified by the particular conservative equilibrium strategy of Tables 2, 3, and 4 do not depend on N and W. Therefore, in these tables, board positions are described by the size compositions of both alignments. It does not matter which alignment is A or C. Therefore, only one of both possibilities is listed. There are 11 types of province distributions (see Table 1) and altogether 56 essentially different decision situations in the game (Tables 2, 3, and 4).

The particular conservative equilibrium strategy described by Tables 2, 3, and 4 is not the only one. Many other conservative equilibrium strategies exist, which deviate from this one with respect to unimportant details. However, in Section 6, we shall argue that under the definition of payoffs as discounted sums of temporary payoffs, r remains as equilibrium strategy if the discount rate is sufficiently small. In this respect, some of these details are not unimportant.

It will be shown that the game is solvable in the sense of 5.1 for $1/3 < F < 3$ and that the strategy described by Tables 2, 3, and 4 is a conservative equilibrium strategy, not only for the game with long run average payoffs, but also for the discounted payoff game if the discount rate is sufficiently small. The result for the game with long run average payoffs will be proved step by step in a sequence of 12 assertions. It will always be assumed that F is in the middle range $1/3 < F < 3$. This will not be explicitly mentioned in Assertions 1 to 12.

5.4 Province Distributions with a Player Owning at least 4 Provinces

If one player owns at least 4 provinces, he can conquer the world and he must do this eventually if he plays an equilibrium strategy. It is clear as far as these province distributions are concerned that Table 2 describes a conservative equilibrium strategy and that values and stability status are uniquely determined as shown in Table 1.

PROVINCE DISTRIBUTION						VALUES				STABILITY STATUS
PLAYERS						4, 5 or 6 provinces	3 provinces	2 provinces	1 province	
1	2	3	4	5	6					
6	-	-	-	-	-	6	-	-	-	stable
5	1	-	-	-	-	6	-	-	-F	transitory
4	2	-	-	-	-	6	-	-F	-	transitory
4	1	1	-	-	-	6	-	-	-F	transitory
3	3	-	-	-	-	-	3	-	-	stable
3	2	1	-	-	-	-	$3\frac{7}{8} - \frac{3F}{16}$	$1\frac{5}{8} - \frac{9F}{16}$	$\frac{1}{2} - \frac{3F}{4}$	transitory
2	2	2	-	-	-	-	-	2	-	stable
3	1	1	1	-	-	-	4	-	$\frac{2}{3} - \frac{F}{2}$	transitory
2	2	1	1	-	-	-	-	2	1	stable
2	1	1	1	1	-	-	-	$1\frac{1}{5}$	$1\frac{1}{5}$	fluid
1	1	1	1	1	1	-	-	-	1	stable

Table 1: The solution for $1/3 < F < 3$.

PROVINCE DISTRIBUTION TYPE	BOARD POSITION according to number of provinces held			DECISION
	A	C	decision maker	
(5,1,0,0,0,0)	-	-	5	attack other player
	-	-	1	non-aligned
(4,2,0,0,0,0)	-	-	4	attack other player
	-	-	2	non-aligned
(4,1,1,0,0,0)	-	-	4	attack one of both other players *)
	-	-	1	non-aligned
	4	1	1	join A
	1	1	4	join A or C *)
(3,3,0,0,0,0)	-	-	3	non-aligned
(3,2,1,0,0,0)	-	-	3	attack size 2 player
	-	-	2	non-aligned
	-	-	1	non-aligned
	3	2	1	join C
	3	1	2	join C
	2	1	3	join C
(2,2,2,0,0,0)	-	-	2	non-aligned
	2	2	2	join A or C *)

Table 2: Equilibrium decisions for province distributions with up to 3 active players and for $1/3 < F < 3$. The local strategies do not depend on N and W.

*) With equal probabilities

PROVINCE DISTRIBUTION TYPE	BOARD POSITION according to number of provinces held A	C	decision maker	DECISION
(3,1,1,1,0,0)	-	-	3	attack one of the other players *)
	-	-	1	non-aligned
	3	1	1	join C
	1	1	3	join A or C *)
	1	1	1	non-aligned
	31	1	1	join A
	11	1	3	join C
	11	3	1	join A
(2,2,1,1,0,0)	-	-	2	non-aligned
	-	-	1	non-aligned
	2	2	1	join A or C *)
	2	1	2	join C
	2	1	1	join C
	1	1	2	join A or C *)
	22	1	1	join A
	21	2	1	join A
	21	1	2	join A
	11	2	2	join A

Table 3: Equilibrium decisions for province distributions with 4 active players and for $1/3 < F < 3$.

*) With equal probabilities

PROVINCE DISTRIBUTION TYPE	BOARD POSITION according to number of provinces held			DECISION
	A	C	decision maker	
(2,1,1,1,1,0)	-	-	2	non-aligned
	-	-	1	attack size-2 player
	2	1	1	join C
	1	1	2	join A or C *)
	1	1	1	non-aligned
	21	1	1	join A
	11	2	1	join A
	11	1	2	join C
	11	1	1	non-aligned
	21	11	1	join A
	11	11	2	join A or C *)
	211	1	1	join A
	111	2	1	join A
	111	1	2	join A
(1,1,1,1,1,1)	-	-	1	non-aligned
	1	1	1	join A or C *)
	11	1	1	join A
	11	11	1	join A or C *)
	111	1	1	join A
	111	11	1	join A
	1111	1	1	join A

Table 4: Equilibrium decisions for (2,1,1,1,1,0) and (1,1,1,1,1,1) and for 1/3 < F < 3.

*) With equal probabilities.

5.5 The Province Distribution (3,3,0,0,0,0)

Values for all other province distributions with at most two active players have been determined already. Conservativeness requires stability of (3,3,0,0,0,0) if this is possible. The choice of "non-aligned" in all pre-attack decision situations connected to (3,3,0,0,0,0) is in equilibrium. The requirements imposed by the definition of a conservative equilibrium strategy are satisfied: As has been pointed out in 4.6, the equilibrium conditions for province distributions with two active players do not exclude the possibility that (3,3,0,0,0,0) is unstable, since each of both active players attacks in the expectation that the other one wants to attack anyhow.

5.6 Province Distributions With Three Active Players

There are two types of province distributions with 3 active players, namely (3,2,1,0,0,0) and (2,2,2,0,0,0). Both of them have to be considered simultaneously, since one can be reached from the other. We first determine values and stability status on the basis of the assumption that the game is solvable. Then, we shall see that the decisions specified by Table 2 for these province distributions meet the requirements imposed by the definition of a conservative equilibrium strategy. In order to shorten formulas, we introduce the following notation:

$$p = V_2((2,2,2,0,0,0),s) \tag{16}$$

$$q_k = V_k((3,2,1,0,0,0),s) \quad \text{for } k = 1,2,3 \tag{17}$$

where s is an equilibrium strategy. Since at most two active players can be eliminated after (2,2,2,0,0,0) has been reached, the sum $3p$ of the values of the three active players in (2,2,2,0,0,0) is at least $6-2F$. In view of $F < 3$, this yields:

$$p > 0. \tag{18}$$

We now prove the following assertion:

Assertion 1: Let s be an equilibrium strategy. Then, (3,2,1,0,0,0) is not viable with respect to s.

Proof: Assume that (3,2,1,0,0,0) is stable. If the size-3 player behaves as specified by Table 2, he can conquer the world with a probability of at least 1/2. In the remaining cases, he always receives something positive, namely 2 if he becomes a size-2 player in a different province distribution of the same type or p if he becomes a size-2 player in (2,2,2,0,0,0). This shows that the size-3 player in (3,2,1,0,0,0) can enforce an expected payoff greater than 3. It follows that $m = (3,2,1,0,0,0)$ is not stable and that $V_3(m,s)$ is greater than 3 for this m. This has the consequence that with positive probability the size-3 player obtains first 4 and eventually 6 provinces. Consequently, (3,2,1,0,0,0) is not viable with respect to s.

Assertion 2: Let s be a conservative equilibrium strategy. If (2,2,2,0,0,0) is stable with respect to s, then for k = 1,2,3 and m = (3,2,1,0,0,0) the values $V_k(m,s)$ agree with those listed in Table 1.

Alignment graph: The proof of Assertion 2 will make use of an "alignment graph" for m = (3,2,1,0,0,0). This graph is shown in Figure 5. The nodes of the alignment graph stand for incompletely described board positions. Only the size compositions of both alignments are indicated, separated by a dash. The lines represent possible changes in the further development of the round. Higher nodes represent earlier situations. The nodes on the bottom describe terminal board positions, without any tokens on "wait for decision". Thus, 3-2 above means, that the size-1 player still can choose to stay non-aligned or to join the size-3 player or the size-2 player. 3-2 below means that the size-1 player has already chosen to stay non-aligned. Below the bottom nodes, the corresponding payoffs are indicated for each number of provinces held in the current round. These payoffs are the expectations of the values obtained for the province distribution of the next round, based on the assumptions of Assertion 2 that a conservative equilibrium strategy is played and that (2,2,2,0,0,0) is stable. This implies p = 2. The assumption that s is a conservative equilibrium strategy permits us to make use of Table 2 as far as province distributions with less than 3 active players or with a size-4 player are concerned.

As an example for the payoff computation, we look at the payoff of the size-3 player for 31-2. With probability 3/4 he obtains 4 provinces worth 6 and with probability 1/4 he becomes the size-3 player in a different province distribution of the type (3,2,1,0,0,0). This yields an expected payoff of $(18+q_3)/4$. Payoffs depend on the unknown variables q_1, q_2, and q_3 which in turn depend on the strategy s being played.

The three arrows next to three lines of the alignment graph do not really belong to the graph, but indicate results of its analysis. As we shall see, changes of the alignments must occur in these directions if an equilibrium strategy is played.

Proof of Assertion 2: If the size-3 player in (3,2,1,0,0,0) behaves as specified by Table 2, he conquers the world with a probability of at least 1/2. The worst which can happen to him in the remaining cases is loss of existence connected to a payoff of -F. In view of F < 3, this yields:

$$q_3 \geq 3 - \frac{F}{2} > \frac{3}{2} \quad . \tag{19}$$

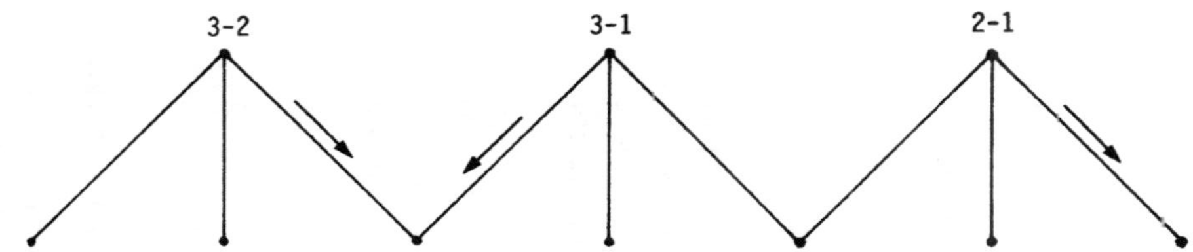

Number of provinces owned	31–2	3–2	3–12	3–1	32–1	2–1	2–31
3	$4\frac{1}{2} + \frac{q_3}{4}$	6	$3\frac{1}{3} + \frac{q_2}{3}$	6	$4\frac{4}{5}$	3	$4\frac{1}{2} + \frac{q_3}{4}$
2	$\frac{q_1}{4} - \frac{3F}{4}$	$-F$	$\frac{1}{3} + \frac{q_3}{3} - \frac{F}{2}$	$-F$	$1\frac{1}{5} - \frac{3F}{5}$	3	$\frac{q_1}{4} - \frac{3F}{4}$
1	$\frac{q_2}{4} - \frac{3F}{4}$	$-F$	$\frac{1}{3} + \frac{q_1}{3} - \frac{F}{2}$	$-F$	$-F$	$-F$	$\frac{q_2}{4} - \frac{3F}{4}$

Figure 5: The alignment graph for (3,2,1,0,0,0).

Consequently, we have

$$4\tfrac{1}{2} + \frac{q_3}{4} > 4\tfrac{7}{8} \ . \qquad (20)$$

It follows by this inequality that in a situation described by the upper node 2-1 of the alignment graph of Figure 5, the size-3 player must join the alignment of the size-1 player. The arrow from 2-1 to 2-31 indicates a necessary change of the alignment situation after a board position corresponding to the upper node 2-1.

Since loss of existence is the worst which can happen, we have:

$$q_2 \geq -F \qquad (21)$$

$$q_1 \geq -F. \qquad (22)$$

In view of $F < 3$, this yields

$$3\tfrac{1}{3} + \frac{q_2}{3} \geq 3\tfrac{1}{3} - \frac{F}{3} > 2\tfrac{1}{3} \ . \qquad (23)$$

Since (3,2,1,0,0,0) is unstable with respect to s, eventually a terminal board position corresponding to one of the bottom nodes of the alignment graph must be reached. We already have shown that the bottom node 2-1 cannot be reached. It follows by (20) and (23) that the payoff of the size-3 player at all the other bottom nodes is greater than 7/3. Consequently, we have:

$$q_3 \geq 3\tfrac{1}{3} - \frac{F}{3} > 2\tfrac{1}{3} \ . \qquad (24)$$

We now show that after a board position corresponding to the upper node 3-1 the alignment situation must move to 3-12 as indicated by an arrow in Figure 5. For this purpose, it is sufficient to show that we have:

$$\frac{1}{3} + \frac{q_3}{3} - \frac{F}{2} > 1\tfrac{1}{5} - \frac{3F}{5} \ . \qquad (25)$$

Obviously, the right-hand side of this inequality is greater than $-F$. In a situation corresponding to the upper node 3-1, it does not pay for the size-2 player to stay non-aligned. With the help of (24) it can be seen that (25) holds for $F > 8/9$. Assume $1/3 < F \leq 8/9$. Then we have:

$$q_3 \geq 3\tfrac{1}{3} + \frac{q_2}{3} \geq 3\tfrac{1}{27}. \qquad (26)$$

This has the consequence that (25) holds for $1/3 < F \leq 8/9$, too. The arrow from 3-1 to 3-12 indicates that the size-2 player must join the alignment of the size-1 player after hostilities between the size-3 player and the size-1 player have been opened.

The next step in the proof of Assertion 2 justifies the third arrow in Figure 5. In order to show that the size-1 player must join the alignment of the size-1 player, after hostilities between both other players have been opened, it is sufficient to establish the validity of the following inequality:

$$\frac{1}{3} + \frac{q_1}{3} - \frac{F}{2} > \frac{q_2}{4} - \frac{3F}{4} . \tag{27}$$

Assume that (27) does not hold. Then, it follows by (22) that the right-hand side of (27) is greater than $-F$. From what we already know, it follows that only the terminal nodes 31-2, 3-12, and 2-31 can be reached in the alignment graph. the left-hand side of (27) is the minimum of the payoffs the size-1 player receives at these nodes. Therefore, we must have:

$$q_1 \geq \frac{1}{3} + \frac{q_1}{3} - \frac{F}{2} . \tag{28}$$

This yields:

$$q_1 \geq \frac{1}{2} - \frac{3F}{4} \tag{29}$$

and consequently:

$$\frac{1}{3} + \frac{q_1}{3} - \frac{F}{2} \geq \frac{1}{2} - \frac{3F}{4} . \tag{30}$$

Since it is assumed that (27) does not hold, this yields:

$$\frac{q_2}{4} - \frac{3F}{4} \geq \frac{1}{2} - \frac{3F}{4} \tag{31}$$

or, equivalently:

$$q_2 \geq 2 . \tag{32}$$

With the help of (32) we obtain:

$$3\frac{1}{3} + \frac{q_2}{3} \geq 4 .$$

This shows that at the terminal nodes not yet excluded the size-3 player receives at least 4. Therefore, we have

$$q_3 \geq 4 \ . \tag{33}$$

All terminal nodes not yet excluded involve a probability of at least 1/2 for the elimination of the size-2 player. In view of (33), his expected number of provinces can be at most 2. This yields:

$$q_2 \leq 2 - \frac{F}{2} \ . \tag{34}$$

Inequality (34) contradicts (32). We can conclude that (27) holds and that the arrow from 3-2 to 3-12 in Figure 5 is justified.

Inequality (27) together with the conclusions expressed by the three arrows in Figure 5 has the consequence that in a pre-attack decision situation it is better for the size-1 player to attack the size-3 player than to attack the size-2 player. As we shall see, it is also better for the size-2 player to attack the size-3 player rather than the size-1 player. Figure 5 shows that this is the case if we have:

$$\frac{1}{3} + \frac{q_3}{3} - \frac{F}{2} > \frac{q_1}{4} - \frac{3F}{4} \ . \tag{35}$$

In view of (24), we must have:

$$\frac{q_1}{4} - \frac{3F}{4} \leq \frac{6-2\frac{1}{3}}{4} - \frac{3F}{4} = \frac{11}{12} - \frac{3F}{4} \ . \tag{36}$$

On the other hand, (24) shows that the following is true

$$\frac{1}{3} + \frac{q_3}{3} - \frac{F}{2} > \frac{10}{9} - \frac{F}{2} \ . \tag{37}$$

Inequalities (36) and (37) yield (35). We can conclude that both the size-1 player and the size-2 player do not attack each other in pre-attack decision situations. This has the consequence that only the terminal node 3-12 can be reached. It follows that q_1, q_2 and q_3 must be the payoffs at this node:

$$q_3 = 3\frac{1}{3} + \frac{q_2}{3} \tag{38}$$

$$q_2 = \frac{1}{3} + \frac{q_3}{3} - \frac{F}{2} \tag{39}$$

$$q_1 = \frac{1}{3} + \frac{q_1}{3} - \frac{F}{2} \,. \tag{40}$$

The equation system (38), (39), and (40) yields the values for $m = (3,2,1,,0,0,0)$ listed in Table 1:

$$q_3 = 3\frac{7}{8} - \frac{3F}{16} \tag{41}$$

$$q_2 = 1\frac{5}{8} - \frac{9F}{16} \qquad ((0U(42$$

$$q_1 = \frac{1}{2} - \frac{3F}{4} \,. \tag{43}$$

We have proved Assertion 2.

<u>Assertion 3</u>: Let s be a conservative equilibrium strategy. Then, for all province distributions with at most three active players values and stability status are as described by Table 1. Moreover, a global strategy r which assigns the local strategies specified by Table 2 to decision situation $D = (m,B,j)$ with $m \in M_{<4}$ satisfies the requirements imposed by the definition of a conservative equilibrium strategy as far as these decision situations are concerned.

<u>Proof</u>: Let r be a global strategy as described above. Consider a round with the province distribution $m = (2,2,2,0,0,0)$ and assume that one of the three active player attacks another one in a pre-attack decision situation and that the other players use r. Then, the local payoff of the attacker is as follows:

$$\frac{1}{2} q_1 + \frac{1}{4} q_2 + \frac{1}{4} q_3 = 1\frac{5}{8} - \frac{9F}{16} \,. \tag{44}$$

Since the right-hand side of (44) is smaller than 2, it is better tho choose "non-aligned". After an attack has happened, the third active player receives the local payoff q_2 for choosing "non-aligned" and the local payoff $(q_2+q_3)/2$ for joining one of both alignments. For him, too, the choice specified in Table 2 is the only local best reply to r.

With the help of Figure 5, it can be seen without difficulty that the local strategies specified by r for decision situations $D = (m,B,j)$ with $m = (3,2,1,0,0,0)$ are local best replies to r. The province distribution (2,2,2,0,0,0) is stable with respect to r. In view of Assertion 1, the province distribution (3,2,1,0,0,0) cannot be viable for any equilibrium strategy. Therefore, the set $P_3(r)$ is maximal in M_3 as required by the definition of conservativeness. This shows that the second part of Assertion 3 holds.

Consider a conservative equilibrium strategy s. If (2,2,2,0,0,0) is not stable with respect to s, then a province distribution of the type (3,2,1,0,0,0) must be reached eventually starting from (2,2,2,0,0,0) if s is played. This has the consequence that the set of active players is reduced with positive probability, which means that (2,2,2,0,0,0) is not viable. If both (2,2,2,0,0,0) and (3,2,1,0,0,0) are not viable, then $P_3(s)$ is a proper subset of $P_3(r)$ and s fails to be conservative. Therefore, (2,2,2,0,0,0) must be stable with respect to s. In view of Assertion 2 together with our earlier results on values and stability status for province distributions with up to 3 active players, it is now clear that Assertion 3 holds.

<u>Comment</u>: In a round with $m = (3,2,1,0,0,0)$, the size-3 player takes the risk to lose one province for an even chance to conquer the world. The two other players form an alignment against him. Other types of military confrontations cannot happen on the equilibrium path, if (3,3,0,0,0,0) and (2,2,2,0,0,0) are stable with respect to the equilibrium strategy played. This result is hardly surprising, but in spite of its plausibility lengthy and tedious arguments were necessary in order to reach the conclusion. This may be due to our failure to find a short proof, but it is also possible that the seemingly obvious is not obvious at all. Difficulties arise from the fact that the values q_3, q_2, and q_1 appear in the payoffs of the alignment graph.

5.7 <u>Province Distributions with Four Active Players</u>

There are two types of province distributions with four active players, namely (3,1,1,1,0,0) and (2,2,1,1,0,0). As we shall see, (3,1,1,1,0,0) cannot be viable and (2,2,1,1,0,0) must be stable with respect to a conservative equilibrium strategy. In this respect, the case of four active players is similar to the case of three active players.

In order to shorten our formulas, we introduce the following notation:

$$u_k = V_k((3,1,1,1,0,0),s) \quad \text{for} \quad k = 1 \quad \text{and} \quad k = 3 \tag{45}$$

$$v_k = V_k((2,2,1,1,0,0),s) \quad \text{for} \quad k = 1,2 \tag{46}$$

where s is an equilibrium strategy.

<u>Transition graph</u>: It will be useful to support our reasoning by a ***transition graph*** which shows the possible transitions from province distributions with four active players to other province distributions with three or four active players. This graph is shown by Figure 6. A node corresponds to the province distribution indicated next to it and a connecting line represents a possible transition in the direction showing the arrow at one of its ends. If an equilibrium strategy s is played, these

transitions occur with the probabilities x, y, z, μ, and η. Near to each connecting line the graph shows the symbol for the corresponding transition probability. (2,2,1,1,0,0) can be reached from (2,2,1,1,0,0) by a round which ends in peace (probability μ) or by a round where a size-1 player wins a province lost by a size-2 player (probability η). The transition graph of Figure 6 is based on the assumption that (3,1,1,1,0,0) is stable. Therefore, no transitions from (3,1,1,1,0,0) to other province distributions are shown.

Assertion 4: Let s be a conservative equilibrium strategy. If (3,1,1,1,0,0) is stable with respect to s, then we have

$$v_2 > 0. \tag{47}$$

Proof: Assume (3,1,1,1,0,0) is stable with respect to s. We first show:

$$v_2 = 2x + \frac{q_2+q_3}{2} y + 2z + \frac{v_1+v_2}{2} \eta + v_2\mu \tag{48}$$

$$v_1 = x + \frac{q_1-F}{2} y + \frac{2-F}{2} z + \frac{v_1+v_2}{2} \eta + v_1\mu . \tag{49}$$

In a transition from (2,2,1,1,0,0) to (3,1,1,1,0,0), one size-2 player wins a province lost by the other size-2 player. Since (3,1,1,1,0,0) is stable and s is symmetric with respect to players, the payoff expectation connected to this transition for a size-2 player is 2 (see Figure 6). This yields the first term $2x$ in (48). Similarly, the second term in (48) is due to the fact that in the transition from (2,2,1,1,0,0) to (3,2,1,0,0,0) a size-2 player becomes a size-3 player or a size-2 player in (3,2,1,0,0,0) with equal probabilities. In this way, all terms in (48) and (49) can be derived easily by looking at the possibilities for the corresponding transitions in Figure 6.

With the help of (48) and (49) we shall show that v_2 must be positive. Consider first the case that we have:

$$x = y = z = 0. \tag{50}$$

If in addition to this $\eta = 0$ holds, (2,2,1,1,0,0) is stable and we have $v_2 = 2$. If $\eta > 0$ holds, then (2,2,1,1,0,0) is fluid and the sum of the values of the active players is 6. This, together with (48) and (49), yields $v_2 = v_1 = 3/2$. Consequently, v_2 is positive in the case (50). From now on, assume that (50) does not hold. Obviously, we must have:

$$x + y + z + \eta + \mu = 1. \tag{51}$$

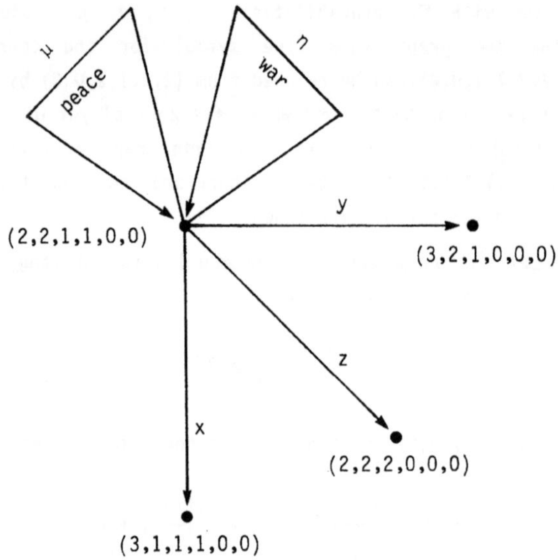

Figure 6: Transition graph for province distributions with four active players if (3,1,1,1,0,0) is stable

Moreover, since (50) does not hold, the following is true:

$$x + y + z > 0. \tag{52}$$

With the help of (41), (42), (43), and (51), addition of (48) and (49) and division by x+y+z yields:

$$v_2 + v_1 = 3 - \frac{5/4y + 1/2z}{x+y+z} F. \tag{53}$$

Similarly, subtraction of (49) from (48) leads to the following conclusion:

$$v_2 - v_1 = \frac{x + \frac{5}{2}y + z + (y+z)\frac{F}{2}}{x + y + z + \eta} . \tag{54}$$

Equation (54) shows that we have $v_2 > v_1$. Therefore $v_2 > 0$ holds for $v_1 + v_2 \geq 0$. In the following, we shall assume $v_1 + v_2 \leq 0$. In view of $F < 3$ and (53) the inequality $v_1 + v_2 \leq 0$ implies

$$\frac{5}{4}y + \frac{z}{2} \geq x+y+z. \tag{55}$$

A transition from (2,2,1,1,0,0) to a different province distribution of the same type cannot happen unless one of the following military confrontations takes place: 2-11 or 21-2 or 21-21 or 211-2. In all these cases the probability of reaching a province distribution of the type (3,1,1,1,0) is at least as great as the probability of reaching a province distribution of the type (2,2,1,1,0,0). Therefore, we have $x \geq \eta$. This together with (54) yields:

$$v_2 - v_1 \geq 1 + \frac{x + y + z + (y+z)\frac{F}{2}}{2x + y + z} . \tag{56}$$

Equation (53) together with $F < 3$ implies $v_1 + v_2 \geq -3/4$. This inequality added to (56) yields the conclusion that v_2 is at least 1/8. Therefore the assertion holds.

Assertion 5: Let s be a conservative equilibrium strategy. Then, (3,1,1,1,0,0) is not viable with respect to s.

Proof: Assume that in a round with the province distribution (3,1,1,1,0,0) the size-3 player behaves as specified by Table 3. Then, he obtains 6 with a probability of at least 1/2. In the remaining cases, he becomes a size-2 player in (2,2,1,1,0,0) or (3,2,1,0,0,0); this means that in the remaining cases he obtains either 2 or v_2. His expected payoff is greater than 3 if v_2 is positive. Therefore, (3,1,1,1,0,0) cannot be stable for $v_2 > 0$. It follows by Assertion 4 that (3,1,1,1,0,0) must be unstable with respect to s.

The instability of (3,1,1,1,0,0) has the consequence that in a round with this province distribution an attack happens with a positive probability if s is played. After an attack, the size-3 player cannot have a higher probability than 1/2 of being at the losing side. Therefore, after an attack one of the size-1 players is eliminated with a probability of at least 1/2. This shows that (3,1,1,1,0,0) cannot be viable with respect to s.

Alignment graphs for (3,1,1,1,0,0): Assume that s is a conservative equilibrium strategy and that (2,2,1,1,0,0) is stable with respect to s. Later, it will be shown that (2,2,1,1,0,0) is actually stable in our solution of the game. It is convenient to split the alignment graph for (3,1,1,1,0,0) into two parts shown in Figures 7 and 8. The graphical conventions are essentially the same as in Figure 5. However, in Figure 8 the situation 1-1 at the start may be changed by a decision of the size-3 player to join one of both alliances or by a decision of the remaining size-1 player to join one of both alliances. These two players may also stay non-aligned. Accordingly, five edges connect 1-1 to lower nodes. In the computation of payoffs, it is now necessary to distinguish between size-1 players in the alignment on the left of the dash (1 left), in the alignment on the right of the dash (1 right) and those which are non-aligned at the end of the round (1 in N).

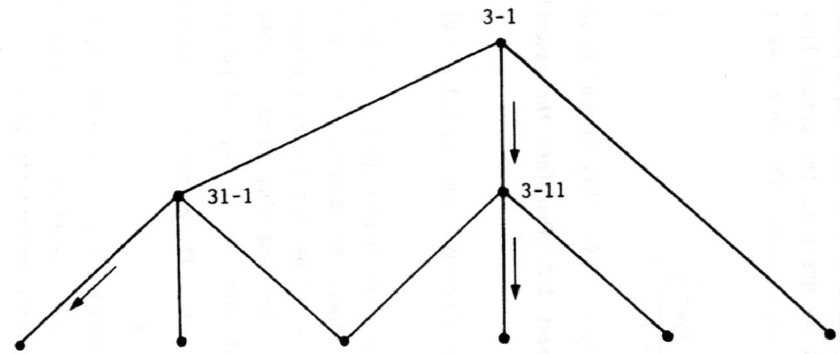

Player type	311–1	31–1	31–11	3–111	3–11	3–1
3	$5\frac{3}{20} - \frac{3F}{40}$	$5\frac{15}{32} - \frac{3F}{64}$	$5\frac{15}{32} - \frac{3F}{64}$	4	6	6
1 left	$\frac{17}{40} - \frac{69F}{80}$	$\frac{13}{32} - \frac{57F}{64}$	$\frac{13}{32} - \frac{57F}{64}$	—	–F	–F
1 right	–F	–F	$\frac{1}{16} - \frac{31F}{32}$	$\frac{2}{3} - \frac{F}{2}$	–F	–F
1 in N	—	$\frac{1}{8} - \frac{15F}{16}$	—	—	–F	–F

Figure 7: Alignment graph for (3,1,1,1,0,0) after 3-1.
It is assumed that (2,2,1,1,0,0) is stable.

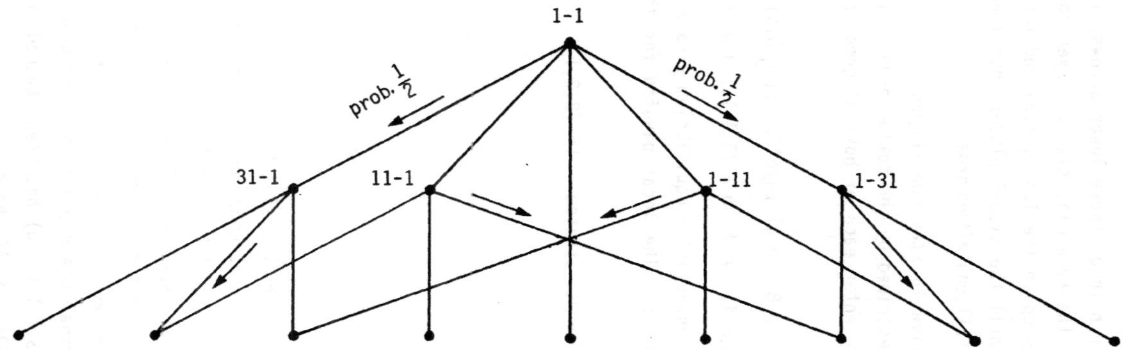

player type	31–1	311–1	31–11	11–1	1–1	1–11	11–31	1–311	1–31
3	$5\frac{15}{32} - \frac{3F}{64}$	$5\frac{3}{20} - \frac{3F}{40}$	$5\frac{15}{32} - \frac{3F}{64}$	$3\frac{7}{8} - \frac{3F}{16}$	$3\frac{7}{8} - \frac{3F}{16}$	$3\frac{7}{8} - \frac{3F}{16}$	$5\frac{15}{32} - \frac{3F}{64}$	$5\frac{3}{20} - \frac{3F}{40}$	$5\frac{15}{32} - \frac{3F}{64}$
1 left	$\frac{13}{32} - \frac{57F}{64}$	$\frac{17}{40} - \frac{69F}{80}$	$\frac{13}{32} - \frac{57F}{64}$	$\frac{17}{16} - \frac{21F}{32}$	$\frac{13}{16} - \frac{25F}{32}$	$-F$	$\frac{1}{16} - \frac{31F}{32}$	$-F$	$-F$
1 right	$-F$	$-F$	$\frac{1}{16} - \frac{31F}{32}$	$-F$	$\frac{13}{16} - \frac{25F}{32}$	$\frac{17}{16} - \frac{21F}{32}$	$\frac{13}{32} - \frac{57F}{64}$	$\frac{17}{40} - \frac{69F}{80}$	$\frac{13}{32} - \frac{57F}{64}$
1 in N	$\frac{1}{8} - \frac{15F}{16}$	—	—	—	$\frac{1}{2} - \frac{3F}{4}$	—	—	—	$\frac{1}{8} - \frac{15F}{16}$

Figure 8: Alignment graph for (3,1,1,1,0,0) after 1-1.
It is assumed that (2,2,1,1,0,0) is stable.

The computation of payoffs is straightforward. It is easy to see that the arrows in Figure 7 represent the only possible equilibrium changes after 3-1. The same can be said on the lower arrows in Figure 8. With the help of these lower arrows, it can be seen immediately that it would be unwise for the remaining size-1 player to join one of both alignments; if he does this, he ends up on the losing side and receives (2-31F)/32. It is better to stay non-aligned until the size-3 player enters one of both alignments in order to join the stronger batallions afterwards.

Obviously, the size-3 player has no incentive to stay non-aligned after 1-1. This would result in a terminal board position described by the bottom node 1-1. In view of the symmetry properties of s, he must enter each of both alignments with probability 1/2 as indicated in Figure 8.

<u>Attacker's payoff</u>: With the help of Figures 7 and 8, local payoffs $L(r_D,D,s)$ for an attack r_D in a pre-attack situation $D = (m,B,j)$ with $m = (3,1,1,1,0,0)$ can be determined. We continue to assume that s is a conservative equilibrium strategy and that (2,2,1,1,0,0) is stable with respect to s. The local payoffs for the 3 possible types of attacks are shown by Table 5.

<u>Assertion 6</u>: Let s be a conservative equilibrium strategy. If (2,2,1,1,0,0) is stable with respect to s, then we have:

$$u_3 = 4 \tag{57}$$

$$u_1 = \frac{2}{3} - \frac{F}{2} . \tag{58}$$

(For the definition of u_1 and u_3 see (45).)

<u>Proof</u>: Table 5 shows that a size-1 player prefers the decision to attack the size-3 player to the decision to attack another size-1 player. Consequently, the situation 1-1 at the start of Figure 8 cannot occur on the equilibrium path. In view of this fact, it can be seen that a confrontation of the type 3-1 must be reached with positive probability in a round with the province distribution (3,1,1,1,0,0), since otherwise (3,1,1,1,0,0) would be stable with respect to s. This is excluded by Assertion 5. It follows that eventually a terminal board position corresponding to the bottom node 3-111 will be reached after (3,1,1,1,0,0) has been reached for the first time. This leads to the conclusion that (58) and (59) hold.

<u>Assertion 7</u>: Let s be a conservative equilibrium strategy. If (2,2,1,1,0,0) is viable with respect to s, then (2,2,1,1,0,0) is stable with respect to s.

<u>Proof</u>: The transition graph of Figure 6 shows which province distributions can be reached from (2,2,1,1,0,0). In the case of a transition to (3,2,1,0,0,0) or (2,2,2,0,0,0), one active player is eliminated. Assertion 5 shows that (3,1,1,1,0,0) is not viable. Therefore, active players are eliminated with positive probability

Number of Provinces		Attacker's local payoff for the attack
of attacker	of attacked player	
3	1	4
1	3	$\frac{2}{3} - \frac{1}{2}F$
1	1	$\frac{17}{64} - \frac{117}{128}F$

Table 5: Attacker's local payoff for an attack in a pre-attack decision situation D = (m,B,j) with m = (3,1,1,1,0,0). It is assumed that (2,2,1,1,0,0) is stable.

after a transition to (3,1,1,1,0,0,0). If (2,2,1,1,0,0) is viable with respect to s, then a province distribution of the same type must be reached with probability 1 at the end of a round with (2,2,1,1,0,0). Assume that (2,2,1,1,0,0) is fluid with respect to s. Then, in a round with (2,2,1,1,0,0) there must be a positive probability for a terminal board position with non-empty alignments A and C. It can be seen without difficulty that no such board position results in a new province distribution of the type (2,2,1,1,0,0) with probability 1. A different province distribution of this type can be reached only if a size-1 player wins a province lost by a size-2 player. In order to exclude other types to be reached, only size-1 players should be on the winning side and only size-2 players should be on the losing side; moreover, the alignment with the size-1 players should be stronger than the other one. This is impossible. Consequently, Assertion 7 holds.

Alignment graphs for (2,2,1,1,0,0): Figures 9, 10 and 11 show alignment graphs for (2,2,1,1,0,0) starting with the three types of possible attacks 2-1, 2-2, and 1-1. The payoffs are computed on the basis of the assumption that (2,2,1,1,0,0) is stable with respect to a conservative equilibrium strategy s. This implies that we assume:

$$v_2 = 2 \tag{59}$$

$$v_1 = 1. \tag{60}$$

Moreover, we can make use of the Eqs. (58) and (59) for u_3 and u_1 and, of course, of the Eqs. (41), (42), and (43) for q_3, q_2, and q_1. The computation of payoffs in Figs. 9, 10, and 11 is tedious, but straightforward. It can be seen without difficulty that for $1/3 < F < 3$ the arrows indicate the possible transitions on the equilibrium path.

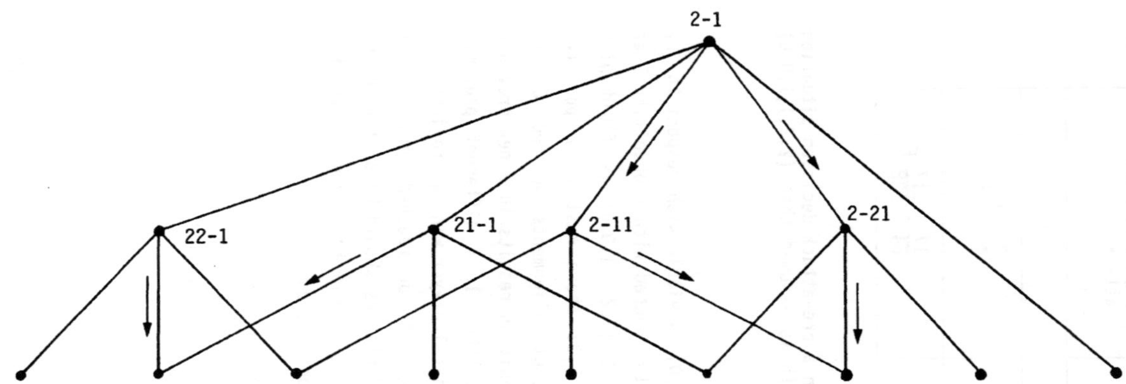

player type	22–1	221–1	22–11	21–1	2–11	21–21	2–211	2–21	2–1
2 left	$2\frac{3}{4} - \frac{3F}{8}$	$2\frac{3}{5} - \frac{3F}{10}$	$2\frac{3}{4} - \frac{3F}{8}$	$3\frac{1}{12} - \frac{F}{8}$	$2\frac{7}{16} - \frac{3F}{32}$	$2\frac{11}{54} - \frac{7F}{36}$	$\frac{5}{6} - \frac{F}{4}$	$\frac{7}{9} - \frac{F}{3}$	$3\frac{7}{8} - \frac{3F}{16}$
2 right	—	—	—	—	—	$2\frac{11}{54} - \frac{7F}{36}$	3	$3\frac{1}{3}$	—
2 in N	—	—	—	$1\frac{3}{4} - \frac{3F}{8}$	$1\frac{13}{16} - \frac{9F}{32}$	—	—	—	$1\frac{5}{8} - \frac{9F}{16}$
1 left	—	$\frac{4}{5} - \frac{3F}{5}$	—	$1 - \frac{F}{2}$	—	$\frac{43}{54} - \frac{17F}{36}$	—	—	$-F$
1 right	$-F$	$-F$	$\frac{1}{4} - \frac{7F}{8}$	$-F$	$\frac{7}{8} - \frac{7F}{16}$	$\frac{43}{54} - \frac{17F}{36}$	$1\frac{1}{12} - \frac{F}{4}$	$1\frac{1}{9} - \frac{F}{6}$	—
1 in N	$\frac{1}{2} - \frac{3F}{4}$	—	—	—	—	—	—	$\frac{7}{9} - \frac{F}{6}$	$\frac{1}{2} - \frac{3F}{4}$

Figure 9: Alignment graph for (2,2,1,1,0,0) after 2-1.

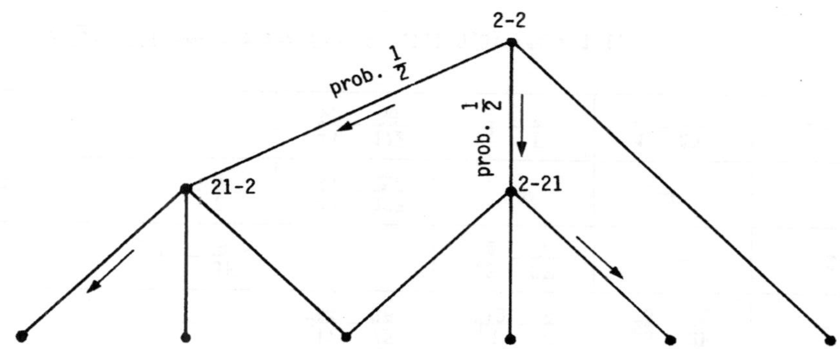

Player type	211–2	21–2	21–21	2–21	2–211	2–2
2 left	3	$3\frac{1}{3}$	$2\frac{11}{54} - \frac{7F}{36}$	$\frac{7}{9} - \frac{F}{3}$	$\frac{5}{6} - \frac{F}{4}$	$2\frac{1}{3} - \frac{F}{4}$
2 right	$\frac{5}{6} - \frac{F}{4}$	$\frac{7}{9} - \frac{F}{3}$	$2\frac{11}{54} - \frac{7F}{36}$	$3\frac{1}{3}$	3	$2\frac{1}{3} - \frac{F}{4}$
1 left	$1\frac{1}{12} - \frac{F}{4}$	$1\frac{1}{9} - \frac{F}{3}$	$\frac{43}{54} - \frac{17F}{36}$	—	—	—
1 right	—	—	$\frac{43}{54} - \frac{17F}{36}$	$1\frac{1}{9} - \frac{F}{3}$	$1\frac{1}{12} - \frac{F}{4}$	—
1 in N	—	$\frac{7}{9} - \frac{F}{3}$	—	$\frac{7}{9} - \frac{F}{3}$	—	$\frac{2}{3} - \frac{F}{2}$

Figure 10: Alignment graph for (2,2,1,1,0,0) after 2-2.

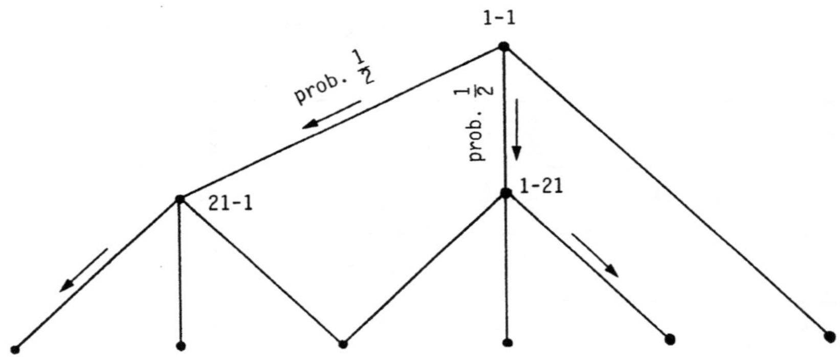

Player type	221–1	21–1	21–21	1–21	1–221	1–1
2 left	$2\frac{3}{5} - \frac{3F}{10}$	$3\frac{1}{12} - \frac{F}{8}$	$2\frac{11}{54} - \frac{7F}{36}$	—	—	—
2 right	—	—	$2\frac{11}{54} - \frac{7F}{36}$	$3\frac{1}{12} - \frac{F}{8}$	$2\frac{3}{5} - \frac{3F}{10}$	—
2 in N	—	$1\frac{3}{4} - \frac{3F}{8}$	—	$1\frac{3}{4} - \frac{3F}{8}$	—	2
1 left	$\frac{4}{5} - \frac{3F}{5}$	$1 - \frac{F}{2}$	$\frac{43}{54} - \frac{17F}{36}$	$-F$	$-F$	$1 - \frac{F}{2}$
1 right	$-F$	$-F$	$\frac{43}{54} - \frac{17F}{36}$	$1 - \frac{F}{2}$	$\frac{4}{5} - \frac{3F}{5}$	$1 - \frac{F}{2}$

Figure 11: Alignment graph for (2,2,1,1,0,0) after 1-1.

Number of Provinces		Attacker's local payoff for the attack
of attacker	of attacked player	
2	2	$1\frac{11}{12} - \frac{F}{8}$
2	1	$\frac{5}{6} - \frac{F}{4}$
1	2	$1\frac{1}{12} - \frac{F}{4}$
1	1	$\frac{2}{5} - \frac{4F}{5}$

<u>Table 6</u>: Attacker's local payoff in a pre-attack decision situation $D = (m,B,j)$ with $m = (2,2,1,1,0,0)$.

<u>Attacker's payoff for (2,2,1,1,0,0)</u>: With the help of Figs. 9, 10, and 11, the entries in Table 6 can be determined easily. It can be seen immediately that for $F > 1/3$ an attack does not pay if nobody else attacks in a pre-attack situation. For $F < 1/3$, a size-1 player would have an incentive to attack a size-2 player, since then the expected payoff for this attack would be greater than 1. For $F < 1/3$, the province distribution (2,2,1,1,0,0) could not be stable with respect to a conservative equilibrium strategy s.

<u>Assertion 8</u>: Let s be a conservative equilibrium strategy. Then, for all province distributions with at most four active players values, and stability status are as described by Table 1. Moreover, a global strategy r which assigns the local strategies specified by Tables 2 and 3 to decision situations $D = (m,B,j)$ with $m \in M_{<5}$ satisfies the requirements imposed by the definition of a conservative equilibrium strategy, as far as these decision situations are concerned.

<u>Proof</u>: Let r be a global strategy as described above. With the help of Figs. 7 to 11, it can be seen that r assigns local best replies to r to all decision situations $D = (m,B,j)$ with $m = (3,1,1,1,0,0)$ or $m = (2,2,1,1,0,0)$. Moreover, (2,2,1,1,0,0) is stable with respect to r. Since (3,1,1,1,0,0) cannot be viable in view of Assertion 5, every conservative equilibrium strategy s must have the property that the set $P_4(s)$ of viable province distributions with respect to s agrees with the set $P_4(r)$ of viable province distributions with respect to r. It follows by Assertion 7 that (2,2,1,1,0,0) is stable with respect to every conservative equilibrium strategy s. This together with Assertions 5 and 6 shows

that as far as (3,1,1,1,0,0) and (2,2,1,1,0,0) are concerned values and stability status are uniquely determined and are correctly presented by Table 1.

5.8 Province Distributions With Five Active Players

There is only one type of province distribution with five active players, namely (2,1,1,1,1,0). As we shall see, these province distributions are fluid (viable but not stable) with respect to conservative equilibrium strategies. We shall use the following notation:

$$w_k = V_k((2,1,1,1,1,0),s) \quad \text{for } k = 1,2. \tag{61}$$

Assertion 9: Let s be a conservative equilibrium strategy. Then, (2,1,1,1,1,0) is not stable with respect to s.

Proof: Assume that s is a conservative equilibrium strategy and that (2,1,1,1,1,0) is stable with respect to s. Then, we have $w_2 = 2$ and $w_1 = 1$. The payoffs in the alignment graphs of Figure 12 are computed on the basis of this assumption. The arrows in Figure 12 show that a size-1 player has an incentive to attack the size-2 player. He can count on all other size-1 players to join him and his local payoff for the attack is 5/4. Consequently, (2,1,1,1,1,0) cannot be stable.

Assertion 10: Let s be a conservative equilibrium strategy. If (2,1,1,1,1,0) is viable with respect to s, then we have:

$$w_2 = w_1 = 1\tfrac{1}{5} \,. \tag{62}$$

Proof: The transition graph of Figure 13 shows which province distributions can be reached from (2,1,1,1,1,0). The graphical conventions are the same as in Figure 6. Assume that s is a conservative equilibrium strategy and that (2,1,1,1,1,0) is viable with respect to s. Let x, y, α, and β be the transition probabilities generated by s belonging to the transitions indicated by the connecting lines next to which the x, y, α, and β resp. are shown. We must have

$$x = y = 0 \tag{63}$$

since otherwise one of the active players would be eliminated with positive probability after (2,1,1,1,1,0) has been reached. In view of Assertion 9, this province distribution is not stable with respect to s. Therefore, α must be positive. Whenever non-empty alignments are formed in a round with (2,1,1,1,1,0), one of the size-1 players wins a province lost by the size-2 player.

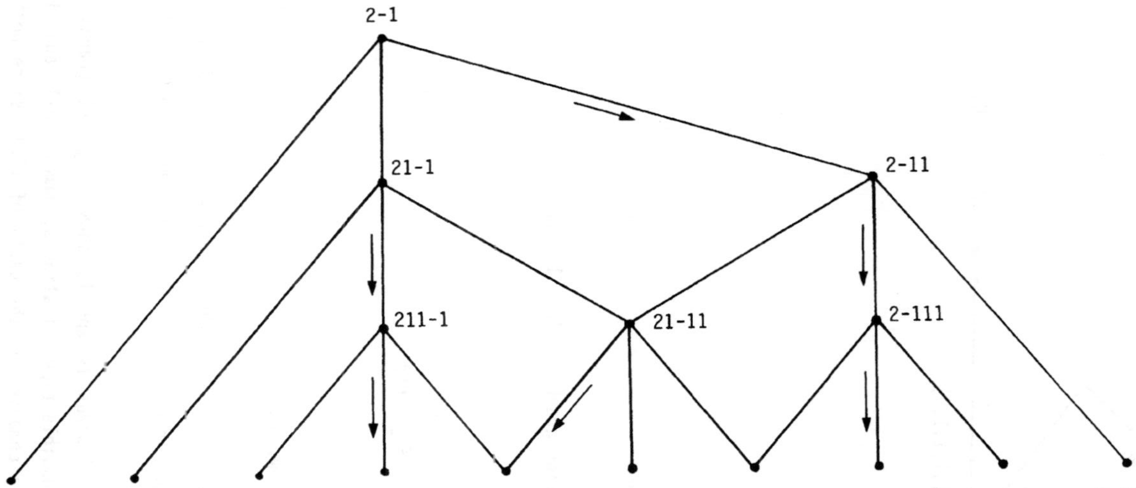

player type	2–1	21–1	211–1	2111–1	211–11	21–11	21–111	2–1111	2–111	2–11
2	4	$3\frac{1}{3}$	3	$2\frac{4}{5}$	3	$3\frac{1}{3}$	$2\frac{1}{3}$	1	1	$2\frac{1}{2}$
1 left	—	$1\frac{1}{9} - \frac{F}{3}$	$1\frac{1}{12} - \frac{F}{4}$	$1\frac{1}{15} - \frac{F}{5}$	$1\frac{1}{12} - \frac{F}{4}$	$1\frac{1}{9} - \frac{F}{3}$	$\frac{8}{9} - \frac{F}{3}$	—	—	$\frac{11}{12} - \frac{3F}{8}$
1 right	$-F$	$-F$	$-F$	$-F$	$\frac{5}{12} - \frac{5F}{8}$	$\frac{7}{18} - \frac{2F}{3}$	$\frac{25}{27} - \frac{5F}{18}$	$1\frac{1}{4}$	$1\frac{1}{3}$	—
1 in N	$\frac{2}{3} - \frac{F}{2}$	$\frac{7}{9} - \frac{F}{3}$	$\frac{5}{6} - \frac{F}{4}$	—	—	$\frac{7}{9} - \frac{F}{3}$	—	—	1	$\frac{5}{6} - \frac{F}{4}$

Figure 12: Alignment graph for (2,1,1,1,1,0) after 2-1 if $w_2 = 2$ and $w_1 = 1$ holds.

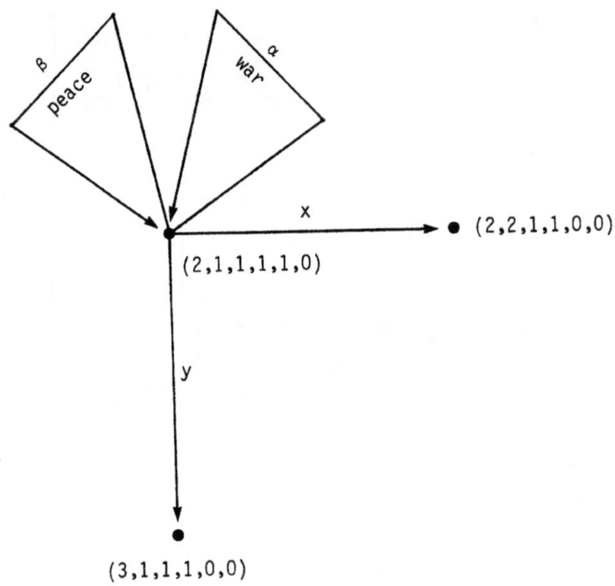

Figure 13: Transition graph for (2,1,1,1,1,0)

Consequently, we have:

$$w_2 = \alpha w_1 + \beta w_2 \tag{64}$$

In view of $\alpha + \beta = 1$, this yields

$$w_1 = w_2. \tag{65}$$

Since (2,1,1,1,1,0) is viable, the sum w_2+4w_1 of the values of all five players is equal to 6, the total number of provinces. This, together with $\alpha > 0$ and Eq. (65), yields (62).

Alignment graphs for (2,1,1,1,1,0): Figures 14 and 15 show alignment graphs for (2,1,1,1,1,0) starting with the two possible types of attacks, namely 2-1 and 1-1. Payoffs in these alignment graphs are computed on the basis of (62), since nothing else is possible if (2,1,1,1,1,0) is viable with respect to a conservative equilibrium strategy s. Of course, the values u_3 and u_1 of (3,1,1,1,0,0) and v_2 and v_1 of (2,2,1,1,0,0) which have been determined earlier are used, too. In order to simplify Figure 15, possible changes also shown by Figure 14 are omitted if in Figure 14 these changes are excluded by the arrows shown there.

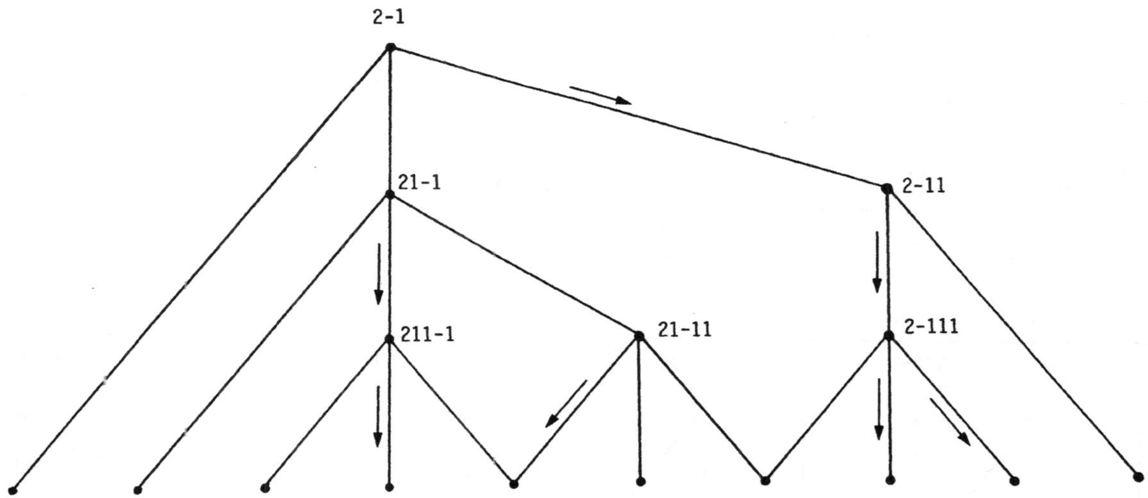

player type	2–1	21–1	211–1	2111–1	211–11	21–11	21–111	2–1111	2–111	2–11
2	4	$3\frac{1}{3}$	3	$2\frac{4}{5}$	3	$3\frac{1}{3}$	$2\frac{2}{5}$	$1\frac{1}{5}$	$1\frac{1}{5}$	$2\frac{3}{5}$
1 left	—	$1\frac{1}{9} - \frac{F}{3}$	$1\frac{1}{12} - \frac{F}{4}$	$1\frac{1}{15} - \frac{F}{5}$	$1\frac{1}{12} - \frac{F}{4}$	$1\frac{1}{9} - \frac{F}{3}$	$\frac{43}{45} - \frac{F}{3}$	—	—	—
1 right	–F	–F	–F	–F	$\frac{5}{12} - \frac{5F}{8}$	$\frac{7}{18} - \frac{2F}{3}$	$\frac{119}{135} - \frac{5F}{18}$	$1\frac{1}{5}$	$1\frac{1}{5}$	$\frac{23}{30} - \frac{3F}{8}$
1 in N	$\frac{2}{3} - \frac{F}{2}$	$\frac{7}{9} - \frac{F}{3}$	$\frac{5}{6} - \frac{F}{4}$	—	—	$\frac{7}{9} - \frac{F}{3}$	—	—	$1\frac{1}{5}$	$\frac{14}{15} - \frac{F}{4}$

Figure 14: Alignment graph for (2,1,1,1,1,0) after 2-1 if $w_2 = w_1 = 1\frac{1}{5}$.

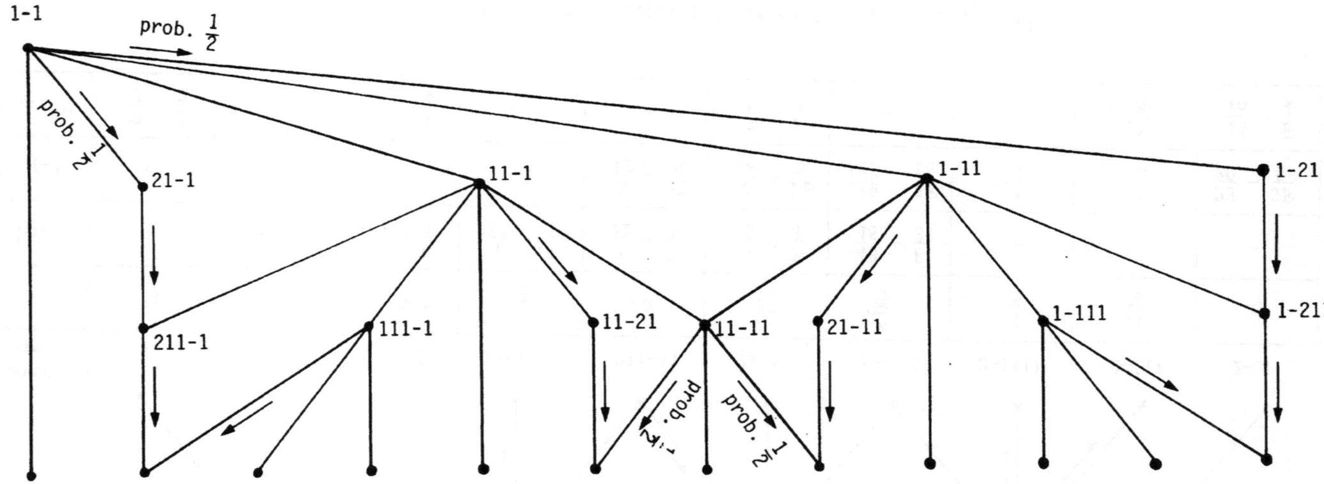

player type	1–1	2111–1	111–21	111–1	11–1	11–211	11–11	211–11	1–11	1–111	21–111	1–2111
2	2	$2\frac{4}{5}$	$2\frac{3}{5}$	2	2	3	2	3	2	2	$2\frac{3}{5}$	$2\frac{4}{5}$
1 left	$1-\frac{F}{2}$	$1\frac{1}{15}-\frac{F}{5}$	$\frac{119}{135}-\frac{5F}{18}$	$1\frac{1}{3}$	$1\frac{1}{2}$	$\frac{5}{12}-\frac{5F}{8}$	$1-\frac{F}{4}$	$1\frac{1}{12}-\frac{F}{4}$	$-F$	$-F$	$\frac{43}{45}-\frac{F}{3}$	$-F$
1 right	$1-\frac{F}{2}$	$-F$	$\frac{43}{45}-\frac{F}{3}$	$-F$	$-F$	$1\frac{1}{12}-\frac{F}{4}$	$1-\frac{F}{4}$	$\frac{5}{12}-\frac{5F}{8}$	$1\frac{1}{2}$	$1\frac{1}{3}$	$\frac{119}{135}-\frac{5F}{18}$	$1\frac{1}{15}-\frac{F}{5}$
1 in N	1	—	—	—	1	—	—	—	1	—	—	—

Figure 15: Alignment graph for (2,1,1,1,1,0) after 1-1 if $w_2 = w_1 = 1\frac{1}{5}$.
Changes which cannot happen according to Figure 14 are omitted.

In Figure 15, it is unwise for a size-1 player to join one of both alignments in the situation 1-1. If he does this, he ends up on the losing side. He should wait until the size-2 player has entered one of both alignments and then join the side of the size-2 player.

<u>Assertion 11</u>: Let s be a conservative equilibrium strategy. Then, for (2,1,1,1,1,0) values and stability status are as described by Table 1. Moreover, a global strategy which assigns the local strategies specified by Tables 2, 3, and 4 to decision situations D = (m,B,j) with $m \in M_{<6}$ satisfies the requirements imposed by the definition of a conservative equilibrium strategy, as far as these decision situations are concerned.

<u>Proof</u>: Let r be a global strategy as described above. With the help of Figures 14 and 15 and Table 7, it can be seen that r assigns local best replies to r to all decision situations D = (m,B,j) with m = (2,1,1,1,1,0). Moreover, (2,1,1,1,1,0) is viable with respect to r. Therefore, the definition of conservativeness requires that (2,1,1,1,1,0) is viable with respect to every conservative equilibrium strategy s. In view of Assertions 9 and 10, the province distributions of the type (2,1,1,1,1,0) are fluid with respect to every conservative equilibrium strategy s and the values for (2,1,1,1,1,0) are uniquely determined by (62).

<u>Remark</u>: Since the definition of an equilibrium strategy has been based on local conditions only, a conservative equilibrium strategy may involve a positive probability of a round with (2,1,1,1,1,0) ending in peace. If this is the case, a player can increase his payoff by a permanent change of his strategy: If in the role of a size-1 player he always attacks the size-2 player in a pre-attack decision situation D = (m,B,j) with m = (2,1,1,1,1,0), then his stationary probability of owning 2 provinces will be greater than 1/5. This results in an expected payoff of more than 6/5.

Number of Provinces		Attacker's local payoff for the attack
of attacker	of attacked player	
2	1	$1\frac{1}{5}$
1	2	$1\frac{1}{5}$
1	1	$\frac{8}{15} - \frac{3F}{5}$

<u>Table 7</u>: Attacker's local payoff in a pre-attack decision situation D = (m,B,j) with m = (2,1,1,1,1,0) for $w_2 = w_1 = 1\frac{1}{5}$.

5.9 <u>The Province Distribution (1,1,1,1,1,1)</u>

There is only one province distribution with 6 active players, namely (1,1,1,1,1,1). Fortunately, it is not necessary to draw an alignment graph in this case.

<u>Assertion 12</u>: Let s be a conservative equilibrium strategy. Then, (1,1,1,1,1,1) is stable with respect to s. Moreover, the strategy r described by Tables 2, 3, and 4 is a conservative equilibrium strategy.

<u>Proof</u>: If in a round with the province distribution (1,1,1,1,1,1) an attack takes place, the next province distribution is of the type (2,1,1,1,1,0). Therefore, all players who survive obtain the same payoff $w_2 = w_1 = 6/5$. This has the consequence that the local strategies specified for decisions situations $D = (m,B,j)$ with $m = (1,1,1,1,1,1)$ are local best replies to r. If follows by Assertion 11 that r is an equilibrium strategy. Obviously, (1,1,1,1,1,1) is stable with respect to r. It follows by the definition of conservativeness together with Assertion 11 that r is a conservative equilibrium strategy and that (1,1,1,1,1,1) is stable with respect to every other conservative equilibrium strategy.

5.10 <u>The Result of the Analysis</u>

The analysis of the game is now complete. We have obtained the following result:

<u>Result 1</u>: The game is solvable for every existence value F in the interval $1/3 < F < 3$. Values and stability status for each basic province distribution are as shown by Table 1. The strategy r described by Tables 2, 3, and 4 is a conservative equilibrium strategy.

<u>Remark</u>: For $F < 1/3$, the province distribution (2,2,1,1,0,0) cannot be stable with respect to a conservative equilibrium strategy. This can be seen with the help of Table 6. Therefore, our result cannot be extended to $F < 1/3$. The assumption $F < 3$ has been used in the proof of Assertion 4. With the help of Assertion 4, we have proved that (3,1,1,1,0,0) cannot be viable with respect to a conservative equilibrium strategy. The assumption $F < 3$ has also been used at other places in our proofs. However, it is possible that nevertheless our results can be extended to $F > 3$ by different proofs.

6. Discounted Payoffs

6.1 <u>The Problem</u>

The definition of an equilibrium strategy in Section 4 has been based on local optimality conditions only. Therefore, a conservative equilibrium strategy does not necessarily have the global equilibrium properties of a subgame perfect equilibrium point. This problem does not arise for a different payoff definition based on discounted sums of temporary payoffs. In this section, we shall show that the strategy described by Tables 2, 3, and 4 remains an equilibrium strategy under the changed payoff definition if the discount rate is sufficiently small. This shows that conservative equilibrium strategies exist which under the changed payoff definition do not only have local but also global equilibrium properties. As the discount rate approaches zero, the local payoffs approach those of the game with long-run average payoffs. The solution derived in Section 5 can be interpreted as a description of the limit of conservative equilibrium strategies under the changed payoff definition obtained for a discount rate approaching zero.

We shall not ask the question whether for sufficiently small discount rates under the changed payoff definition all conservative equilibrium strategies are approximately described by the solution derived in Section 5. It is the purpose of this section to show that this is the case for at least one conservative equilibrium strategy.

6.2 The Discounted Payoff Game

In the following, we shall change the theoretical model introduced in Section 4 by the introduction of a different payoff definition. Everything else will remain the same.

The payoff definition depends on a parameter ρ, called the *discount factor*. The discount factor satisfies the following inequality

$$0 < \rho < 1. \tag{66}$$

We are mainly interested in the case that ρ is very near to 1. The following notation will be used

$$\eta = 1-\rho. \tag{67}$$

We refer to η as the *discount rate*.

As in 4.3, assume that we have $m(t) = m$ and that in rounds $t, t+1, \ldots$ the same global strategy s is always played by all players. A player's expected payoff $H_{\rho i}(m,s)$ is now defined as follows

$$H_{\rho i}(m,s) = \eta\, E[\sum_{r=t}^{\infty} \rho^{r-t}\, h_i(m_i(r))] \tag{68}$$

where $m(t), m(t+1), \ldots$ is a realization of the stochastic process over M described in 4.3 and where E is the expectation operator; of course, $m_i(t)$ is player i's component in $m(t)$. The expected discounted payoff sum is multiplied by the discount rate η in order to make the payoff comparable to a one-round payoff.

For example, if $m = (m_1,\ldots,m_6)$ is stable with respect to s, we obtain $H_{\rho i}(m,s) = m_i$. The *payoff vector* $H_\rho(m,s)$ is defined analogously to (8) and *local payoffs* are defined as in (9) by

$$L_\rho(D,r_D,s) = E(H_{\rho j}(m(t+1),s)) \tag{69}$$

with $m(t) = m$ and $D = (m,B,j)$. We refer to the changed model as the *discounted payoff game* with the discount factor ρ. Local best replies and equilibrium strategies are defined as before with L_ρ instead of L. Conservativeness is also defined as before.

6.3 Local and Global Optimality

Without going into too much detail, we want to indicate why in the case of the discounted game local equilibrium conditions are sufficient for subgame perfect equilibrium in the usual sense. Consider an equilibrium strategy s for the discounted payoff game and assume that player i plays a behavior strategy b_i (not necessarily a stationary one) whereas all other players always play s. The definition of payoffs for this situation is straightforward and will not be given here. Assume that player i obtains a better payoff for b_i than for s if the other players use s.

For every $T > 1$, we construct a *truncated game* which ends after T rounds, but otherwise has the same structure as the discounted payoff game. The payoffs in this truncated game are those which would be obtained in the discounted payoff game, if in all rounds $T+1,T+2,\ldots$ the players always play s. It can be seen without difficulty that for sufficiently great T player i's payoff for using b_i in rounds $1,\ldots,T$ must be greater than player i's payoff for using s in rounds $1,\ldots,T$ if all other players play s in rounds $1,\ldots,T$. However, the truncated game is finite and the restriction of s to the truncated game satisfies the local equilibrium conditions for this finite game; this implies that the restriction of s to the truncated game being played by all players is a subgame perfect equilibrium point of the truncated game, contrary to our assumption. Therefore, in the discounted game local equilibrium conditions are sufficient for subgame perfect equilibrium in the usual sense.

6.4 The Strategy Described by Tables 2, 3, and 4

Let r be the strategy described by Tables 2, 3, and 4. In the following, we shall show that for sufficiently small discount rate η the strategy r is an equilibrium strategy of the discounted payoff game; in other words, a $\rho_0 < 1$ exists such that for $1/3 < F < 3$ and $\rho_0 < \rho < 1$ the strategy r is an equilibrium strategy for the discounted payoff game with the discount factor ρ.

Values: We shall use the notation

$$V_{\rho k}(m) = H_{\rho i}(m,r) \quad \text{with} \quad m_i = k > 0 \tag{70}$$

where m_i is player i's component of m. If m is stable with respect to r, then $V_{\rho k}(m)$ agrees with the corresponding value listed in Table 1. For those province distributions which are not stable with respect to r, it can be seen without difficulty that we have:

$$\lim_{\rho \to 1} V_{\rho k}(m) = V_k(m,r) \tag{71}$$

wherever values are defined.

Province distributions where one player has at least 4 provinces: Obviously, a player with at least 4 provinces optimizes his payoff in the discounted payoff game if he behaves according to r. The same is true for weaker players who face a player with at least 4 provinces in a pre-attack decision situation. If the player with at least 4 provinces is in N, "non-aligned" is the only local best reply; if he is in W, it does not hurt to choose "non-aligned". Only one post-attack decision situation arises for a weaker player faced by a player with 4 or more provinces: after an attack of the size-4 player in (4,1,1,0,0,0), the remaining size-1 player has to decide whether he wants to join A or C or whether he wants to stay non-aligned. The expected temporary payoffs for the decision maker are shown by Table 8. It is

DECISION	EXPECTED TEMPORARY PAYOFFS			
	ROUNDS			
	t	t+1	t+2	>t+2
join A	1	$1\frac{1}{5}$	$\frac{1-4F}{5}$	-F
join C	1	$\frac{1-F}{2}$	-F	-F
non-aligned	1	1	-F	-F

Table 8: Expected temporary payoffs for the remaining size-1 player after an attack of the size-4 player in a round t with the province distribution (4,1,1,0,0,0).

clear that for every discount factor ρ with $0 < \rho < 1$ the decision "join A" yields the highest expected payoffs.

<u>Local payoff comparisons</u>: In view of (71), local payoff comparisons in the long-run average payoff case based on strong inequalities remain valid in the discounted payoff case, if ρ is sufficiently near to 1. Similarly, equalities between local payoffs for different choices in the long-run average case are preserved in the discounted payoff case for ρ sufficiently near to 1, if they are due to symmetry or to the same outcome being reached at the end of the round.

Additional strong inequalities which are not present in the long-run average payoff case must be considered in the discounted payoff case. A player whose value for the current province distribution is lower than his number of provinces should stay non-aligned in a pre-attack decision situation where all other players are in N. A player whose value for the current province distribution is higher than his number of provinces should attack with probability 1 rather than with a smaller positive probability. These requirements of payoff maximization in the discounted payoff game are satisfied by the strategy r of Tables 2, 3, and 4.

<u>Province distributions with up to 4 active players and with at most 3 provinces for each player</u>: In view of what has been said on local payoff comparisons, it can be seen with the help of Figures 5, 7 to 11 and Tables 5 and 6 that for ρ sufficiently near to 1 the local strategies specified by r are local best replies to r in the discounted payoff game at decision situations with such province distributions.

<u>The province distribution (2,1,1,1,1,0)</u>: For $k = 1,2$, let w_k be the value of a size-k player for (2,1,1,1,1,0) and r in the discounted payoff game. Since the outcome of a round with (2,1,1,1,1,0) is 2-1111 if r is played (see Figure 12), the following equations hold:

$$w_2 = 2\eta + \rho w_1 \tag{72}$$

$$w_1 = \eta + \rho(\tfrac{3}{4}w_1 + \tfrac{1}{4}w_2). \tag{73}$$

This yields:

$$w_2 = \frac{6}{5} + \frac{4\eta}{5 + \frac{5}{4}\rho} \tag{74}$$

$$w_1 = \frac{6}{5} - \frac{\eta}{5 + \frac{5}{4}\rho} . \tag{75}$$

For every discount factor ρ the value w_2 is greater than w_1. The arrows in Figures 14 and 15 are based on strong inequalities between local payoffs. In view of what has been said on local payoff comparisons, it can be seen with the help of Figures 14 and 15 and Table 7 that the local strategies specified by r for decision situations D = (m,B,j) with m = (2,1,1,1,1,0) are local best replies to r in the discounted payoff game for every discount factor ρ with $0 < \rho < 1$.

The province distribution (1,1,1,1,1,1): In the long-run average payoff case, all survivors after an attack in a round with (1,1,1,1,1,1) receive the same expected payoff 6/5. In view of $v_2 > v_1$, the expected payoff of a survivor in the discounted payoff game depends on his probability of obtaining 2 provinces at the end of the round; the greater this probability is, the greater is his expected payoff. After an attack in a round with (1,1,1,1,1,1), the local strategies specified by r maximize the probability of obtaining 2 provinces at the end of the round. For every ρ with $0 < \rho < 1$, the local strategies specified by r for decision situations D = (m,B,j) with m = (1,1,1,1,1,1) are local best replies to r. We have obtained the following result:

Result 2: A discount factor ρ_0 with $0 < \rho_0 < 1$ exists such that the following is true for $\rho_0 < \rho < 1$ and $1/3 < F < 3$: The strategy described by Tables 2, 3, and 4 is an equilibrium strategy for the discounted payoff game with the discount factor ρ.

7. The Interpretation of the Solution

In Section 5, we have derived a unique solution for our theoretical model with 6 provinces and the assumption that the existence value is in the middle range $1/3 < F < 3$. In the following, we shall look at the reasons for stability and instability in this solution.

Figure 16 shows the transitions among types of province distributions permitted by the solution. Types of province distributions are represented by nodes, and transitions are represented by connecting lines. Arrowheads indicate the direction of the transition. The numbers written next to the connecting lines are the probabilities of the corresponding transitions if the equilibrium strategy described by Tables 2, 3, and 4 is played. Circles surrounding nodes indicate stable province distribution types and the fluid province distribution type is graphically distinguished by a square surrounding its node. Figure 16 provides an overview over important qualitative features of the solution.

Five of the eleven types of province distributions are stable. We shall now look at each of these types separately in order to discuss the reason for its stability.

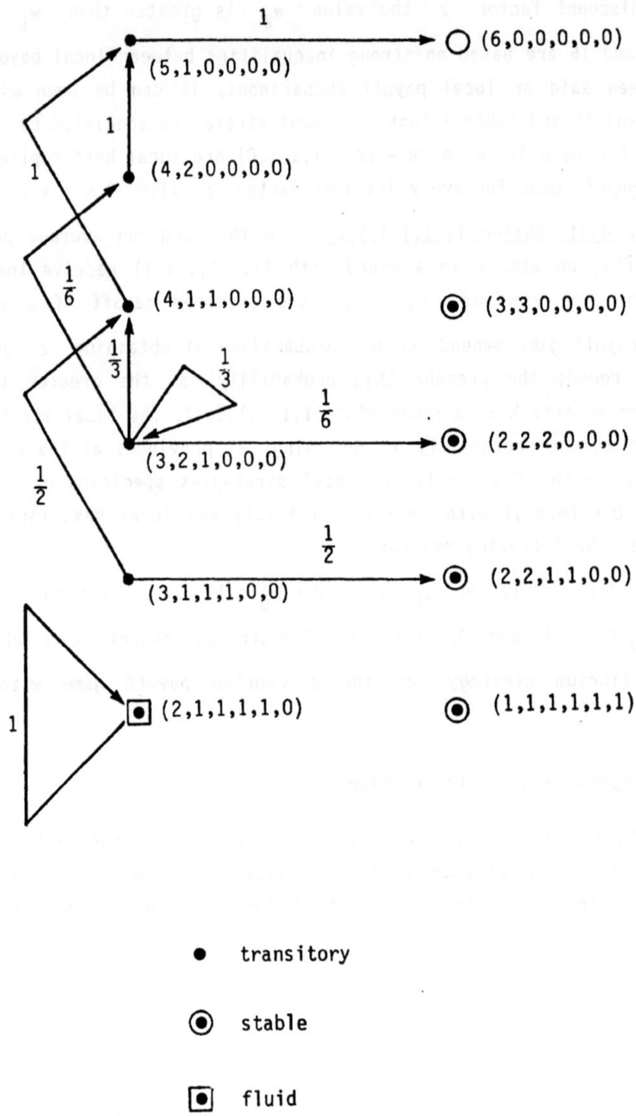

Figure 16: Transitions among types of province distributions for the equilibrium strategy described by Tables 2, 3, and 4.

The province distribution type (6,0,0,0,0,0): This type is stable for the trivial reason that wars cannot arise if there is only one power left.

The province distribution type (3,3,0,0,0,0): Here, a potential attacker is deterred by half a chance to lose his existence which in view of $F > 0$ has a greater weight than half a chance to conquer the world. As has been pointed out, the expectation of stability is necessary for the stabilization of (3,3,0,0,0,0) in the case of long-run average payoffs. This is different in the discounted payoff game. There, "non-aligned" is the only possible equilibrium choice of one of the action players if the other one has already chosen "non-aligned". This makes it optimal to choose "non-aligned" in all pre-attack decision situations in a round with (3,3,0,0,0,0).

The province distribution type (2,2,2,0,0,0): Here, all three active players hold the same number of provinces. This has the consequence that after an attack of one active player on another the remaining active player must join both sides with equal probability. Therefore, an attacker has half a chance to be on the losing side. If there were a riskless way to build up a coalition of two active players against the third one, such a coalition would be formed. However, the symmetry properties of global strategies exclude this possibility. After an attack, both sides must be treated symmetrically by the remaining player. One may say that symmetry is a source of stabilizing uncertainty on the formation of alignments.

The province distribution type (2,2,1,1,0,0): It is of crucial importance for the stability of this type that the existence value is assumed to be greater than 1/3. For $F < 1/3$, a size-1 player would have an incentive to attack a size-2 player (see Table 6). A size-1 player who attacks a size-2 player can be sure to be on the winning side (see Figure 9). However, with probability 1/2 he becomes a size-1 player in an unstable province distribution of the type (3,2,1,0,0,0). This creates a risk of being eliminated. We may say that the danger of instability in the next round is a stabilizing force.

The province distribution type (1,1,1,1,1,1): Here, an attacker has half a chance to be on the losing side, since after an attack both sides are treated equally by the next decision maker. In this case, too, symmetry is a source of stabilizing uncertainty on the formation of alignments.

Reasons for stability and instability: The discussion of the five types of stable province distributions has revealed several possible reasons for stability. In the world ownership case (6,0,0,0,0,0), the reason for stability is trivial. the *bipolar case* (3,3,0,0,0,0) is stabilized by mutual deterrence. the discussion of the three stable province distributions with at least three active players has revealed two stabilizing forces:

(a) Symmetry as a source of uncertainty on the formation of alignments.

(b) The danger of instability in the next round.

In the solution of our model, transitory instability is always connected to an attempt of a player with at least 3 provinces to conquer the world. However, other reasons of instability are possible. For an existence value smaller than 1/3, the stability of (2,2,1,1,0,0) breaks down, because a weak player, who can be sure to be on the winning side, has an incentive to attack one of the stronger players. The danger of instability in the next round and the associated risk of losing existence destroy this incentive if the existence value is greater than 1/3.

Coalitions against actors who threaten to become too strong: Informal balance of power theories convey the impression that the most important stabilizing force is a tendency to form coalitions against actors who threaten to become too strong. Our solution does not seem to support this view. Coalitions against actors who threaten to become too strong are formed in the cases of the two unstable types of province distributions (3,2,1,0,0,0) and (3,1,1,1,0,0). These coalitions are desperate attempts with at best half a chance of success to prevent the strong player from reaching world ownership and to establish a balance of power among the existing active players. It is true that the coalitions against the strong player in rounds with (3,2,1,0,0,0) and (3,1,1,1,0,0) decrease the incentive for the stronger players to attack in a round with (2,2,1,1,0,0), but they also increase the incentive for an attack by a weaker player, and as we have seen, an attack of a weaker player is the real danger to the stability of (2,2,1,1,0,0).

The fluid province distributions of the type (2,1,1,1,1,0): Here, we can speak of a balance of power even if these province distributions are not stable. Peace is never achieved, but none of the actors must fear to become eliminated. Whenever a war occurs, a coalition against the strongest actor forms, but not in order to prevent him to become too strong, but rather for the purpose of taking one province away from him. However, the impossibility of building up a successful alliance including the current owner of two provinces may be attributed to the danger that this player threatens to become too strong. This suggests that in some situations exploitative coalitions against the strongest player are facilitated by the fact that other coalitions are too risky for the weaker players who must fear that the strongest player becomes too strong.

Limitations of the analysis: It is not difficult to analyze analogous games with a smaller number of provinces. 6 is the minimum number which yields an interesting case. It must be expected that additional phenomena emerge for greater numbers of provinces.

Our analysis is based on assumptions of stationarity and symmetry. Experiences with the parlor game suggest that human players do not play stationary strategies. Informal coalitions are built up which may last for many rounds. Expectations on the behavior of other players may crucially depend on past history. Informal agreements can destroy the inherent symmetry of a situation like (1,1,1,1,1,1). This province distribution must be expected to be unstable in a game played by human players.

Our observations of plays of the parlor game are not yet sufficiently extensive to permit the construction of a descriptive theory. It is clear that a descriptive theory must take account of the limited rationality of human players. Unfortunately, a precise general theory of limited rationality did not yet emerge from empirical evidence. Therefore, a game theoretical approach based on strong rationality assumptions must serve as a substitute for a more realistic approach. Moreover, the contrast between game theoretical analysis and observed behavior provides insights into the problem of limited rationality.

References

Aumann, R.J., and M. Maschler (1964). The Bargaining Set for Cooperative Games. In: Dresher et al. (eds.), Advances in Game Theory. Princeton, N.J.: Princeton University Press.

Bernholz, P. (1985). The International Game of Power. Berlin: Mouton.

Clausewitz, C. von (1963). Vom Kriege. Leck (Schleswig): Clausen und Bosse.

Feller, W. (1968). An Introduction to Probability Theory and its Applications, Vol. I, 3rd ed. New York: Wiley.

Harsanyi, J.C., and R. Selten (1988). A General Theory of Equilibrium Selection in Games. Cambridge, Mass.: MIT Press.

Kaplan, M.A. (1957). System and Process in International Politics. New York: Wiley.

Kautilya, A.-S. (1957). Excerpts. In: S. Radakrishan and Ch.A. Moore: A Source Book of Indian Philosophy. Princeton: Princeton University Press.

Luce, D., and H. Raiffa (1957). Games and Decisions. New York: Wiley.

Niou, E.M. and P.C. Ordeshook (1986). A Theory of the Balance of Power in International Systems. J. Conflict Resol. **30(4)**: 685-715.

Niou, E.M., and P.C. Ordeshook (1987). Preventive War and the Balance of Power. J. Conflict Resol. **31(3)**.

Selten, R., and R. Stoecker (1986). End Behavior in Sequences of Finite Prisoner's Dilemma Supergames. J. Econ. Beh. Organ. **7**: 47-70.

Wagner, H. (1986). The Theory of Games and the Balance of Power. World Politics **37(4)**: 546-576.

Zinnes, D.A. (1967). The Analytical Study of the Balance of Power Theories. J. Peace Res. **4**: 270-328.

MODELLING ALLIANCE FORMATION: A NONCOOPERATIVE APPROACH

by

Roy Gardner and Werner Güth

Abstract: Instead of constructing a simple but complex alliance formation game, we analyse three different extensive game models illuminating special aspects of alliance formation. In the first model, various alliances can form from an initial state of nonalignment. It helps to answer questions like: will nonalignment continue, or are bipolarity or balance-of-power configurations more likely to develop? Next, we model arms races such that the effects on arms expenditures and alliance formation can be analyzed. The third model considers how an alliance containing one superpower and a potential periphery of smaller countries can develop if the superpower is only incompletely informed about its potential partners. Throughout the paper we try to relate our results to actual political observations.

I. Introduction

Ours is a time of great ferment in international relations, both theoretically and practically. Policitical scientists such as Snidal (1985) have called for the development of better game models of international relations, and considerable work has been done on cooperative game models of alliance formation (Ordeshook and Niou, 1986, 1987). On the practical side, new attitudes in Moscow and Washington have led to the greatest relaxation of tension along the Central European front since the formation of NATO and WTO over three decades ago. We feel that noncooperative game theory also has something to say about these developments, and in this spirit of exploration offer the following sequence of models of alliance formation. Indeed, if one has a reasonably good theory of alliance formation, then one can better predict when an alliance loses its raison d'être.

In the first model, we consider the formation of various alliances from an initial situation of nonalignment. This model allows one to predict in a broad way whether continued nonalignment, bipolarity, or balance-of-power configurations are the most likely to develop. The second model attempts to model arms races which are basically bipolar. These models allow one to predict arms expenditures, at least qualitatively, in bipolar situations such as that of present day Europe. The third model considers the problem of formation of an alliance containing a superpower and a potential periphery of other countries, all of whom face a fixed external threat. The theoretical complication in this case is that the superpower has only incomplete information about its potential alliance partners, so that the games studied here involve incomplete information and probabilistic alliance formation.

The common thread uniting this entire sequence of models is the approach of noncooperative game theory. It seems to us that a cooperative game theory, which presupposes the formation of coalitions, begs the question of which coalitions -- here alliances -- actually form, and that the only logically defensible approach starts with an initial position of noncooperation. In such an approach, cooperation must be sustained as an equilibrium of the noncooperative game. Since not all equilibria are equally reasonable, we adopt the position of Harsanyi and Selten (1988) that only perfect equilibria are worthy of our attention. We also make considerable use of the method of equilibrium scenarios (Güth, 1984). Using this method, we are able to show what sorts of outcomes are possible as perfect equilibria of alliance formation games on nondegenerate regions of the parameter space on which the game is defined.

The models we propose are far from the most general imaginable. Even in their present specificity, however, they shed light on questions such as ideological cohesion of alliances, and burden sharing within alliances, that is simply not available from either a non-game theoretic approach or a cooperative game theoretic approach.

II. Alliance Formation and Ideological Relatedness

Alliances often claim that they share a common worldview which distinguishes them from other alliances. In the present bipolar world, one alliance stands for democracy and markets, the other for socialism and planning. Another important linking factor besides political-economic ideology can be religion, as reflected by the Arab league. In this section, we do not formally distinguish between the varieties of possible worldview. What we assume is that each country can be located at a point on the unit interval, representing its ideal point. The closer two countries' ideal points are, the closer the two countries' ideological or other cultural climates are.

Let countries $1,\ldots,n$ $(n \geq 2)$ be indexed so that

$$0 \leq t_1 < t_2 \ \ldots \ < t_n \leq 1, \tag{1}$$

where t_i denotes the ideal point of country i. (1) reflects the intuition that no two countries are exactly alike. Denote the set of countries by N. An alliance M^j satisfies

$$M^j \subset N, \quad |M^j| \geq 1. \tag{2}$$

A country which belongs to no larger alliance, we term *nonaligned*. We define the tension of an alliance for member i, $t_i(M^j)$

$$t_i(M^j) = (|t_i - t_k|)_{\substack{k \neq i \\ k \in M^j}} \tag{3}$$

to be the vector of ideal point differences between i and its alliance partners. From (1) it follows that the tension vector is positive. We assume that within an alliance less tension is preferred to more.

Tension within an alliance can be compensated for by the security an alliance offers its members. To measure security, consider the set $\mathcal{M}$ of all possible alliance configurations with typical member M

$$M = (M^j) \tag{4}$$

where the collection (M^j) forms a partition of N.

The military strength of country i is represented by its weight $w_i > 0$, normalized so that

$$\sum_{i=1}^{n} w_i = 1. \tag{5}$$

We assume that military strength is additive, so that the military strength of alliance M^j is given by that of its members:

$$w(M^j) = \sum_{i \in M^j} w_i \, . \tag{6}$$

The interaction of military strength and ideological position determines the threat potential of an alliance, $t(M^j)$:

$$t(M^j) = \sum_{i \in M^j} w_i t_i \Big/ \sum_{i \in M^j} w_i \, , \; i \in M^j \, . \tag{7}$$

The idea here is that $t(M^j)$ measures what an alliance has both the will and the power to realize. If $t(M^j)$ differs greatly from $t(M^k)$, then an alliance M^k has reason to feel threatened. On the other hand, if $t(M^j)$ and $t(M^k)$ are not too different, then they might even be potential partners in a grand alliance between them.

For a given alliance configuration M, let $w(M)$ be the vector of military strength $w(M^j)$ and $t(M)$ be the vector of threat potentials $t(M^j)$. Country i in alliance M^j feels more secure as $w(M^j)$ increases and $|t(M^j) - t(M^k)|$ decreases. Indeed, a country would feel most secure in the grand alliance $M = \{N\}$. However, the tension of the grand alliance usually will prevent it from forming.

To see this, consider the following recursive procedure of alliance formation. Start with the initial alliance configuration M_1 consisting of all nonaligned countries. At time t, $t = 1,\ldots,T$, any member i of alliance M_t^j of the current alliance configuration M_t can propose that M_t^j form a larger alliance with another member of M_t. If member i wants to preserve the status quo, he makes no proposal. Two alliances M_t^j and M_t^k merge at time t if all members of both alliances propose this joint alliance. In this case, $M_t^j \cup M_t^k$ is a member of M_{t+1}. The alliance formation process terminates at time T if no alliance has merged, that is $M_{T-1} = M_T$. What we claim then is that only under very special conditions will $M_T = \{N\}$.

Let s_i denote a (pure) behavior strategy for country i, namely a unique proposal for country i at each possible stage of the alliance formation game. The vector of strategies $s = (s_1,\ldots,s_n)$ determines uniquely an ending alliance configuration $M(s)$. Here we assume that countries' preferences are determined by the final alliance configuration that results, and not by the path by which that configuration is reached. Thus, country i's payoff, hence preference between strategies s and s', depends only on the comparison between $M(s)$ and $M(s')$. We express this by the utility function $u_i(s)$ for country i

$$u_i(s) = u_i(t_i(M^j(s)), w(M(s)), t(M(s))) \tag{8}$$

where $t_i(M^j(s))$ is the tension i feels in the coalition $M^j(s)$ to which it belongs at the end of the alliance formation game, and $w(M(s)), t(M(s))$ are the military strength and threat vectors respectively of the final alliance configuration. Utility is decreasing in within alliance tension, outside alliance threat potential and outside alliance military strength, but increasing in within alliance military strength. A specific functional form embodying these assumption is the following:

$$u_i(s) = \alpha_i \prod_{\substack{k \neq j \\ M^k(s) \in M(s)}} \left[\frac{w(M^j(s))}{w(M^j(s)+w(M^k(s))}\right]^{|t(M^j(s)-t(M^k(s))|} - \sum_{\substack{h \in M^j(s) \\ h \neq 1}} |t_i - t_h|^{\beta_i}. \tag{9}$$

We shall sometimes use this specific functional form later on. Note that $\alpha_i > 0$ and $\beta_i > 1$ in (9).

The alliance formation game described above can be summarized by the vector P

$$P = (t,w,u) \tag{10}$$

consisting of individual country ideal points, military strengths, and utility functions. In the event of specification (9) on utility, then the vector of utility functions can be denoted by the preference parameter vectors (α,β).

When applying game theory to a game such as P, one usually tries to determine all strategy vectors s which satisfy certain rationality requirements, such as subgame perfect equilibrium (Selten, 1975). When this does not by itself lead to definite answers, one may apply even stronger rationality requirements, for example those embodied in equilibrium selection theory (Harsanyi and Selten, 1988). Unfortunately, doing so can be very difficult in games as complicated as the alliance formation games studied here. Often in applications, and particularly in questions of international relations, the basic question is whether a certain outcome is consistent with any rationality requirements. In Hegel's terms (1831), is what is actual really rational? The method of equilibrium scenarios is useful in answering such questions. Suppose the scenario is that of a bipolar world. Then one uses game theory to ask whether bipolarity is a subgame perfect equilibrium, say, of the alliance formation game. If not, then we can infer that bipolarity is in fact strategically unstable: if countries play rationally, the next time the question of alliance membership is opened, one would not expect such an outcome to recur. For further discussion and applications of the method of equilibrium scenarios see Güth (1984) and Avenhaus, Güth and Huber (1988, Part II).

The method of equilibrium scenarios can be employed in two ways. First, one can show that a certain outcome can never be an equilibrium, or can only be an equilibrium under very unlikely parameter constellations of P. Second, and more positively, one can determine non-degenerate regions of the parameter space where an outcome is an equilibrium. We now proceed to illustrate both these employments in the question of alliance formation.

Results of Some Equilibrium Scenarios

We begin with the most extreme outcome, namely when M(s) is the configuration of all nonaligned countries, and when M(s) is the grand alliance.

Proposition 1. There are non-degenerate regions of P such that the alliance formation game has subgame perfect equilibria with outcomes M(s) consisting of all nonaligned countries and the grand alliance, respectively.

Proof. To get M(s) to consist of all nonaligned countries, it suffices to take all α_i small and all β_i large in (9). Then no country i will gain by joining any alliance. Hence, the strategy vector s consisting of no proposals at any round is subgame perfect. For general P, it suffices to pick utility functions in the neighborhood of (9).

To get $M(s)$ to be the grand alliance, pick α_i very large and β_i close to 1 for all countries i. Then the strategy consisting of the proposals (s_i = the grand coalition at every stage) is a subgame perfect equilibrium. ■

Proposition 1 illustrates the fact that without special assumptions, our model of alliance formation does not exclude any alliance configurations a priori. Furthermore, it illustrates the potential coordination problem facing countries. For example, suppose that the grand alliance is the best possible equilibrium point. However, if countries do not coordinate their proposals, they may wind up at a strictly inferior equilibrium point, but still subgame perfect. Although one intuitively thinks of alliance formation as basically a bargaining problem, our model exposes a coordination aspect that the popular intuition does not usually capture. It is also important to realize that the process of alliance formation can be quite crucial to the final outcome. Indeed, there may be only one way to reach an outcome $M(s)$, as the following shows:

<u>Proposition 2</u>. There are non-degenerate regions of P such that for all corresponding games, if s and s' are subgame perfect equilibria with $M(s) = M(s')$, then $s = s'$.

<u>Proof</u>. Consider the specification (9). The result is trivial for $n = 2$. For $n = 3$, take t_1 close to 0, t_2 close to .5, and t_3 close to 1; also, w_1 and w_3 close to 0, w_2 close to 1. It is easy to find values of α_i, β_i such that: (1) country 1 does not want 3 to join $\{1,2\}$ once it forms because $|t_1-t_3|$ is large, (2) 2 and 3 want the grand alliance to form (take α_2 and α_3 large and β_2 and β_3 small), and (3) country 3 prefers joining the alliance with 2 to being left nonaligned, since 2 has such large military strength. Clearly, $M(s)$ = the grand alliance is an equilibrium outcome. However, the only way to arrive at this outcome by equilibrium play is first to form $\{2,3\}$ and then form the grand alliance. For $n > 3$, one constructs the desired example by embedding the above construction for $n = 3$. ■

The above example also shows that subgame perfection by itself is not enough to determine the outcome of the game uniquely.

<u>Proposition 3</u>. There are non-degenerate regions of P with multiple subgame perfect equilibria.

<u>Proof</u>. Consider the situation described in the previous proposition. $M(s) = \{\{1,2\},\{3\}\}$ is a subgame perfect equilibrium outcome. ■

This shows that the basic data of the game, P, may be consistent with many different equilibrium outcomes. This is a serious drawback in applying game theory immediately to international relations. On the other hand, it points out the importance of history in the formation or critique of naturally occurring alliance patterns. Ultimately, at some point in the game, only one equilibrium is still available, and to a large extent, it may be historical forces which determine which such point of no-return is in fact reached.

A very important bias that psychologists have uncovered is hindsight bias -- the notion that because something happened, it had to happen, was either inevitable or extremely likely to happen (Camerer, Loewenstein and Weber, 1989). The alignment of the great powers in 1913 might be considered as such a sure-thing outcome. But it seems to us that if one goes back to, say, 1871, the only really certain feature of this alignment is the placing of France and Germany in opposite alliances. Every other detail of 1913 -- especially the placing of Austria-Hungary with Germany, and England with France -- seems far from inevitable.

For a more contemporary illustration, consider the present alignment of European countries in either the North Atlantic Treaty Organization (NATO), the Warsaw Treaty Organization (WTO), or nonaligned powers (as, for instance, Switzerland, Austria, Sweden, Finland, Albania, Yugoslawia). Let the unit interval represent a political-economic-geographical index, with 0 = West and 1 = East. Then the following relationship of ideal points is indicated:

The current alignment then minimizes alliance tensions within the two large alliances, while it is not in the interest of any of the nonaligned countries to join either of the large alliances. (Indeed, Yugoslavia is an obvious defector from the potentially larger WTO.) What we would then have is a configuration which is separating, with all countries below $\underline{t}$ belonging to one alliance, all countries above $\bar{t}$ belonging to another alliance, and all countries in-between nonaligned. We now demonstrate that such a configuration could indeed be an equilibrium:

<u>Proposition 4</u>. There are non-degenerate regions of P with subgame perfect equilibrium alliance configurations which are separating with a buffer in between.

<u>Proof</u>. Let $t_1,\ldots,t_i$ all be close to 0 and t_j $(j>i),\ldots,t_n$ all be close to 1. Then it is not hard to see why $\{1,\ldots,i\}$ and $\{j,\ldots,n\}$ form, since their tension is minimal. Now, for k between i and j, take α_k and β_k very large. Then any such k has no incentive to join either bloc. The proposed alliance configuration is then an equilibrium. ∎

Proposition 4 shows the potential explanatory power of the noncooperative game approach to alliance formation. However, it also raises the question of whether there might also be nonseparating equilibria, such as the following:

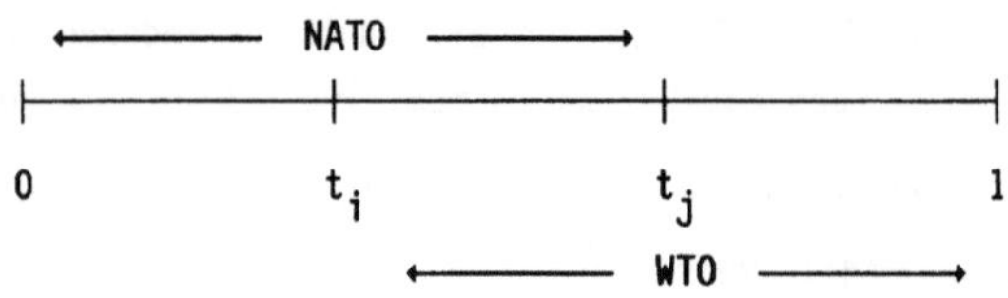

Here, country j belongs to NATO, country i belongs to WTO, and j is east of i. Countries in between can belong either to NATO or WTO or be nonaligned. Indeed, we cannot rule out such a case as pathological (the fear of the Soviet Union that Poland might defect from the WTO and realign itself with France and the UK as in 1939 is well known).

Proposition 5. There are non-degenerate regions of P where the alliance formation game has non-separating equilibria.

Proof. A simple possibility is when $n = 4$. Take t_1 close to 0, t_4 close to 1, t_2 and t_3 close to .5. Suppose further that w_1, w_2, w_3 and w_4 are nearly equal. Let $\{1,3\}$ form first. Then the threat potential of $\{1,3\}$ is sufficient for 2 to join 4 instead of them. The resulting alliance configuration is then an equilibrium outcome. ∎

In an instance such as that of Proposition 5, one might say that balance of power considerations outweigh those that are purely ideological. Indeed, as Bernholz (1985) shows, the current bipolar division of Europe along ideological lines is a very recent development. For centuries before ours, the historical norm was one of balance of power alliances not exclusively along political, economic, or religious lines. The great changes now taking place in the foreign policies of both Washington and Moscow may portend a return to the historical norm in Europe before the end of this century. Such a development, at least, cannot be precluded by game theoretic considerations.

III. Military Technology and Alliance Formation

Ever since World War II, the two antagonistic alliances, NATO and WTO, have tried to develop new weapons yielding each at least a short run advantage. This has led to an arms race with enormous expenditures on both sides. There are three reasons why an arms race is linked to alliance formation. First, an alliance may be more efficient, since it can exploit international division of labor, pool resources, and avoid duplication of effort. Second, a member of an alliance may be able to acquire a technology otherwise unavailable from one of its alliance partners. Third, decision making in an alliance may be better -- more cautious -- than that of a national government in a crisis situation or when contemplating exploiting a temporary advantage. Here we describe and analyze a game model incorporating some of these linkages between arms races on the one hand and alliance formation on the other.

As before, let N be a finite set of countries indexed by i, which are engaged in an arms race. Each country has a positive scalar r_i of resources which can be devoted to weapons research and development. For any alliance M^j the total resources are given by $r(M^j)$ where

$$r(M^j) = \sum_{i \in M^j} r_i . \qquad (11)$$

The total resources that an alliance M^j actually commits to the arms race, $y(M^j)$, is constrained:

$$0 \leq y(M^j) \leq r(M^j). \qquad (12)$$

As before, let $M = (M^j)$ be an alliance configuration, a partition of N, and let $y(M) = (y(M^j))$ be the corresponding vector of alliance expenditures. Given the overall pattern of expenditures, we assume that the probability that M^j is the first alliance to achieve technological breakthrough is given by $p(M^j,M)$, where

$$p(M^j,M) = y(M^j)/r(N). \qquad (13)$$

The rationale behind (13) is that $p(M^j,M)$ is the product of M^j's probability of being first, given that a breakthrough occurs $(y(M^j)/\Sigma_{M^k \in M} y(M^k))$, multiplied by the probability of observing a breakthrough $(\Sigma_{M^k \in M} y(M^k)/r(N))$.

In the interest of simplicity, we assume that each country's utility u_i depends only on its alliance's probability of making the breakthrough and on its own resources not used in the arms race, so that

$$u_i = u_i(p(M^j,M), r_i - y_i) \qquad (14)$$

where $\Sigma_{k \in M^j} y_k = y(M^j)$ and no y_i is allowed to exceed r_i. A simple version satisfying (14) is

$$u_i = p(M^j,M)(r_i - y_i), \qquad (15)$$

a Cobb-Douglas utility function.

We assume that countries play the same alliance formation game as in the previous section. Once a final alliance configuration M_T has been achieved, an expenditure decision stage game is played to determine payoffs. For alliance M^j, actual expenditures $y(M^j)$ depend on the proposed expenditures $\hat{y}_i$ of its members via the relation

$$y_i = \frac{r_i}{r(M^j)} \hat{y}(M^j) \tag{16}$$

with $\hat{y}(M^j) = \Sigma_{k \in M^j} \hat{y}_k$. Once the pattern of expenditures $(y(M_T))$ is chosen, payoffs are determined via (14). This completes the description of the game model.

Since we assume that there is an arms race underway, the formation of the grand alliance in this model will not be meaningful. We therefore add the restriction that M_T consists of at least two members, M' and N/M', with all allowable alliances subsets of these two (M' is fixed a priori). As we will see below, bipolarity is a very real possibility in this kind of model.

Analysis of the Game Model

Rather than conduct an analysis of the utmost generality, we restrict our investigation of arms races and alliance formation to the utility specification (15). More general specifications will add to the complexity and notational burden of the analysis, without adding greatly to its qualitative explanatory power.

We begin our analysis with the expenditure stage. At this stage, a country's payoff depends on the vector of expenditure proposals $(\hat{y}_i)$ and the alliance configuration M. Taking M as given, then from (13), (15), and (16), we can write utility solely as a function of the strategy vector $\hat{y} = (\hat{y}_1, \ldots, \hat{y}_n)$:

$$u_i(\hat{y}) = \frac{r_i}{r(M^j)r(N)} \hat{y}(M^j) \; (r(M^j) - \hat{y}(M^j)). \tag{17}$$

Maximizing u_i with respect to $\hat{y}_i$, one has

$$\hat{y}(M^j) = r(M^j)/2 \; . \tag{18}$$

(18) is an equation with infinitely many solutions $(\hat{y}_h)$. However, this multiplicity of solutions is payoff irrelevant, since all solutions to (18) lead to the same

achieved utility u_i, namely

$$u_i = r_i r(M^j)/4r(N). \tag{19}$$

We have established

<u>Proposition 6</u>. In the expenditure game with utilities given by (17), every alliance will devote half of its available resources to the arms race. Infinitely many expenditure proposal vectors support this result, all of which are payoff equivalent.

It follows from Proposition 6 that a country must always spend half of its resources on arms, regardless of which alliance M^j it belongs to. This gives country i a strong incentive to be in an alliance whose probability of making a breakthrough is as large as possible. Since this probability is large if the ratio of M^j's resources to aggregate resources is large, a country would like to join an alliance with a preponderance of resources. In particular, such incentives could lead to a bipolar configuration, with alliances M^1 and N/M^1 confronting one another. This is the content of

<u>Proposition 7</u>. There exist subgame perfect equilibria inducing the bipolar alliance configuration.

<u>Proof</u>. Since the only possible alliances M^j are subsets of M^1 or N/M^1, there is an upper bound to the resources of an alliance. By monotonicity, the payoff to a member country increases with the resource base of the alliance. Then the alliance formation strategies, with all members of M^1 proposing M^1, and all members of N/M^1 proposing N/M^1 at every stage, is subgame perfect (compare the proof of Proposition 1). ■

Due to the unanimity rule required for the merging of disjoint alliances, it is possible to induce other alliance configurations besides that of Proposition 7. For instance, there exist subgame perfect equilibria which induce the configuration of all nonaligned countries. In such a case, however, one is very attempted to apply arguments from equilibrium selection theory (see Harsanyi and Selten, 1988) to discard such unintuitive, if not unreasonable, equilibria.

We now turn to the study of utility functions more general than (17). Let $c_i = r_i - y_i$, so that utility can be expressed as

$$u_i = f_i(p(M^j,M),c_i) \tag{20}$$

where f_i is strictly increasing in both its arguments. Assuming that the sufficient conditions for an interior maximization of f_i with respect to $\hat{y}_i$ hold, one has the following analogue to (18)

$$\frac{f_i^1}{f_i^2} = \frac{r_i \cdot r(N)}{r(M^j)} \tag{21}$$

where f_i^1, f_i^2 denote the partial derivatives of f_i with respect to its first (respectively second) argument and country i belongs to alliance M^j. Another way of expressing (21) is that marginal rate of substitution in utility space for each member country i is proportional to its share of coalition resources. Interpreted politically, this says that each country faces the trade-off between the alliance goal of making a breakthrough and the member country goal of husbanding its own resources. This trade-off represents the politically sensitive issue now facing both NATO and WTO, that of burden sharing within each alliance. Both Washington and Moscow are complaining now more than ever that their contributions to their respective alliances are a growing burden which is no longer being shared by the other member countries. To the extent that these charges are true, this suggests that an equilibrium of the sort given by (21) no longer exists. What is clear from the data on expenditure at any rate is that the proportional expenditure by the superpowers is substantially greater than that of their alliance partners (in the case of the USSR, there is almost an order of magnitude difference). Thus, although the specification (17) with its attendant one-half expenditure rule seems too special, a specification as flexible as (20) has testable empirical content. Even a model as simple as that proposed here may eventually shed considerable light on the growing controversy over burden sharing within the two alliances dividing Europe.

IV. Alliance Formation With Superpowers: An Incomplete Information Game

In this section, we consider a game where a superpower S and a set of potential alliance partners $N = \{1,...,n\}$ face a fixed external threat, posed by another superpower, possibly also with allies. An immediate example is the fixed threat faced by the USA in Europe, where the potential alliance partners are all states not belonging to the WTO. We assume that the superpower S has the best military technology to counter the external threat. If the potential partners were to provide the same level of protection by themselves, it would require considerably more resources than if S is the provider.

An efficient solution to the alliance formation problem would have S providing protection for all of its potential partners. Suppose S offers to provide such protection, provided that the partners share in S's cost. Then S faces the problem that if it demands too much in the way of burden sharing, then some potential members may not be willing to join the alliance. This problem facing S is especially delicate when S has only incomplete information about the security needs of $1,\ldots,n$. In this event, S is not sure how much another country would be willing to contribute to the security burden.

Güth (1988) has studied a similar situation in a complete information context. In that allocation game with alliance formation, the superpower knows the security needs of its two potential allies, and the protection S offers them comes in the form of a public good -- a deterrence umbrella over Europe, if you will. Here, we relax both the complete information condition, and the public good condition. S can, if it chooses, defend only its alliance partners and not those countries that choose to remain nonaligned.

For each country i, let t_i denote i's type. Type can be interpreted as the positive amount which i would have to spend in order to reach the security level offered by S. Each country i knows its own type, whereas all other countries have a common prior over i's type represented by the probability density $f_i(t_i)$. We assume that the collection of priors over type $f_i(t_i)$ is everywhere positive on the positive real numbers, with the f_i mutually independent, and all of this common knowledge.

The incomplete information aspect of alliance formation can be handled as follows:

Stage 0 (type determination). Chance chooses the vector $t = (t_1,\ldots,t_n)$ of actual types according to the probability density

$$f(t) = \prod_{i=1}^{n} f_i(t_i) . \tag{22}$$

At this stage, t_i is revealed only to i. From (22), one can induce country j's beliefs about its type via conditional probability, while (22) expresses exactly S's beliefs about its potential partners types.

Stage 1 (offer stage). Armed only with its beliefs, S chooses an offer to its potential partners. Since S does not know who will join, its offer must be contingent upon joining. Denote by S' any subset of $\{S\} \cup N$ which contains S. An offer is defined on alliances S', which are the only ones considered here. Let C(S') be cost to S of protecting S'. By budget balance, costs not borne by outside members are borne by S. A cost-sharing rule s(S') is described by

$$s(S') = (c_i(S'),s_i(S')) \tag{23}$$

where for all members of S including S itself $c_i(S')$ is a nonnegative constant and $s_i(S')$ is a nonnegative share of costs which i pays if S' forms and i belongs to S'. Note that these shares are relative to what S does not itself cover. (23) thus expresses a two-part cost-sharing scheme, namely a fixed part and a variable part where member countries may have to pay even if no protection is provided. At the offer stage then, S announces a cost-sharing rule for each possible alliance S'.

Stage 2 (entry stage). Each outside country i must now decide simultaneously with the rest whether to enter an alliance with the offered cost-sharing rule ($d_i = 1$) or to remain nonaligned ($d_i = 0$). Given the total vector of entry decisions $d = (d_1,\ldots,d_n)$, the alliance S' has now been formed. If d is the zero vector, the game is over. Otherwise, play continues with

Stage 3 (type revelation stage). Each outside member country i announces a type $\hat{t}_i$, which may differ from its true type t_i. S computes the security needs based on these announcements according to

$$c(S') = k \sum_{i \in S'} \hat{t}_i \,. \tag{24}$$

Here, k is a positive scalar, but less than one, which measures S's relative efficiency in providing security over that of its alliance partners. Of course, $C(S') = \Sigma\, t_i$, if the alliance partners all underreport their true types. On the other hand, if the partners greatly overstate their types, then serious cost overruns might easily result, even when S is the provider. At any rate, cost shares are computed according to (23) and the game ends.

To complete the specification of the game, it remains to define payoffs. Let $s = (s(S),s_1,\ldots,s_n)$ denote a strategy vector for the 3-stage game, where $s(S)$ is the superpower's strategy, the announcement of a cost-sharing rule for each possible alliance, and s_i is a strategy for country i. s_i can be written

$$s_i = s_i(t_i) = (d_i(s(S),t_i),\hat{t}_i(s(S),S',t_i)) \tag{25}$$

since i's strategy is type-dependent, the decision d_i to join depends on the cost-sharing rule and on type, as does announced type.

The payoff function for the superpower $u_S(s)$ is given by

$$u_S(s) = u(|S'|) - c_S(S') - s_S(S') \max \left\{k \sum_{\substack{i \in S' \\ i \neq S}} \hat{t}_i - \sum_{i \in S'} c_i(S'),0\right\}. \tag{26}$$

The function in $(|S'|)$ is assumed to be increasing in the size $|S'|$ of the alliance S'. (26) expresses the fact that the superpower is interested in a larger alliance and that all costs not covered by the allies are covered by the superpower $(c_S(S'), s_S(S'))$.

Since the superpower does not observe the true type vector t, its payoff will be the expected value of (26) relative to its prior (22).

Similarly for any country i, payoffs are given by

$$u_i(s) = (d_i-1)t_i - d_i\left[c_i(S')+s_i(S')\ \max\left\{k \sum_{\substack{j\in S' \\ j\neq S}} \hat{t}_j - \sum_{j\in S'} c_j(S'),0\right\} + \max\{t_i-\hat{t}_i,0\}\right]. \quad (27)$$

In (27), we suppress the notation that the participation decision d_i depends on type and cost-sharing rule. According to (27), i has to provide for its own security if it does not participate, and gets the security level it announces, t_i, paying its computed cost share, if it does participate. Again, i's payoff will be the expected value of (27) relative to its prior (22) under incomplete information.

Game Theoretic Analysis

We argue backwards from stage 3, the type revelation stage. Differentiating the expected value of (27) with respect to announced type, one has that the left and right-hand derivatives are negative for $s_i(S')$ positive, whence $\hat{t}_i = t_i$ is the best response in this case. Thus, we get the revelation principle at the end of the game: there is an equilibrium supporting truthful revelation of types when cost shares are positive. If a cost share is zero, then country i doesn't have to pay and so has an incentive to announce the highest possible type, $\hat{t}_i = \infty$. Given revelation in the final stage of the game, then at any subgame perfect equilibrium players anticipate revelation. Thus, at stage 2, the participation stage, i will join $(d_i = 1)$ if the expected value $EC_i(S')$ of

$$C_i(S') = c_i(S') + s_i(S')\left[k \sum_{\substack{j\in S' \\ j\neq S}} t_j - \sum_{j\in S'} c_j(S')\right] \quad (28)$$

is smaller than t_i. This condition simply says that it pays i to join, given true revelation by all countries joining the alliance. An analogous condition with membership costs greater than t_i holds for all those not joining.

We are now back to the first stage of the game, the offer stage. If the superpower wants country i to join the alliance, its cost-sharing proposal must satisfy (28). However, since S does not know t_i, it cannot guarantee the formation of any specific S'. S can achieve an alliance with at least one country i if it announces $s_i(S') = 1$ and $c_i(S') = 0$ (the other countries clearly don't get such offers). In this event, $EC_i(S') < t_i$ reduces to

$$kt_i < t_i \tag{29}$$

which is automatically satisfied by the comparative cost advantage condition. We have

Proposition 7. The superpower can always guarantee a 2-member alliance by offering exactly one country i the full cost advantage. However, such an offer excludes the certainty of any other alliance forming.

The question naturally arises whether the superpower can guarantee the formation of larger alliances than those given by Proposition 7. The answer is in general negative, if the superpower is unwilling to subsidize the outside members, i.e. $s_S(S') = 1$ or $c_S(S') > 0$. The problem is with satisfying $EC_i(S') < t_i$ with certainty for more than a single country i. Once the superpower has to induce two countries to join with certainty, there is positive probability on the event that all countries except country i, say, have types very close to zero, for none of whom condition $EC_i(S') < t_i$ can be satisfied. The logic of the situation is that if the superpower wants to guarantee a larger coalition, then it has to subsidize the alliance partners to guarantee participation and subsequent true revelation of type.

More positive results are available if one asks only that an alliance form with some positive probability less than one. Indeed, for this less demanding criterion on alliance formation one has

Proposition 8. For any possible alliance S' there exists a parameter region where S' forms a perfect equilibrium at no cost to the superpower, i.e. $s_S(S') = 0 = c_S(S')$.

Proof. The parameter regions available are (k,f_i). Pick t^* such that $EC_i(S') > t_i$ is satisfied for any $t_i > t^*$. In particular, take $c_i(S') = c_S(S') = 0$, and $s_i(S') = 1/|S'|$, so that $s_S(S') = 0$ and the superpower is offering no subsidy. Now, consider f_i that put a large amount of probability on the event t_i greater than t^*. Then one has a positive probability of forming coalitions of more than two members.
∎

Propositions 7 and 8 taken together allow one to draw certain important conclusions regarding alliance formation under incomplete information. If the superpower wants to ensure a large alliance on its periphery, then it must offer subsidies to its alliance partners. Otherwise, it cannot guarantee that such a large alliance will form. If the superpower is willing to settle for only probabilistic formation of a large alliance, then it can do so without offering any subsidy. Finally, if it is willing to settle for a certain alliance with a single partner, all it has to do is offer that one partner full cost offset.

Our own impression of the events surrounding the formation of both NATO and WTO is that the respective superpowers chose the middle course. They succeeded in getting several alliance partners, but now find themselves with a considerable problem of burden sharing. Moreover, any effort now to change the cost sharing rules could lead the entire equilibrium to unravel, especially if the cost advantage parameter k in the last 30 years has approached one. Given a reduction in the fixed threat facing either alliance (generated, for instance, by bickering within or even defection from the other), the imperative to form an alliance with a superpower is greatly reduced. Thus, the presence of incomplete information in the alliance formation game also points in the direction of a possible return to a balance of power regime within the European theater.

V. Conclusion

In this paper, we have studied three aspects of alliance formation from the standpoint of noncooperative game models. Given the exploratory nature of our undertaking, this seems to have been the most reasonable approach. Another approach is also tempting, however, expecially if one finds the results obtained so far intriguing. That approach would be to build a single very complex model incorporating all the phenomena studied here, and then investigate its equilibria. Such an undertaking, although challenging, offers especial payoffs as well. By allowing for rather subtle linkages between various strategic questions surrounding alliance formation and dissolution, one can approach an even greater reality in both the model and its application.

Another quite intentional limitation of the present paper is our emphasis on equilibrium scenarios, rather than a complete characterization of the equilibrium set, or even more ambitiously, the application of equilibrium selection theory to find the unique rational equilibrium for the countries to play. Even characterizing the equilibrium set for all possible parameter constellations seems a daunting task. Moreover, given the current state of equilibrium selection theory, actually selecting the right equilibrium from a very rich set of equilibrium candidates is beyond the capabilities of current computational procedures. Until the necessary algorithms are up and working, it seems premature to apply equilibrium selection theory to alliance formation models. The writing of efficient programs for equilibrium selection theory is a task very much worth doing.

We have taken the approach of noncooperative game models as basic to our approach. This is not to deny that other approaches are not worthwhile or appealing. As cited earlier, cooperative game models have also had considerable success in addressing some international political questions. Again, all our games have been in extensive form; however, worthwhile results have also been obtained using the normal form, for instance by Brams (1985) and Brams and Kilgour, 1988. Progress in game theory and its applications has always depended on attacking a strategic problem from several forms of games, then comparing the results, the most notable case being the theory of market games. There is no reason to take a dogmatic approach a priori to the research techniques to use in the study of alliance formation. We hope to have shown the utility of extensive game models in our present work.

The reason for doing any theoretical work is to better organize existing data and predict successfully new data. There exists by now a vast historical record of balance of power, bipolar, and hegemonic international systems. We are confident that the application of game theoretical methods, particularly those using extensive games, will prove very useful in organizing at least some of this data. Much new data in the social sciences is now being created using subjects in behavior laboratories under controlled conditions. To the best of our knowledge, the established methodology of experimental game theory (with applications to economics and political science) has not yet been applied to the questions of alliance formation. This we consider to be an urgent task for experimental investigation. For example, by creating a bipolar initial situation in the laboratory which theory predicts will unravel, and then observing the subsequent play of the international relations game, we can tell just how good such a theory is. Given the great uncertainties surrounding the possibility of disengagement of the superpowers in Central Europe, with all the risks that there attend, the cost of running and analysing the results of such experiments is far from prohibitive. Experimental results have advanced theory in the recent past, and for questions such as these the promise of advancement is great indeed.

References

Avenhaus, R., W. Güth and R.K. Huber (1988). Implications of the Defense Efficiency Hypothesis for the Choice of Military Force Structures. Part II: A Sequential Game Including the Possibility of Restructuring Forces. This volume.

Bernholz, P. (1985). The International Game of Power, Past, Present and Future. Berlin.

Brams, S. (1985). Superpower Games: Applying Game Theory to Superpower Conflict. New Haven CT: Yale University Press.

Brams, S. and M. Kilgour (1988). Game Theory and National Security. New York: Basil Blackwell.

Camerer, C., G. Loewenstein and M. Weber (1989). The Curse of Knowledge in Economic Settings: An Experimental Analysis. Journal of Political Economy, forthcoming.

Güth, W. (1984). Egoismus und Altruismus - Eine spieltheoretische und experimentelle Analyse -. In: H. Todt (ed.), Schriften des Vereins für Socialpolitic, N.F. **141**: 35-58.

Güth, W. (1988). The (Stability of the) Western Defense Alliance - A Game Theoretic Analysis. This volume.

Harsanyi, J.C. and R. Selten (1988). A General Theory of Equilibrium Selection in Games. Cambridge/Mass.: MIT Press.

Hegel, G.W.F. (1831). Rechtsphilosophie. Frankfurt/M.: Suhrkamp Verlag.

Ordeshook, P. and E. Niou (1986). A Theory of the Balance of Power in International Systems. Journal of Conflict Resolution **30**: 685-715.

Ordeshook, P. and E. Niou (1987). Preventive War and the Balance of Power: A Game-Theoretic Approach. Journal of Conflict Resolution **31**: 387-419.

Selten, R. (1975). Reexamination of the Perfectness Concept for Equilibrium Points in Extensive Games. Intern. Journal of Game Theory **4**: 25-55.

Snidal, D. (1985). The Game Theory of International Politics. Journal of World Politics **38**: 25-57.

THE (STABILITY OF THE) WESTERN DEFENSE ALLIANCE
- A GAME THEORTIC ANALYSIS -

by

Werner Güth

Abstract: We discuss the problems of forming a Western defense alliance (NATO) and try to analyse them by two special game models with player A (USA) and two stylized Western European players F and G resembling French and German defense policy. In the first model F and G are two possible types of Western Europe expressing A's possible beliefs about Western Europe's true security interests. The second model with complete information and two Western European countries F and G concentrates on the instability of an alliance due to the fact that one can enjoy its military protection without being its member. By comparing our theoretical results with the actual state of NATO we derive new interesting restrictions for the main political parameters.

I Introduction

Our main intention is to study the problems of forming a Western defense alliance in spite of the different security interests in North America and Western Europe. This will be done with the help of two special game models with player A (the United States of America) representing North America and two independent players F (standing for France or its policy) and G (standing for the Federal Republic of Germany or its policy) representing Western Europe. Of course, in reality the Western defense alliance N (the North Atlantic Treaty Organization / NATO) has more potential member nations. But our models aim more at analysing basic aspects of international political interaction than at representing the world in the most realistic way. In our view, some of the most crucial aspects of international military policy can be demonstrated very clearly by relatively simple and abstract game models (see BRAMS, 1985, for an inspiring selection). Furthermore, a very realistic game model of the Western defense alliance would be very complex and possibly much more difficult to analyse.

Another important goal of our study is to demonstrate some methodological tools of modern game theory which till now are not frequently used by political scientists. So we will not always rely on complete information but also allow for information deficits concerning other countries' political intentions, strategies and information conditions. Furthermore, it will be demonstrated that one can derive exact policy conclusions in spite of the fact that the game has different equilibria with conflicting implications.

Before describing the two game models let us first discuss some basic advantanges and problems of forming a Western defense alliance. The most straightforward reason is, of course, that each member country has to *sacrifice less resources* to balance the military power of the East if more Western countries combine their military potentials. At least for the smaller countries in Western Europe a satisfactory security against foreign attacks could not have been reached without a Western defense alliance.

Another positive aspect of such an alliance is that a *war between two member countries* of the same alliance *becomes unlikely* if not impossible. So, for instance, a war between Greece and Turkey may have been avoided since due to their common NATO-membership military experts in both countries were aware that no country can gain by attacking the other. The danger of an attack within a defense alliance can obviously be further reduced by having mixed armies, i.e. all armies on all territories contain soldiers from all member countries. In such a situation attacking foreign territory within the alliance means to risk the life of one's own soldiers. This shows that mixed armies may not exclude a war within an alliance but will make it less attractive and more risky.

On the other hand, all organizational problems of voluntary cooperation will typically arise if independent countries are supposed to form a defense alliance. Here the *direct costs of forming the alliance* as the costs of communicating, negotiations, adjustment of military equipment etc. seem to be less important when compared to the increase in security achievable by cooperation. In our view a more serious problem is that the individual potential members will have greatly different security interests. So, for instance, the security interests in North America and Western Europe can hardly be compared. Whereas Western European countries are seriously endangered by most nuclear missiles based in Eastern Europe only long range missiles have to be feared in North America. Furthermore, Western Europe is seriously threatened by a conventional military invasion whereas a conventional invasion of North America can hardly be imagined. Even Western European countries will have different fears and therefore different desires for security. Whereas any European conventional war will most likely take place on German territory, a conventional military invasion of France by Eastern European countries is less likely. As a consequence, the national members of a defense alliance will mostly disagree on the necessary measures to protect all member countries, we refer to this as the *problem of coordinating defense policy.*

Since the security provided by the alliance will be a public good for all its national members they, furthermore, will often try to strategically hide their true desires for security against foreign attacks in order to reduce the own defense expenditures by making other countries pay more. The public good property of security against foreign attacks may even induce some countries not to join the alliance at all. So it could, for instance, be argued that the neutral countries in Northern and Central Europe like Finland, Sweden, Switzerland, and Austria are freeriders enjoying at least partly NATO's protection without paying for it. The stability problem of a Western defense alliance as caused by the public good property of its main product "security against Eastern attacks" and the incentives for strategic misrepresentation of different national interests for security will be called the *public good problem* of a Western defense alliance.

Another reason why a country may refrain from joining a defense alliance is that this necessarily means to give up some of its autonomy and maybe national pride. It is often argued that this, for instance, was the decisive reason for France to leave the Western defense alliance NATO. We will refer to this aspect as the *problem of sacrificing national autonomy*.

There are other reasons influencing the decision whether a country will want to participate in a defense alliance or not. So the pure fact that the Western defense alliance NATO at least claims to stand up for democracy and market economies might be a reason for a country to join or not to join it. Here we will not investigate further such additional aspects of a Western defense alliance. When interpreting our parameter assumptions later on we will refer only to the positive or negative aspects mentioned above.

The remainder of this paper is organized as follows. In Section II we introduce a game with incomplete information where A is trying to form an alliance with a Western European country E which can be of type F or G representing F(rench) or G(erman) defense policy. Thus the main attempt of the 'Incomplete Information about Western Europe's Security Interests' - game is to study the strategic interaction of A and Western Europe where A is not completely aware of the true security interests of Western European countries. A major shortcoming of this model is, of course, that it does not consider the fact that Western Europe is composed of many autonomous countries with their own specific interests. The 'Allocating the Cost for American Protection of Western Europe' - game tries to include this aspect by distinguishing two Western European countries F(rance) and G(ermany). There are three possible ways to share the cost of American protection of Western Europe which A can propose. An important aspect of the game is that one Western European country can leave the overall Western defense alliance and avoid to pay for American protection without endangering the existence of a Western defense alliance including A and the other Western European country, i.e. countries in Western Europe may be freeriding on American protection for Western Europe. By comparing our theoretical results with the actual state of the Western defense alliance NATO, we will be able to derive important and interesting restrictions for the parameters of the game. We conclude the paper by summarizing its main results and by indicating some ways how to incorporate further aspects into the strategic analysis of the (stability of the) Western defense alliance.

II Incomplete Information about Western Europe's Security Interests

The following game model does not distinguish different countries in Western Europe but assumes that Western Europe is an indivisible entity represented by a unique player E. Nevertheless, for the North American player A Western Europe's policy is difficult to predict since A is not completely sure about Western Europe's security interests or its strategic possibilities or its information conditions. In the terminology of game theory this is usually described by saying that A is incompletely informed about Western Europe E's true type, i.e. a type is a complete description of a player's objectives, his strategic possibilities and information conditions. As shown by HARSANYI (1967/68) the differences between two possible types of a player can be remodelled as different payoff functions of the player. We can therefore assume without loss of generality that player A, standing for the United States of America, is only incompletely informed about E's true payoff function describing Western Europe's political evaluation of all possible international defense measures. For the sake of simplicity we will distinguish only between two types of Western Europe reflecting the two possible payoff functions of E expected by player A.

a) The game model

Apart from the information deficit of player A the 'Incomplete Information about Western Europe's Security Interests' - game is a rather simple normal form game where both players A and E must independently decide whether they want to form the Western defense alliance consisting of A and E or not. Denote by N player A's decision to form the alliance and by $\bar{N}$ to refrain from such a measure. For Western Europe E, which can choose between n (to form the alliance) and $\bar{n}$ (not to form it), the situation is more complicated since, at least in view of its potential partner A, Western Europe E can be of two possible types F and G with different payoff functions representing A's expectations about what Western Europe is heading for. The decisive trick (HARSANYI, 1967/68) to capture player A's information deficit concerning E's true type and obtain a nevertheless informationally closed system (a game with complete information) is to assume a fictitious initial chance move taking place before the actual game begins. By this chance move it is determined whether E is of type F or of type G. The probabilities for F and G are, furthermore, determined according to player A's subjective beliefs whether E is of type F or G.

In Figure II.1 we have graphically illustrated the extensive form of the game with complete information which results if one includes the fictious initial chance move described above. Here it is, of course, assumed that player E, respectively his

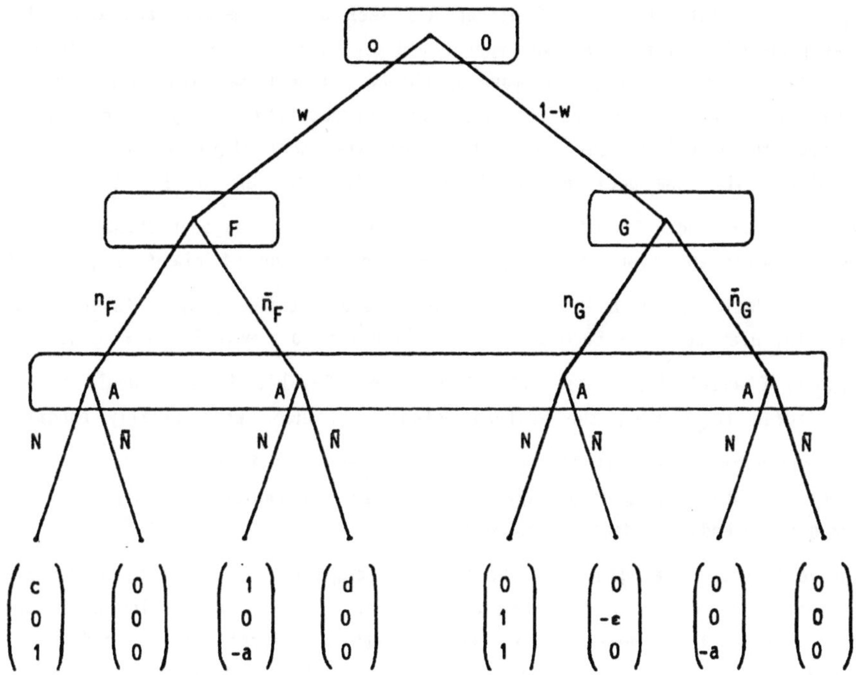

Figure II.1: The 'Incomplete Information about Western Europe's Security' - Game Γ (the game starts with chance player 0's move determining Western Europe E's type F or G with probability w or 1-w, respectively; knowing his own type Western Europe E's type F or G chooses between n_F and $\bar{n}_F$, or n_G and $\bar{n}_G$, respectively; knowing neither E's type nor the decision of E player A finally decides between N and $\bar{N}$; information conditions are graphically illustrated by encircling all decision nodes which the player, who is going to move, cannot distinguish; a play is a sequence of moves from the origin o to an endpoint, which is a node where the game ends; the evaluation of a play is determined by the payoff vector attached to the respective endpoint; at each endpoint payoffs are given in the order F, G, A where we use the convention that the non-present type of E evaluates all plays with "0").

true type, is completeley aware of player A's lack of information and knows A's subjective probabilities for the two types F and G. The play starts at the origin o (the top of the game tree) with the move of chance player 0 determining E's type F with probability w where 1>w>0 and its type G with probability 1-w. By encircling the decision node for F and the one for G it is indicated that player E knows his own type. To distinguish the decision of E's type F from the one of E's type G, we indicate by a lower index F or G whether the decision $\bar{n}$ or n has been chosen by F or G. Without knowing the type chosen for E nor the decision of player E player A finally has to decide for N or $\bar{N}$. To give an example assume that player E's true type is F. Then the play would be such as if it started with 0's move F. Knowing his own type F player E would then choose between n_F or $\bar{n}_F$. Say his decision would be n_F, namely to agree to forming the Western defense alliance. When finally choosing between N and $\bar{N}$ player A neither knows that E's true type is F (all what A knows about E are his subjective probabilities w and 1-w for the two possible types F and G of E) nor does he know the decision n_F of E.

Observe that the real game with incomplete information is defined by player E's true type, i.e. it is that part of the game tree in Figure II.1 starting at F's or G's decision node. This shows that in order to obtain informational closedness we usually have to solve a much more complex game. As a matter of fact when solving the game of Figure II.1 the true type of E is of no importance at all.

b) <u>Restrictions on the parameters of the game</u>

We want to analyse the class of games illustrated by Figure II.1 which is defined by the following conditions:

$$1 > w > 0 \;,\quad a > 0 \;,\quad \varepsilon > 0 \;,\quad d > 0 \;,\quad 1 > c \;. \qquad \text{(II.1)}$$

Condition 1 > w > 0 guarantees that player A is not completely sure about Western Europe E's type. Before trying to justify the other parameter conditions in (II.1) we want to present the strategic interaction of players A and E with the help of familiar 2×2-bimatrix games. If A knew that E's type is F, players A and E would obviously play the game Γ_F of Table II.1 whereas the strategic interaction of A and E would be described by Γ_G if A knew that E's type is G. From our parameter conditions (II.1) follows that Γ_F has as unique equilibrium point (a strategy vector from which no individual player will want to deviate), namely the strategy combination $(\bar{n}_F,\bar{N})$ according to which both players do not want to form the Western defense alliance. For the game Γ_G the corresponding vector $(\bar{n}_G,\bar{N})$ is also in equilibrium but according to (II.1) the game Γ_G has another equilibrium point (n_G,N) leading to the formation of the Western defense alliance N.

Γ_F	N	$\bar{N}$
n_F	c, 1	0, 0
$\bar{n}_F$	1, $-a$	d, 0

Γ_G	N	$\bar{N}$
n_G	1, 1	$-\varepsilon$, 0
$\bar{n}_G$	0, $-a$	0, 0

Table II.1: The 2×2-bimatrix games resulting from Figure II.1 if player A is aware of E's type, i.e. Γ_F or Γ_G is the game if E is of type F or G, respectively (player E is the row and player A the column player; player E's (A's) payoff appears in the upper left (lower right) corner of the bimatrix element determined by E's and A's move).

Whereas E knows whether they are going to play Γ_F or Γ_G player A expects Γ_F to be played with probability w and Γ_G with probability 1-w. If this is known to player E we have a game which is adequately represented by Figure II.1 where player A's incomplete information about E's type is transformed into imperfect information resulting from a partly unobservable chance move.

Let us now discuss the other conditions of (II.1). Here it should be kept in mind that the Western defense alliance will only be formed if both players agree to it, i.e. only the strategy vectors (n_F,N) and (n_G,N) will lead to the formation of N. For A the formation of the defense alliance is obviously a desirable result. But if Western Europe does not want to form N, i.e. if F (G) chooses $\bar{n}_F$ ($\bar{n}_G$), also A prefers not to vote for N. The respective parameter condition $-a<0$ expresses the loss of North America's political influence and the reduced popularity and prestige of its political leaders resulting from an unsuccessful attempt to form a Western defense alliance. Because of $-\varepsilon<0$ the situation of player E's type G is similar to the one of player A. Type G also wants the alliance to be formed but will also refrain from voting for it if A chooses $\bar{N}$.

Different from G Western Europe E's type F always prefers not to vote for the formation of the alliance regardless of the strategy chosen by A, i.e. the choice of $\bar{n}_F$ is always better than the one of n_F. This is a consequence of our parameter restrictions $d>0$ and $1>c$. Thus only for A and G the positive aspects of the alliance, as discussed in our Introduction, will dominate the direct costs of forming N, the problem of coordinating defense policy, the public good problem, the loss of national autonomy etc. For Western Europe's type F the alliance does not yield a sufficient security increase to outweigh these negative aspects. The reason can be that for E's type F national autonomy is considerably more important than for G. When comparing

French and West German foreign policy one often has the impression that for F(rance) national prestige and autonomy is fundamentally more important than for G(ermany). But observe that this can only justify the condition $l>c$ since only in case that A chooses N type F determines whether the alliance will form or not. The condition $d>0$ expresses again that, if the other player votes against N, it is better also not to vote for it in order to avoid a loss of political influence and prestige.

Thus the strategic situation in the two games Γ_F and Γ_G of Table II.1 is fundamentally different. $(\bar{n}_F,\bar{N})$ is the obvious solution of Γ_F since it is the only equilibrium point of Γ_F. For Γ_G the likely result is the equilibrium point (n_G,N) since it payoff dominates the other equilibrium point $(\bar{n}_G,\bar{N})$, i.e. both, A and G, prefer (n_G,N) to $(\bar{n}_G,\bar{N})$. If E's type is known to be F, no Western defense alliance will form, if it is known to be G, North America and Western Europe will both want to vote for it. By solving the game of Figure II.1 we will determine what happens if A is not sure whether Western Europe is of type F or G. Due to player A's information deficit about E the game of Figure II.1 has essentially three players, namely player A and the types F and G of Western Europe. Even for F or G who know that the other type is not present it is nevertheless essential to consider the other type's behavior since it influences player A's decision.

c) The solution of the game

The pure strategy equilibrium points of the game Γ illustrated by Figure II.1 and specified by the parameter restrictions (II.1) are $(\bar{n}_F,\bar{n}_G,\bar{N})$ and $(\bar{n}_F,n_G,N)$ where the latter equilibrium is strict only if assumption

$$1/(1+a) > w \tag{II.2}$$

is satisfied. An equilibrium is called strict if every player looses from deviating unilaterally. If inequality (II.2) is reversed, $(\bar{n}_F,\bar{n}_G,\bar{N})$ is the only equilibrium point of the game. Observe that due to $a>0$ the left-hand side of inequality (II.2) is a positive number smaller than 1. Thus a first result is that all games Γ satisfying (II.1) and

$$1/(1+a) < w \tag{II.3}$$

have the solution $(\bar{n}_F,\bar{n}_G,\bar{N})$. Thus all three players, player A and the two types F and G of Western Europe, will vote against the Western defense alliance if North America expects Western Europe to be of type F with high probability in the sense of condition (II.3). Another possibility for interpreting condition (II.3) is to consider the range of parameter values $a>0$ satisfying (II.3) for a given probability w. Remember that the parameter a measures A's loss of political influence and the reduced popularity of its political leaders due to an unsuccessful attempt to form a Western defense alliance. If a is rather small the left side of inequality (II.3) is only slightly smaller than 1. This shows that for given probability w an unsuccessful attempt to form an alliance in the sense of $(\bar{n}_F,N)$ or $(\bar{n}_G,N)$ has to imply quite a political disaster in order to satisfy condition (II.3).

Although, in our view, our assumption a>0 is rather obvious we do not think that a is large given the normalization rule of A's cardinal utility according to which A evaluates the formation of the alliance by 1 and the bilateral vote against it by 0. Remember that the Western defense alliance implies considerable differences in the international military constellation and in national resource allocation. In all three possible strategy combinations $(\bar{n}_i,N)$, $(\bar{n}_i,\bar{N})$, $(n_i,\bar{N})$ for i=F and i=G leading to no alliance of A and E there are no such essential differences. The three cases only differ with respect to who is responsible for the nonformation of the alliance. It is reasonable to assume that this responsibility differences are far less important than primary security and resource allocation effects. Thus, since a is rather small, situations satisfying (II.2) seem to be more likely than those satisfying inequality (II.3).

Given the more relevant case (II.2) we face a game with two pure strategy-equilibria, namely $(\bar{n}_F,\bar{n}_G,\bar{N})$ and $(\bar{n}_F,n_G,N)$ and one equilibrium point $(\bar{n}_F,q_G,q_A)$ in partly mixed strategies according to which G votes for the alliance with probability

$$q_G(n_G) = \frac{a}{(1-w)(1+a)} \tag{II.4}$$

and A uses his strategy N with probability

$$q_A(N) = \varepsilon/(1+\varepsilon). \tag{II.5}$$

The other pure strategies $\bar{n}_G$ and $\bar{N}$ are chosen with the residual probabilities by G and A, respectively. Due to $\varepsilon>0$ and a>0 both probabilities in (II.4) and (II.5) are positive and smaller than 1. The payoffs for F, G, and A implied by the equilibrium point $(\bar{n}_F,q_G,q_A)$ are given by the payoff vector

$$\left(\frac{d+\varepsilon}{1+\varepsilon},0,0\right) . \tag{II.6}$$

According to all three equilibria F chooses $\bar{n}_F$ whereas G's and A's behavior crucially depend on which of the three equilibria is selected as the solution of the game.

Now the pure fact that a game has many equilibria with different political implications does not mean that the political results of a game theoretic analysis must be ambiguous. In modern game theory we can distinguish two main approaches to select among equilibria. Whereas refinement concepts (e.g. KREPS and WILSON, 1982, MYERSON, 1978, KALAI and SAMET, 1984, KOHLBERG and MERTENS, 1986, for a survey see VAN DAMME, 1987) try to rule out equilibria not satisfying some stronger rationality requirement than just the equilibrium property equilibrium selection theories (HARSANYI and SELTEN, 1988, and GÜTH and KALKOFEN, 1989) try to resolve strategic uncertainty completely by selecting a unique equilibrium as the solution to the game. The problem of most refinement concepts is that often results remain ambiguous. For instance, most refinement concepts will not rule out a strict equilibrium point in the sense that every player looses by deviating unilaterally from the equilibrium. Thus in case of (II.2) they will not enable us to distinguish between the two pure strategy-equilibria $(\bar{n}_F,\bar{n}_G,\bar{N})$ and $(\bar{n}_F,n_G,N)$. This illustrates that the refinement approach is only a half-hearted attempt to resolve strategic uncertainty. We therefore will apply equilibrium selection theory in order to derive definite political conclusions in the case (II.2) with multiple equilibria.

Now applying equilibrium selection theory to games illustrated by Figure II.1 and satisfying (II.1) and (II.2) is trivial since F has a unique undominated strategy $\bar{n}_F$ and since, furthermore, the two players G and A, who use different strategies in the three equilibria, prefer the equilibrium point $(\bar{n}_F,n_G,N)$ to the two other equilibria. Due to the fact that $\bar{n}_F$ is the unique undominated strategy of F only G and A are active players in the restricted game for the comparison of two solution candidates. Now HARSANYI and SELTEN (1988) assume priority of payoff over risk dominance and, following HARSANYI and SELTEN, also GÜTH and KALKOFEN (1989) impose priority of payoff dominance over resistance dominance. Thus in case of (II.2) the solution is $(\bar{n}_F,n_G,N)$ regardless which of the two equilibrium selection theories one applies.

In Figure II.2 we have graphically illustrated the solution of the games described by Figure II.1 which satisfy the parameter conditions (II.1). In the area $(\bar{n}_F,n_G,N)$ below the $1=(1+a)w$-line where $(\bar{n}_F,n_G,N)$ is the solution the Western defense alliance NATO will form if Western Europe E is of type G but it will not result in case of type F of Western Europe although North America is trying to realize it. In the area with the solution $(\bar{n}_F,\bar{n}_G,\bar{N})$ there is no attempt at all to realize the Western defense alliance NATO.

For a given value $a>0$ it is rather natural that A will try to realize NATO if w is small and will refrain from this if $w \in (0,1)$ is large. Similarly, the greater a the more disastrous it will be for A to meet F. A will therefore prefer not to vote

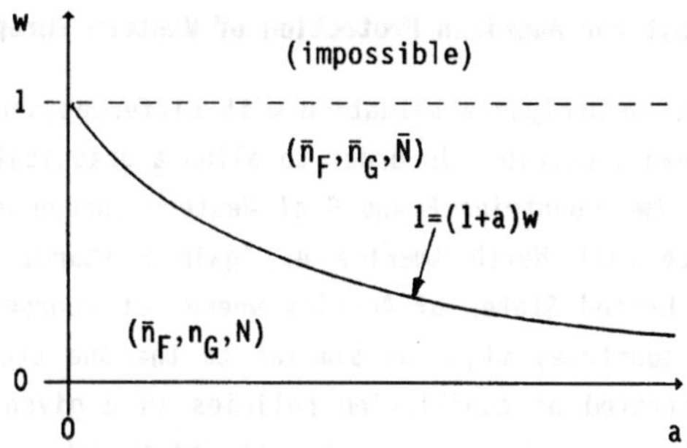

Figure II.2: The solution of games described by Figure II.1 and condition (II.1) in the (a,w)-diagram (below the $1=(1+a)w$-line the solution is $(\bar{n}_F,n_G,N)$, above this line it is $(\bar{n}_F,\bar{n}_G,\bar{N})$).

for the Western defense alliance if for given w the value of a becomes large. In other words: In case of (II.3) A's vote for N is very likely to become a political failure with all its embarrassments. Player A will therefore not dare to propose the Western defense alliance. As a matter of fact no equilibrium will support such a vote of player A.

For small values of w in the sense of (II.2) the situation is different. Here both choices of players G and A are supported by equilibrium points. None the less, both players will want to realize the Western defense alliance although A is not sure that Western Europe is of type G. This shows that only equilibrium selection theory, which resolves strategic uncertainty completely, can distinguish clearly parameter constellations implying that the Western defense alliance is reached from those discouraging its formation.

Obviously F and G stand for F(rench) and G(erman) international defense policy. Our approach here was not to consider countries with different policies but to show that depending on the evaluation of political results both kinds of international defense policies could become the prevailing policy in a Western European country. This was done by studying the strategic interaction of A and two possible types of his potential ally with very different political preferences. In this sense the 'Incomplete Information about Western Europe's Security Interests'-game concentrates on the problem of conflicting ideas about Western Europe's true security interests at the cost of neglecting the different military needs of different countries in Western Europe.

III On Allocating the Cost for American Protection of Western Europe

In the following we want to analyse a situation with different security interests of different Western European countries. In order to allow a graphical representation of the game we assume only two countries F and G of Western Europe which try to form a Western defense alliance with North America A. Again F stands for France, G for Germany, and A for the United States of America where, of course, the situation of two other neighbouring countries might be similar to the one studied here. Whereas the former model concentrated on conflicting policies in a given Western European country, this model tries to capture the different and thereby possibly conflicting security interests of different countries. Naturally the public good problem of a Western defense alliance, which has been neglected so far, will become very crucial. If one Western European country does not become a member of the Western defense alliance, this does not imply, that there will be no defense alliance containing A and other Western European countries. This, of course, opens the possibility of freeriding with all its consequences for the existence and stability of the alliance referred to as its public good problem.

a) The game model

In Figure III.1 the 'Allocating the Cost for American Protection of Western Europe'-game is graphically illustrated in the usual way. The game starts at the origin o with player A's initial decision for h, m, or l. Here h(igh) means that the cost C (> 0) for American protection of Western Europe has to be paid by Western Europe. The decision m(iddle) means that only the part p with $0 < p < 1$ has to be covered by the Western Europeans whereas l(ow) says that A himself pays C. Knowing A's decision $s_A \in \{h, m, l\}$ players F und G then decide independently whether to participate in the Western defense alliance with the cost burden determined by s_A (the decision $n_F^{s_A}$ and $n_G^{s_A}$, respectively) or not (the decision $\bar{n}_F^{s_A}$ and $\bar{n}_G^{s_A}$, respectively). Formally the extensive form requires a time order of the moves by player F and player G (in Figure III.1 player F decides before player G). But since, due to our information conditions as indicated by the information sets in Figure III.1, G does not observe F's decision, the time order of Figure III.1 could be reversed without changing the results of the game. The decision for one specific time order of moves in case of independent decisions is obviously an overspecification in the description of the game required by the extensive form.If at least one Western European country joins the Western defense alliance proposed by A, the game is over. If the alliance includes all its potential members A, F, and G, the security effect expressed in utility terms is denoted by S_i for i = A, F, G. If only country $j \in \{F, G\}$ fails to join the alliance, the corresponding security effects are S_i^{-j}. In all the cases leading to the formation of a Western defense alliance comprising A and at least one Western European country, the payoff H_i of player i is determined by the

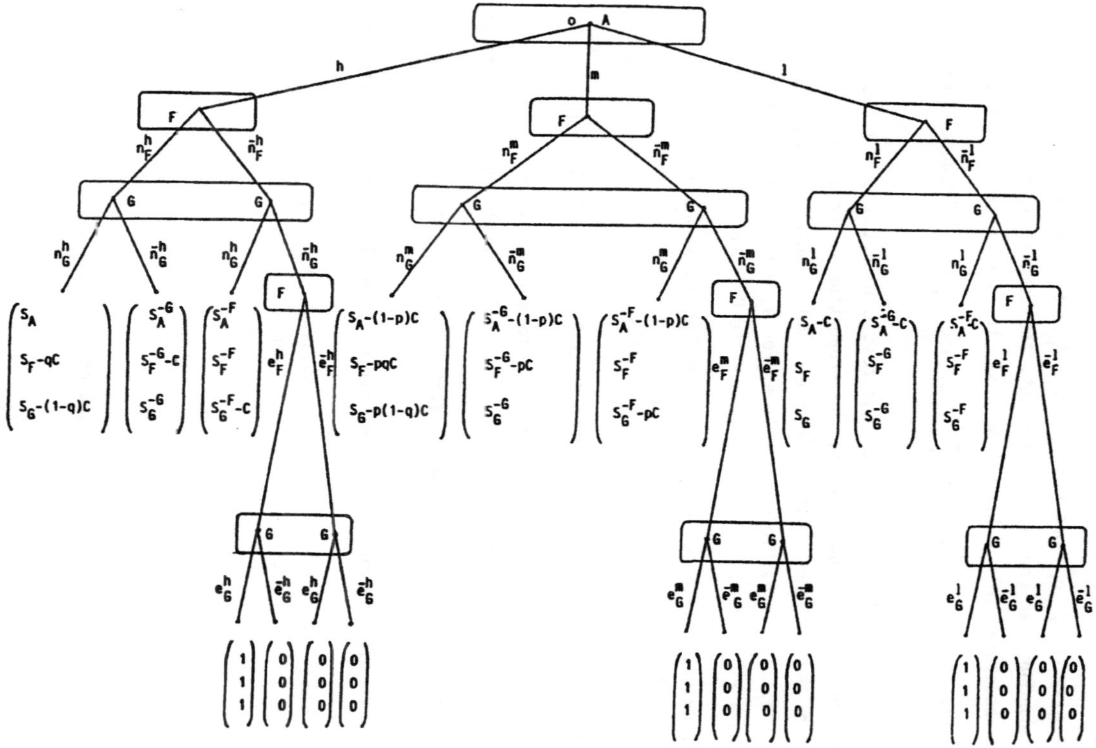

Figure III.1: The 'Allocating the Cost for American Protection of Western Europe'-game (for $i = A, F, G$ the parameter S_i expresses the security level (in utility units) implied by the overall Western defense alliance with member countries A, F, and G; the parameters S_i^{-G} and S_i^{-F} are the analogous security levels for the situations where G or F, respectively, is not a member of the alliance; C (>0) is the cost of American protection for Western Europe; p and q with $0<p<1$ and $0<q<1$ are shares determining the allocation of C).

difference of the security effect defined above and the cost for American protection to be covered by country i. When F and G form an alliance with A, the cost burden of Western Europe is allocated in the proportion q for F and 1-q for G, where $0 < q < 1$. Thus in case of the play (h, n_F^h, n_G^h) one has $H_A = S_A$, $H_F = S_F - qC$, $H_G = S_G - (1-q)C$. If one of the Western European countries drops out, the remaining one has to cover all the cost burden according to A's initial move s_A. If, for instance, the play is $(h, n_F^h, \bar{n}_G^h)$, the payoffs are $H_A = S_A^{-G}$, $H_F = S_F^{-G} - C$, and $H_G = S_G^{-G}$ because G has not joined the alliance.

In case of $s_A = m$ only the share pC of the cost C with $0 < p < 1$ has to be covered by the Western European countries in the proportion q for F and 1-q for G when they both join the alliance. Otherwise the country which became the only European member of the alliance has to pay pC alone. For $s_A = 1$ the cost C are solely paid by A regardless whether the alliance comprises F and G or only one of them.

If no Western defense alliance including North America A has formed, i.e. in case of $(s_A, \bar{n}_F^{s_A}, \bar{n}_G^{s_A})$, the two Western European countries can form a purely European defense alliance. Formally this alliance is induced by the independent choice $e_F^{s_A}$ of player F and $e_G^{s_A}$ of player G. If one of them fails to make this vote, not even a purely European defense alliance has formed. It is assumed that all countries prefer a purely European defense alliance induced by $(e_F^{s_A}, e_G^{s_A})$ to not having any defense alliance protecting Western Europe. Since cardinal utilities are only uniquely defined up to positive affine transformations, there is no loss of generality in assuming that players i = A, F, G evaluate all situations $(s_A, \bar{n}_F^{s_A}, \bar{n}_G^{s_A}, e_F^{s_A}, e_G^{s_A})$ with $s_A = h, m, 1$ by $H_i = 1$ and the situation with no alliance at all by $H_i = 0$. This completes the description of the 'Allocating the Cost for American Protection of Western Europe' - game described in Figure III.1.

In the following we restrict ourselves to games which satisfy the following parameter conditions:

$$S_i > S_i^{-j} \qquad \text{for all } j = F, G \text{ and } i = F, G \,, \tag{III.1}$$

$$(S_F - S_F^{-F})/q < pC < S_F^{-G} - 1 < C \,, \tag{III.2.a}$$

$$(S_G - S_G^{-G})/(1-q) < pC < S_G^{-F} - 1 < C \,, \tag{III.2.b}$$

$$\frac{S_F^{-G} - pC - 1}{S_F^{-F} - S_F + pqC} \neq \frac{S_G^{-F} - pC - 1}{S_G^{-G} - S_G + p(1-q)C} \,. \tag{III.3}$$

Inequality (III.1) expresses that a more comprising defense alliance is more effective in guaranteeing the security of its member countries. Observe that we do not impose any restrictions on player A's payoff function although one might think of some intuitive requirements for the parameters determining A's evaluation of plays. So one might, for instance, assume that North America A does not want to grant American protection for Western Europe just as a gift, i.e. in case F and G do not contribute to the cost C, player A prefers the purely European alliance to the overall alliance. In the past American politicians haven often expressed views along these lines, especially shortly before bargaining with G(ermany) about how to share the cost for American protection of Western Europe.

Whereas condition (III.1) is rather immediate, assumptions (III.2.a) and (III.2.b) seem to be more debatable. Inequalities (III.2.a) and (III.2.b) would, for instance, not be satisfied if A could produce military security at very low cost, i.e. if C is nearly 0, or if at least one European country prefers to go along with the other European country instead of having an alliance with A, i.e. if $S_F^{-G}<1$ or $S_G^{-F}<1$. At least for F(rance) the latter could be true.

By $S_F^{-G}-1 <C$ and $S_G^{-F}-1 < C$ it is assumed that neither F nor G wants to bear all the cost C of American protection of Western Europe, i.e. the purely European alliance is preferred if neither A nor the other European country is willing to pay at least part of the cost C.

According to (III.2.a) and (III.2.b) neither F nor G wants to pay for American protection of Western Europe if he can drop out without endangering the existence of a Western defense alliance including one European country beside A. The public good problem of American protection for Western Europe is therefore adequately represented due to $(S_F-S_F^{-F}) < qpC$ and $S_G-S_G^{-G}) < (1-q)pC$. Observe that these two conditions require that Western Europe's cost share p cannot be too low and that the allocation parameter $q \in [0,1]$ deciding on how the cost pC is allocated to F and G is not too extreme. It might be questionable whether, due to G's specific location at the Iron Curtain, G really would like to be in the outsider position. Here this has not been excluded since this is the more delicate case to be analysed.The game would have a unique perfect equilibrium point (SELTEN, 1975) if an outsider advantage existed for F(rance) only.

On the other hand, the remaining conditions in (III.2.a) and (III.2.b) imply that both, F and G, prefer to accept the share p of the cost C if their decision not to participate ($\bar{n}_F^m$ or $\bar{n}_G^m$) means that there will be no American protection of Western Europe. At least for G(ermany) the assumption $S_G^{-F}-1 > pC$ seems to reflect the political views of the main political groups (Christian Democrats, Social Democrats, Liberals) except for some parts of the Green Movement. The analogous assumption S_F^{-G} -

pC > 1 is more debatable. What we know from reality is that F(rance) prefers the outsider position with its security level S_F^{-F} to the overall Western defense alliance. What we do not know is whether F(rance) would prefer to go along with G only to form a purely European defense alliance.

Here we have imposed the condition $S_F^{-G} - pC > 1$ since this is the more difficult case to be analysed. As a matter of fact the game could be easily solved by backward induction if condition $S_F^{-G} - pC > 1$ would be reversed. But whereas in case of (III.2.a) the result can be derived only by selecting among equilibria, reversing condition $S_F^{-G} - pC > 1$ would mean that the game has a unique perfect equilibrium point (SELTEN, 1975). We therefore simply concentrate on the more difficult case. The analysis and the results of the simpler case are obvious.

Whereas all previous assumptions treat F and G in a similar way, condition (III.3) excludes the special case of identical interests in F(rance) and G(ermany). The extremely special case of equality in (III.3) has obviously no political relevance since the security of different countries are significantly influenced by their different geographical location with all its strategic consequences.

b) Perfect equilibrium points

To solve the 'Allocating the Cost for American Protection of Western Europe'-game satisfying conditions (III.1) to (III.3) we first determine its perfect equilibrium points (SELTEN, 1975). To derive the perfect equilibria for the game at hand it suffices to know that no perfect equilibrium will prescribe the choice of an inferior strategy and that perfect equilibrium points are subgame perfect (SELTEN, 1975). For the subgames following the decisions $(s_A, \bar{n}_F^{s_A}, \bar{n}_G^{s_A})$ this implies that $(e_F^{s_A}, e_G^{s_A})$ is the only perfect equilibrium point (the decision vector $(\bar{e}_F^{s_A}, \bar{e}_G^{s_A})$ is also an equilibrium point but an imperfect one). Knowing this these subgames can be substituted by the payoff vector (1,1,1) implied by their unique perfect equilibria. With the help of this substitution the subgames following the decisions $s_A \in \{h, m, l\}$ of player A can be described as 2×2-bimatrix games as shown in Table III.1. Since player A is not active in these subgames his payoff function is irrelevant for analysing these subgames.

Because of (III.2.a) and (III.2.b) one has $S_F - qC < S_F^{-F}$ and $S_G - (1-q)C < S_G^{-G}$ and also $S_F^{-G} - C < 1$ and $S_G^{-F} - C < 1$. Thus in the subgame following $s_A = h$ (the left bimatrix of Table III.1) the strategies n_F^h and n_G^h are always dominated by $\bar{n}_F^h$ and $\bar{n}_G^h$, respectively, which shows that $(\bar{n}_F^h, \bar{n}_G^h)$ is the unique (perfect) equilibrium point of this subgame.

$s_A=h$	n_G^h	$\bar{n}_G^h$
n_F^h	S_F-qC $S_G-(1-q)C$	$S_F^{-G}-C$ S_G^{-G}
$\bar{n}_F^h$	S_F^{-F} $S_G^{-F}-C$	1 1

$s_A=m$	n_G^m	$\bar{n}_G^m$
n_F^m	S_F-pqC $S_G-(1-q)pC$	$S_F^{-G}-pC$ S_G^{-G}
$\bar{n}_F^m$	S_F^{-F} $S_G^{-F}-pC$	1 1

$s_A=l$	n_G^l	$\bar{n}_G^l$
n_F^l	S_F $S_G f$	S_F^{-G} S_G^{-G}
$\bar{n}_F^l$	S_F^{-F} S_G^{-F}	1 1

Table III.1: The subgames following the decisions $s_A \in \{h, m, l\}$ of player A in Figure III.1 where the subgames following $(s_A, \bar{n}_F^{s_A}, \bar{n}_G^{s_A})$ are substituted by their solution payoff vector (1,1,1) (the upper (lower) payoff is the one of F (G); player A's payoff has been neglected since he is not active in these games).

Due to (III.2.a) and (III.2.b) the subgame following $s_A = m$ has three equilibria, namely $(n_F^m, \bar{n}_G^m)$, $(\bar{n}_F^m, n_G^m)$, and the mixed strategy vector $q^m = (q_F^m, q_G^m)$ with

$$q_F^m(n_F^m) = \frac{1 + pC - S_G^{-F}}{1 + pC - S_G^{-F} + S_G - S_G^{-G} - p(1-q)C} \tag{III.4}$$

and

$$q_G^m(n_G^m) = \frac{1 + pC - S_F^{-G}}{1 + pC - S_F^{-G} + S_F - S_F^{-F} - pqC} \tag{III.5}$$

which are all perfect. Because of (III.2.a) and (III.2.b), respectively, the right-hand side in (III.4) and (III.5) is positive and smaller than 1. If an equilibrium point is strict, i.e. every player looses by deviating unilaterally, or completely mixed, i.e. every pure strategy is chosen with positive probability, then it is also perfect.

Because of (III.1) as well as (III.2.a) and (III.2.b) the subgame following s_A = l has a unique (perfect) equilibrium point (n_F^l, n_G^l) since n_F^l dominates $\bar{n}_F^l$ and n_G^l

dominates $\bar{n}_G^1$. Thus, if player A chooses $s_A = 1$, he can be sure that the overall Western defense alliance will form whereas in case of $s_A = h$ he can anticipate that no European country wants to go along with him and that the purely European defense alliance is formed. With $s_A = m$ he cannot guarantee the overall Western defense alliance but at least according to the two pure strategy equilibria there will always exist an alliance providing American protection of Western Europe.

Knowing this player A's dilemma has become obvious. If A tries to avoid that he himself has to contribute to the cost for American protection of Western Europe, then there will be no American protection of Western Europe. But if A wants the overall alliance with members A, F, and G to form, then he has to accept all its costs. On the other hand, sharing these costs means to split up Western Europe (at least with positive probability) which definitely is an undesirable result. The reason for this is that the compromise $s_A = m$ suffers from the public good-problem which is implied by costly American protection for Western Europe.

According to our results above player A will get 1 if he chooses $s_A = h$ and $S_A - C$ in case of $s_A = 1$. When he uses $s_A = m$, his payoff is $S_A^{-F} - (1-p)C$ if $(\bar{n}_F^m, n_G^m)$ is the solution of the bimatrix game with $s_A = m$ in Table III.1 whereas it is $S_A^{-G} - (1-p)C$ in case of $(n_F^m, \bar{n}_G^m)$ and

$$
\begin{aligned}
&H_A(q_F^m, q_G^m) \\
&= S_A \cdot q_F^m(n_F^m) \cdot q_G^m(n_G^m) + S_A^{-G} \cdot q_F^m(n_F^m) \cdot (1-q_G^m(n_G^m)) + S_A^{-F} \cdot (1-q_F^m(n_F^m)) \cdot q_G^m(n_G^m) \\
&\qquad + (1+(1-p)C) \cdot (1-q_F^m(n_F^m)) \cdot (1-q_G^m(n_G^m)) - (1-p)C \qquad \text{(III.6)}
\end{aligned}
$$

for (q_F^m, q_G^m) where $q_F^m(n_F^m)$ amd $q_G^m(n_G^m)$ are determined by (III.4) and (III.5), respectively.

Although a complete solution of the game will require to solve uniquely also the subgame following $s_A = m$, it is clear that player A will choose $s_A = h$ if 1 is greater than $S_A - C$, $S_A^{-F} - (1-p)C$, $S_A^{-G} - (1-p)C$ or $H_A(q_F^m, q_G^m)$ and that he will choose $s_A = 1$ in case that $S_A - C$ is greater than 1, $S_A^{-F} - (1-p)C$, $S_A^{-G} - (1-p)C$ or $H_A(q_F^m, q_G^m)$. If A uses $s_A = h$, he apparently believes that a purely European alliance is better than to grant protection for Western Europe as a gift and to split up Western Europe. In case of $s_A = 1$ player A prefers to grant his protection freely in order to avoid that not all Western European countries join the alliance. But none of the two cases seems to reflect reality since there is no purely European defense alliance and since North America A has caused its European allies to bear at least some of the cost for American protection of Western Europe. Thus the most relevant situation seems to be the one where $S_A^{-F}-(1-p)C$, $S_A^{-G}-(1-p)C$ or $H_A(q_F^m,q_G^m)$ is greater than 1 and greater than S_A-C, i.e. where

$$\max\{S_A^{-F}, S_A^{-G}\} + pC > \max\{S_A, 1+C\} \tag{III.7}$$

or

$$H_A(q_F^m,q_F^m) > \max\{1, S_A-C\}\ . \tag{III.8}$$

If either (III.7) or (III.8) is true the solution of the subgame following $s_A = m$ must be determined in order to learn whether A will choose $s_A = m$ and what will happen afterwards in case he does so. In the following we will determine a unique solution for the subgame following $s_A = m$ by applying the theory of equilibrium selection.

c) The solution of the subgame following $s_A = m$

Since the subgame following A's initial move $s_A = m$ is a 2×2-bimatrix game with two strict equilibrium points, both equilibrium selection theories (the one of HARSANYI and SELTEN, 1988, as well as the related ESBORA-concept of GÜTH and KALKOFEN, 1989) determine the same dominance relationship between the two strict equilibria. In their pioneering approach Harsanyi and Selten have shown that for the special class of 2×2-bimatrix games risk dominance between strict equilibria is uniquely implied by the three requirements: Invariance with respect to isomorphisms, best reply invariance and payoff monotonicity (HARSANYI and SELTEN, 1988, Chapter 3).

Two games are isomorphic if the rules of one game can be mapped into the rules of the other game by renaming players or strategies and by positive affine transformations of the players' utilities. A dominance relation between strict equilibria is invariant with respect to isomorphisms if it determines - up to ismorphisms - the same dominance relations for isomorphic games.

Best reply invariance requires that the dominance relations are not influenced by payoff transformations which do not change the best replies to the other player's strategies. A dominance relation is payoff monotonic if an increase of one player's equilibrium payoff will make this equilibrium dominant whenever it was not dominated before.

Since for the special class of 2×2-bimatrix games the same three axioms also characterize the resistance dominance relation between two strict equilibria (GÜTH and KALKOFEN, 1989, Part I) both equilibrium selection concepts determine the same dominance relationship between the two strict equilibria of the subgame following A's move $s_A = m$. In the following we will simply use the three requirements in order to determine whether $(\bar{n}_F^m,n_G^m)$ dominates $(n_F^m,\bar{n}_G^m)$ or vice versa.

	n_G^m	$\bar{n}_G^m$
n_F^m	0 0	$S_F^{-G}-pC-1$ $S_G^{-G}-S_G+p(1-q)C$
$\bar{n}_F^m$	$S_F^{-F}-S_F+pqC$ $S_G^{-F}-pC-1$	0 0

	n_G^m	$\bar{n}_G^m$
n_F^m	0 0	1 $\frac{S_G^{-G}-S_G+p(1-q)C}{S_G^{-F}-pC-1}$
$\bar{n}_F^m$	$\frac{S_F^{-F}-S_F+pqC}{S_F^{-G}-pC-1}$ 1	0 0

Table III.2: Auxiliary games to determine the dominance relation between $(\bar{n}_F^m, n_G^m)$ and $(n_F^m, \bar{n}_G^m)$; the upper game is obtained from the subgame after $s_A = m$ by using best reply invariance (for each pure strategy of opponent j one substracts i's payoff for his worse pure reply strategy from i's payoff for his best reply), the lower game is derived from the upper game by dividing each player i's payoff by his equilibrium payoff for n_i^m and $\bar{n}_j^m$, i.e. by using invariance with respect to isomorphisms.

The upper game of Table III.2 is obtained from the subgame after $s_A = m$ by subtracting for each row (or column) player G's (F's) non-equilibrium payoff from his equilibrium payoff for that row (or column). F's payoff $S_F^{-F} - S_F + pqC$, for instance, is nothing else than the difference of his equilibrium payoff S_F^{-F} for $(\bar{n}_F^m, n_G^m)$ and his non-equilibrium payoff S_F - pqC for (n_F^m, n_G^n) corresponding to the column n_G^m of the subgame following $s_A = m$, described in Table III.1. Since in 2×2-bimatrix games adding a constant to player i's payoff for a given strategy of his opponent does not change player i's best replies to his opponent's behavior (see HARSANYI and SELTEN,

1988, Chapter 3), we can conclude: For $i = 1, 2$ player i's best reply to any of his opponent's strategies in the upper game of Table III.2 is the same as the one in the game $s_A = m$ of Table III.1. This shows that the upper game of Table III.2 is obtained from the subgame after $s_A=m$ by using the requirement of best reply invariance.

The lower game of Table III.2 is derived from the upper game of Table III.2 by the following positive linear transformation of both utility functions: For $i = 1, 2$ player i's utility is divided by his positive equilibrium payoff for n_i^m and $\bar{n}_j^m$ $(j \neq i)$. Clearly, transforming the upper game into the lower game in Table III.2 is nothing else than using invariance with respect to isomorphisms. Thus we have shown: To determine the dominance relation between $(\bar{n}_F^m, n_G^m)$ and $(n_F^m, \bar{n}_G^m)$ for the subgame after $s_A = m$, one can use the lower game of Table III.2 whenever the dominance relation is best reply invariant as well as invariant with respect to isomorphisms.

Consider now the lower game of Table III.2 for the special case of equality in (III.3) which has been excluded. Since invariance with respect to isomorphisms implies symmetry invariance, none of the two strict equilibria can dominate the other. Therefore payoff monotonicity implies that $(\bar{n}_F^m, n_G^m)$ dominates $(n_F^m, \bar{n}_G^m)$ whenever

$$\frac{S_F^{-F} - S_F + pqC}{S_F^{-G} - pC - 1} > \frac{S_G^{-G} - S_G + p(1-q)C}{S_G^{-F} - pC - 1} \tag{III.9}$$

whereas in case of the opposite inequality

$$\frac{S_F^{-F} - S_F + pqC}{S_F^{-G} - pC - 1} < \frac{S_G^{-G} - S_G + p(1-q)C}{S_G^{-F} - pC - 1} \tag{III.10}$$

the dominance relation between the two strict equilibria is reversed. Since the lower game of Table III.2 is obtained from the subgame following $s_A = m$ by using best reply invariance and invariance with respect to isomorphisms, this proves the folowing result: According to any dominance relation, which satisfies invariance with respect to isomorphisms, best reply invariance and payoff monotonicity, $(\bar{n}_F^m, n_G^m)$ dominates $(n_F^m, \bar{n}_G^m)$ if (III.9) holds whereas $(n_F^m, \bar{n}_G^m)$ dominates $(\bar{n}_F^m, n_G^m)$ in case of (III.10).

It should be mentioned that according to the Harsanyi-Selten theory of equilibrium selection the dominance relation, implied by the three requirements, is only relevant if none of the two strict equilibria payoff dominates the other. Obviously, conditions (III.2.a) and (III.2.b) together imply that neither $(\bar{n}_F^m, n_G^m)$ payoff dominates $(n_F^m, \bar{n}_G^m)$ nor vice versa. For the ESBORA-concept of Güth and Kalkofen it is, furthermore, necessary that at least one of the two strict equilibria payoff dominates the mixed strategy equilibrium $q^m = (q_F^m, q_G^m)$ described in (III.4) and (III.5). Because of (III.2.a) the equilibrium payoff S_F^{-F} of F for $(\bar{n}_F^m, n_G^m)$ is F's highest payoff at all. Now, the condition that G's payoff for $(\bar{n}_F^m, n_G^m)$ is greater than the one for q^m is equivalent to

$$S_G - S_G^{-F} + pqC < 0 \tag{III.11}$$

and therefore not satisfied. From an analogous condition it follows that also $(n_F^m, \bar{n}_G^m)$ does not payoff dominate q^m. Thus the mixed strategy equilibrium point cannot be excluded purely by considerations of payoff dominance among solution candidates.

To prove that the ESBORA-solution is $(\bar{n}_F^m, n_G^m)$ in case of (III.9) and $(n_F^m, \bar{n}_G^m)$ in case of (III.10) one has to explore the resistance dominance relationships among all three perfect equilibrium points of the subgame following s_A = m. Since the mixed strategy equilibrium point is weak (i.e. not strict), one, furthermore, would have to determine the resistance dominance for the so-called perturbed games where every pure strategy must be chosen with a positive minimum probability and where payoffs are perturbed such that all perfect equilibria become strict (see GÜTH and KALKOFEN, 1989).

Here we do not want to focus our attention on situations where (q_F^m, q_G^m) is the ESBORA-solution of the game. After all it seems difficult to imagine that countries are gambling when determining whether to join an alliance or not. Even if it looks as if another country would be randomizing in order to determine its foreign policy, we prefer to capture this by introducing incomplete information about the other country's type. If the possible types behave differently, this will look as if the other country is using a mixed strategy as shown in Section II.

It has been shown (KALKOFEN, 1989) that q^m can never be the ESBORA-solution of the subgame after s_A = m, whenever one of the two strict equilibria resistance dominates the other. In case of (III.9) player A will obtain the payoff S_A^{-F} - (1-p)C whereas his payoff is S_A^{-G} - (1-p)C in case of (III.10). This shows how by applying equilibrium selection theory one can derive a unique payoff of A for his choice of s_A=m. Depending on whether s_A = m or s_A = h or s_A = l yields the highest payoff one can now easily determine the solution of the whole game of Figure III.1 and the play which it implies. If, for instance, condition (III.9) holds and if S_A^{-F} - (1-p)C is greater than 1 and greater than S_A-C, player A will choose s_A = m and players F and G the strategies $s_F = (\bar{n}_F^h, e_F^h, \bar{n}_F^m, e_F^m, n_F^l, e_F^l)$ and $s_G = (\bar{n}_G^h, e_G^h, n_G^m, e_G^m, n_G^l, e_G^l)$, respectively. The corresponding solution play is given by the move sequence m, $\bar{n}_F^m$, n_G^m implying the solution payoffs S_A^{-F}-(1-p)C, S_F^{-F}, and S_G^{-F}-pC for A, F, and G, respectively. This shows how our model offers ways to explain the actual political situation of the Western defense alliance.

Here we do not want to analyse the solution for all possible constellations of payoff parameters satisfying our restrictions (III.1) to (III.3). What we want to do instead is to analyse those conditions which have to be satisfied for the solution play $(m,\bar{n}_F^m,n_G^m)$ resembling the present political situation. The conditions for $(m,\bar{n}_F^m,n_G^m)$ to be the solution play are (III.9) and

$$S_A^{-F} - (1-p)C > \max \{1, S_A-C\} . \qquad \text{(III.12)}$$

Condition (III.12) simply expresses that North America A must prefer the alliance with G only to the purely European alliance and also to having to bear all the cost C of the overall alliance with members A, F, and G. According to (III.9) one has to compare the ratio

$$\frac{S_i^{-i} - S_i + p\Pi_i C}{S_i^{-j} - pC - 1} \qquad \text{(III.13)}$$

of payoff differences for i = F and i = G where $j \neq i$ and $\Pi_F = q$ and Π_G = 1-q. The denominator of (III.13) can be described as i's incentive to form an alliance with A instead of one with j. The numerator is player i's incentive to leave the overall alliance, it could be called the outsider advantage. Condition (III.9) requires that the ratio (III.13) is greater for i = F than for i = G. One way to explain (III.9) is by arguing that 1 - q is much smaller than q, i.e. if F(rance) would have to pay a much higher share than G(ermany) in the overall alliance. In our view, such a biased cost sharing parameter does not appear very convincing. Thus our explanation of (III.9) has to rely on arguments that either S_G^{-F} is greater than S_F^{-G} or that $S_G - S_G^{-G}$ is greater than $S_F - S_F^{-F}$. Since due to its geographical location G(ermany) is much more endangered by the threat of an Eastern military invasion, the assumption $S_G^{-F} > S_F^{-G}$ is very convincing. American protection is obviously much more vital for G than for F. The other possibility to justify (III.9), namely by

$$S_G - S_G^{-G} > S_F - S_F^{-F} , \qquad \text{(III.14)}$$

is also convincing since, due to the reasons mentioned above, one will naturally assume that $S_G > S_F$ and $S_G^{-G} < S_F^{-F}$ which implies (III.14). If our model captures the essential aspects of forming a Western defense alliance, one thus should have expected that only G(ermany) is willing to cooperate with A in forming a Western defense alliance to balance the military threat of the East.

d) <u>Endogenous determination of the cost sharing parameter</u>

An obvious way to generalize the game is to endogenize p by substituting A's initial move set {h, m, l} by the continuum [0,1] of moves $s_A = p$ satisfying $0 \leq p \leq 1$, i.e. instead of only three cost sharing parameters all possible ways to share the cost C among North America and its European allies can be proposed. The payoff function of the new game would be determined as for $s_A = m$ with the only difference that now p is not exogeneously given but player A's initial move.

Let $\underline{p}$ be the highest cost sharing parameter $p \in [0,1]$ such that for all $p < \underline{p}$ both European countries will join A according to the solution of the game. Similarly, $\bar{p}$ denotes the highest cost sharing parameter such that for all $p < \bar{p}$ at least one European country will join A if both European countries behave as prescribed by the solution. For the sake of simplicity let us assume that for $p = \underline{p}$ both European countries enter the alliance and that at least one of them does so in case of $p = \bar{p}$.

Clearly, A will choose either $s_A = \underline{p}$ to ensure the overall alliance with lowest cost, or $s_A = \bar{p}$ to guarantee American protection for Western Europe at lowest cost, or a prohibitively high cost parameter p inducing the purely European alliance. Clearly, the decision will be $s_A = \underline{p}$ if $S_A - S_A^{-i}$ for i = F or i = G is greater than $(\bar{p}-\underline{p})C$ and if $S_A - (1-\underline{p})C$ is greater than 1. For $s_A = \bar{p}$ to be optimal one must have that $(\bar{p}-\underline{p})C$ exceeds $S_A - S_A^{-i}$ and that $S_A^{-i} - (1-\bar{p})C$ is greater than 1. If neither $s_A = \bar{p}$ nor $s_A = \underline{p}$ is optimal, player A obviously prefers the purely European alliance to having himself to pay for American protection of Western Europe.

This shows that one can easily include all possible ways of sharing the cost of American protection for Western Europe. Here we have not started with the analysis of the more general game model since the graphical representation of extensive form games requires finite move sets.

IV Conclusions

The strategic problems of forming a Western defense alliance including North America and Western Europe have been studied in a rather abstract way by assuming just one North American country and at most two Western European countries.

The first model concentrates on the uncertainty in judging the true security interests of Western Europe. Player A's incomplete information about Western Europe's true type can, for instance, be explained by the uncertain results of democratic elections in Western European countries where often enough the competing parties have very different views how national security should and can be achieved. We have distinguished two possible types F and G of Western Europe resembling French and German international defense policy. Of course, for a given Western European country there may be more than just two competing ideas of how to achieve national security, e.g. the idea that neutrality is advantageous. But, in our view, including more types of Western Europe would have complicated the game without adding much to the conclusions of strategic analysis.

An advantage of modelling conflicting views about international policy as games with incomplete information is that it helps to yield very new explanations of why different countries choose different policies or why different parties of the same country follow different strategies. So the 'Incomplete Information about Western Europe's Security Interests'-game specifies parameter conditions for which both, French and German defense policy, are in equilibrium. The parameters include those reflecting probabilistic views about other countries' political behavior as well as those reflecting the political evaluation of certain international military constellations, of resource implications or national autonomy.

Whereas F's policy can be described as one of high respect for national autonomy, G is more afraid that he himself may not be strong enough to prevent the East from any attempt to change the territorial constellation in Europe. A's political problem is that he would like to cooperate with Western European countries in order to counter the military threat of the East but that he must fear that Western Europe prefers to remain independent since any Western defense alliance including A will be dominated by the United States of America. Our model allows us to specify exact parameter conditions for situations where A will try to form an alliance. By comparing the actual political results with the parameter conditions for such a political constellation one can derive important conclusions about political preferences. It is in this sense that strategic analysis of international political problems can help to reveal political preferences.

The 'Allocating the Cost for American Protection of Western Europe'-game neglects incomplete information about other countries' political interests but tries to capture the different security needs of different Western countries. Of course, one might prefer to analyse a model which includes both aspects. Here this has not been done since we wanted to have models which can still be graphically illustrated and whose analytical tractability is not too demanding. The most simple model with two Western European countries which both can be of two possible types would have required already a very complex system of rules. Furthermore, the solution of such a game will often enough rely on a complicated case distinction which is difficult to interprete politically.

If American protection of Western Europe is not provided as a free good, Western European countries might want to freeride on American protection if an alliance of A with other Western European countries is sufficient to guarantee the territorial status quo in Europe. In our view, this public good property of American protection explains why so many countries in Central Europe are not members of NATO although their territorial integrity is as endangered by the East as those of countries which have joined NATO.

Another aspect of the 'Allocating the Cost for American Protection of Western Europe'-game is that A, when trying to reduce his own cost for granting Western Europe's security, must fear to find no ally at all or to split up Western Europe, as it is presently the case, in NATO-member countries and in non-NATO-member countries. By comparing the actual political situation with the corresponding parameter conditions for such a constellation to be the solution of the game we are again able to find important parameter restrictions. Thus both models demonstrate how strategic analysis can be used to reveal political preferences.

Another motivation of our study is to illustrate the use of some new methodological tools of modern game theory. Although we do not know of many applications of games with incomplete information to the political sciences, many political scientists are probably aware of this important development. We hope that our study will demonstrate that one can include incomplete information without loosing analytic tractability. The other methodological message is to use equilibrium selection theory whenever one cannot derive unambiguous results by just requiring the equilibrium property. Both models illustrate that one can resolve strategic uncertainty completely in situations where one cannot make definite conclusions by just demanding that the solution is an equilibrium point. In our view, some of the most obvious game models of international political problems will have many equilibrium points with conflicting political conclusions (see, for instance, the game models with multiple equilibria studied in BRAMS, 1985). Equilibrium selection theory allows to study these game models without giving up the attempt to derive unambiguous results. If one does not want to apply equilibrium selection theory, one will have to restrict oneself to very special subclasses of games instead of analysing the most natural game models for the situation at hand.

References

Brams, Steven J. (1985): Superpower Games - Applying Game Theory to Superpower Conflicts -, Yale University Press, New Haven - London.

Güth, Werner and Brigitte Kalkofen (1989): Unique Solutions for Strategic Games - Equilibrium Selection Based on Resistance Avoidance -, Lecture Notes in Economics and Mathematical Systems 328, Springer, Berlin Heidelberg.

Harsanyi, John C. (1967/68): Games with Incomplete Information played by 'Bayesian' Players, Parts I-III, Management Science 14, 159-182, 320-334 and 486-502.

Harsanyi, John C. and Reinhard Selten (1988): A General Theory of Equilibrium Selection in Games, M.I.T. Press, Cambridge Mass.

Kalai, Ehud and Dov Samet (1984): Persistent Equilibria in Strategic Games, International Journal of Game Theory 13, 129-144.

Kalkofen, Brigitte (1989): Gleichgewichtsauswahl in strategischen Spielen - Theorie und Anwendungen, University of Frankfurt/Main.

Kohlberg, Elon and Jean-Francois Mertens (1986): On the Strategic Stability of Equilibria, Econometrica 54, 1003-1037.

Kreps, David and Robert Wilson (1982): Sequential Equilibria, Econometrica 50, 863-894.

Myerson, Robert B. (1978): Refinement of the Nash Equilibrium Concept, International Journal of Game Theory 7, 73-80.

Selten, Reinhard (1975): Reexamination of the Perfectness Concept for Equilibrium Points in Extensive Games, International Journal of Game Theory 4, 25 55.

van Damme, Eric E. C. (1983): Refinements of the Nash Equilibrium Concept, Lecture Notes in Economics and Mathematical Systems 219, Springer, Berlin Heidelberg.

IMPLICATIONS OF THE DEFENSE EFFICIENCY HYPOTHESIS FOR THE CHOICE OF MILITARY FORCE STRUCTURES. PART I: GAMES WITH AND WITHOUT COMPLETE INFORMATION ABOUT THE ANTAGONIST'S INTENTIONS[1]

by

Rudolf Avenhaus, Werner Güth, and Reiner K. Huber

Abstract: The so-called Defense Efficiency Hypothesis (DEH) says that conventional forces designed specifically for defensive operations and, thus incapable of any major offensive operations can exploit the intrinsic defense advantage more efficiently than forces designed to operate in all combat modes including offensive operations. If this were true, conventional stability between two antagonistic parties could be brought about by a mutual adoption of defensive force structures.

However, with a view to the uncertainty about the true strategic intentions of the opponent, the soundness of such a force conversion policy is frequently called into question. For this reason, an attempt is made to shed some light on the issue by studying a simple game-theoretic model of defense policy interactions between two antagonistic parties. In particular, the equilibria are determined in a series of games in which the antagonistic parties are free to choose between offensive and defensive force structures given the DEH were true.

First, it is assumed that both parties know each other's intentions - expressed by the payoffs of the antagonists. Thereafter, it is assumed that one party knows about the other's intentions, but not vice versa, and finally that both parties only have subjective probabilities about the other party's intentions.

1. Defense Efficiency Hypothesis

At present, heavy armor and its associated combat support equipment are considered to be the quintessential tools of modern conventional forces. However, since armor is the backbone of offensive operations as well, it follows that a so-called "worst case" military assessment would very likely advise the leadership of two antagonistic parties to consider the opponent's military plans as offensive, irrespective of what his true strategic objectives and political intentions may be. In a crisis, such perceptions could become a powerful motor of instability because either side might be tempted to preempt an expected attack by the other. Short of more or less complete disarmament, mutual arms reductions may even exacerbate this instability problem. This is because, if surprised, a defense thinned by the negotiated force reductions would face considerably more unfavorable local force ratios at the points where the

[1]The two parts deal with related, but structurally different game models and can be read independently.

attacker pursues his main thrusts. In addition, the reductions would deprive the attacker of some of his operational reserves. Thus, he must be expected to rely even more on a quick initial success by means of surprise.

Thus, with regard to crisis stability there is clearly a dilemma that cannot be solved by the classical approach to conventional arms control as, for example, practiced in the Mutual Balanced Force Reductions (MBFR)-talks between NATO and the Warsaw Pact (WP). They have been aimed at symmetrical reductions of troop strengths that were to leave the basic force structures essentially intact. For this reason, several experts have argued that, before sizeable reductions are implemented, the structures of the conventional forces need to be changed in a manner that they may not be perceived by either opponent as being capable of major offensive operations. Most of the respective proposals recommend to limit force mobility and replace armor by anti-armor systems of some kind. Indeed, if both sides were incapable of offensive operations, the military situation between them would obviously be highly stable.

However, based on a conceptual transarmament model HUBER (1987) has shown that there is no possibility for a transition to that ultra-stable situation without a temporary deterioration of crisis stability unless the Defense Efficiency Hypothesis (DEH) proposed by HUBER and HOFMANN (1984) is true. That hypothesis says that, in a conventional war, "reactive" military forces designed specifically for defensive operations and incapable of major offensive actions can exploit the intrinsic defense advantage[2] more efficiently, i.e., at higher cost-effectiveness, than the traditional "active" forces that are designed to operate in all combat modes.

Thus, if the Defense Efficiency Hypothesis were true, it is obvious that some degree of replacement of active by reactive units would improve the defense potential of a force while simultaneously reducing its offensive capabilities. For a given operational depth available for defensive operations, the optimum degree of replacement that maximizes the defense potential would depend, among other things, primarily on the configuration of the reactive force elements and on the degree to which the opponent has replaced active by reactive units in his forces.

Whether and under which circumstances the Defense Efficiency Hypothesis might hold has not yet been convincingly demonstrated.[3] But assuming it were true, there remains the question whether the adoption of a largely reactive, i.e., defensive force structure would be a rational policy, especially with a view to the

---

[2]The fact that a well organized defense has an inherent advantage over the attacker has been substantiated by many a historical battle. It has led VON CLAUSEWITZ to state *"that the defensive form of warfare is intrinsically stronger than the offensive"* (On War, Book Six, Chapter 2). See also Quester (1977).

[3]A major part of the research at the institute of the first and third named authors is devoted to testing the DEH by means of combat simulation experiments (see HUBER (1987) and HOFMANN, HUBER, STEIGER (1986)).

uncertainties about the opponent's true political and strategic intentions. In order to somewhat illuminate that issue, an attempt is made to determine equilibrium[4] points in a series of games between two antagonistic parties which are free to choose between offensive and defensive structures and the strategic intentions of which may be either defensive or offensive.

In this first part of the paper we consider situations of independent choice, i.e., both parties are assumed to decide on their force structures simultaneously. The first case is characterized by both parties being fully informed about each other's strategic situations which are expressed in terms of the payoff-functions of the game. Thereafter, the case is considered where one party knows about the intentions of the other, but not vice versa, and finally that both parties are uncertain about the opponent's intentions, but have subjective probabilistic beliefs about them.

In the first case, explicit use will be made of payoffs and probabilities for successful attack and defense. It will be shown which equilibria result if the Defense Efficiency Hypothesis were true. In the two cases with *incomplete information* only expected payoffs will be considered. With the help of the result obtained for the first case, the equilibria that follow from the Defense Efficiency Hypothesis are determined.

In a second part the analysis will be extended to address situations in which both antagonists had previously chosen offensive force structures and, after learning that this has led to a highly unstable military constellation rather than to a situation in which they may coerce their opponent, may consider to restructure their forces from offensive to defensive ones. This is done by formulating the choice situation in terms of a fictituous multi-stage game.

2. Completely Informed Players

In this chapter we describe the problem of two antagonistic states choosing one of two alternative force structures with the help of a set of four noncooperative 2×2-two-person games. We assume complete information on both sides and determine among others, those equilibrium points which exist as a consequence of the Defense Efficiency Hypothesis. In case of multiple equilibria, we apply the equilibrium selection theory by Harsanyi and Selten (1988).

---

[4]It should be noted that, in the game-theoretic context of this paper, the term equilibrium does not necessarily imply a numerical force balance. Rather, it refers to a situation that represents a rational choice for both sides in the sense that any other individual decision would result in less favorable outcomes for the deviator.

2.1. Definition of the Games

Given the two antagonistic parties X and Y may choose between the two alternative force structures o and v, where

o is the attack oriented (offensive) force structure, and
v is the defensive one.

Let P be the probability that the attacking party is successful, and W the probability that the defending party is successful. The probabilities P_{ij}^X, W_{ij}^X, P_{ij}^Y, W_{ij}^Y are defined as given in Table 1. The upper index refers to the State whose probabilities for successful attack or defense one is considering. The lower index combination ij indicates the strategy constellation $s = (s_X, s_Y)$ with $(s_X, s_Y) \in \{o,v\}$ to which the respective probability refers. Note that in case $(s_X, s_Y) = (v,v)$ war cannot occur.

With respect to these probabilities we make the following

Assumption 1

1.1 For (o,o) we have

$$P_{oo}^X = W_{oo}^X, \quad P_{oo}^Y = W_{oo}^Y, \tag{2-1a}$$

i.e., there is no advantage for one side to attack first ('crisis stability'). Furthermore, we have

$$P_{oo}^X + P_{oo}^Y = W_{oo}^X + W_{oo}^Y = 1, \tag{2-1b}$$

which means that in case of war one party will win.

X\Y	o	v
o	P_{oo}^X, W_{oo}^X P_{oo}^Y, W_{oo}^Y	P_{ov}^X — — W_{ov}^Y
v	— W_{vo}^X P_{vo}^Y —	—

Table 1: Probabilities for successful attack (P) and defense (W).

1.2 For (o,v) and (v,o) we have

$$P^Y_{vo} + W^X_{vo} = 1, \quad P^X_{ov} + W^Y_{ov} = 1, \tag{2-2}$$

which means again that in case of war one party will win.

1.3 The Defense Efficiency Hypothesis (DEH) is given by

$$W^X_{oo} < W^X_{vo}, \quad W^Y_{oo} < W^Y_{ov}, \tag{2-3}$$

when we assume, as mentioned in the introduction, that the total effort spent for military forces is kept at the same level by both parties.

As can be taken from Table 1, we use the probability P only for those players who have an offensive force structure, whereas we use the probability W in those cases where the antagonist has such a structure. This means that we assume that a party with a defense oriented force structure cannot attack.

Let the payoffs to the two players X and Y under the condition that war will occur, and in case of the two strategies $(i,j) \in \{o,v\}$ – $(i,j) \neq (v,v)$ since in this case war cannot occur – be given

$$\begin{array}{llll} \tilde{a}_{ij} & & \text{victory} & \\ -\tilde{b}_{ij} & \text{for} & \text{defeat} & \text{for } i,j = v,o \\ 0 & & \text{no war} & \end{array} \tag{2-4a}$$

and for Y by

$$\begin{array}{llll} \tilde{c}_{ij} & & \text{victory} & \\ -\tilde{d}_{ij} & \text{for} & \text{defeat} & \text{for } i,j = o,v \\ 0 & & \text{no war.} & \end{array} \tag{2-4b}$$

Furthermore, let for $i,j = o,v$ be p_{ij} the probabilities that war will occur. Since we assumed already implicitly $p_{vv} = 0$, we obtain the expected payoffs to both players as given in Table 2.

If we now define the (conditional) expected payoffs

$$\tilde{a}_{ij}p_{ij} =: a_{ij}, \; \tilde{b}_{ij}p_{ij} =: b_{ij}, \; \tilde{c}_{ij}p_{ij} =: c_{ij}, \; \tilde{d}_{ij}p_{ij} =: d_{ij} \tag{2-4}$$

for $i,j = o,v$, then we may write the expected payoffs to the two players as given in Table 3.

X\Y	o	v
o	$[(\tilde{a}_{oo}+\tilde{b}_{oo})\cdot p^X_{oo}-\tilde{b}_{oo}]\cdot p_{oo}$ $[(\tilde{c}_{oo}+\tilde{d}_{oo})\cdot p^Y_{oo}-\tilde{d}_{oo}]\cdot p_{oo}$	$[(\tilde{a}_{ov}+\tilde{b}_{ov})\cdot p^X_{ov}-\tilde{b}_{ov}]\cdot p_{ov}$ $[(\tilde{c}_{ov}+\tilde{d}_{ov})\cdot p^Y_{ov}-\tilde{d}_{ov}]\cdot p_{ov}$
v	$[(\tilde{a}_{vo}+\tilde{b}_{vo})\cdot p^X_{vo}-\tilde{b}_{vo}]\cdot p_{vo}$ $[(\tilde{c}_{vo}+\tilde{d}_{vo})\cdot p^Y_{vo}-\tilde{d}_{vo}]\cdot p_{vo}$	0 0

Table 2: Two-person game in normal form.

X\Y	o	v
o	$(a_{oo}+b_{oo})\cdot p^X_{oo}-b_{oo}$ $(c_{oo}+d_{oo})\cdot p^Y_{oo}-d_{oo}$	$(a_{ov}+b_{ov})\cdot p^X_{ov}-b_{ov}$ $(c_{ov}+d_{ov})\cdot p^Y_{ov}-d_{ov}$
v	$(a_{vo}+b_{vo})\cdot p^X_{vo}-b_{vo}$ $(c_{vo}+d_{vo})\cdot p^Y_{vo}-d_{vo}$	0 0

Table 3: Two-person game in normal form with redefined payoffs.

Now, we distinguish two types of players, namely a peaceful (f-)type and an imperialistic (i-)type of party. In general, a type is a complete description of a player including his strategic possibilities, his information conditions, his evaluation of results, his beliefs about the corresponding characteristics of other players including their beliefs about his own characteristics, etc. As shown by Harsanyi (1967/68) all these characteristics can be adequately represented by the payoff function of this player. Thus the possible types of a player can be simply defined as the possible payoff functions of this player as expected by any of his opponents.

Assumption 2

2.1 For both types of both players we assume the losses in case of defeat to be positive;

$$b_{ij} > 0 \quad \text{and} \quad d_{ij} > 0 \quad \text{for} \quad i,j = o,v. \tag{2-5a}$$

For the i-type State we assume the gains in case of victory to be positive,

$$a_{ij} > 0 \quad \text{and} \quad c_{ij} > 0 \quad \text{for} \quad i,j = o,v, \tag{2-5b}$$

whereas for the f-type State we assume these gains to be negative,

$$a_{ij} < 0 \quad \text{and} \quad c_{ij} < 0 \quad \text{for} \quad i,j = o,v. \tag{2-5c}$$

2.2 For an i-type State X we assume the losses in case of defeat to be larger than the gains in case of victory,

$$b_{oo} > a_{oo}, \; b_{ov} > a_{ov}, \; b_{vo} > a_{vo} \tag{2-5d}$$

and furthermore,

$$b_{vo} > a_{oo}, \; a_{oo} > a_{vo}, \; b_{oo} > b_{vo} \; . \tag{2-5e}$$

The analogous assumptions are made for an i-type State Y.

For an f-type State X we assume the losses in case of defeat to be larger than those in case of victory,

$$b_{oo} > -a_{oo}, \; b_{ov} > -a_{ov}, \; b_{vo} > -a_{vo} \; , \tag{2-5f}$$

and furthermore,

$$b_{vo} > -a_{oo}, \; -a_{oo} > -a_{vo}, \; b_{oo} > b_{vo} \; . \tag{2-5g}$$

The analogous assumptions are made for an f-type State Y.

From the Assumptions 2.2 we get the following partial ordering of the payoff parameters:

$$b_{oo} > b_{vo} > a_{oo} > a_{vo} > 0 \quad \text{for i-type } X$$

$$d_{oo} > d_{ov} > c_{oo} > c_{ov} > 0 \quad \text{for i-type } Y$$

$$b_{oo} > b_{vo} > -a_{oo} > -a_{vo} > 0 \quad \text{for f-type } X$$

$$d_{oo} > d_{ov} > -c_{oo} > -c_{ov} > 0 \quad \text{for f-type } Y. \tag{2-5h}$$

1. Remark

The very plausible assumption $b_{vo} > a_{oo}$ permits to establish a complete order of the parameters b_{oo}, b_{vo}, a_{oo}, a_{vo}; however, it is not used in the subsequent analysis. The same is true for $b_{ov} > a_{ov}$.

2. Remark

The same letters have been used to designate the payoffs of both types of X and of Y. However, this does not mean that the values as such are the same for both types; it also does not mean that the payoffs are the same for both types of the antagonistic parties. On the contrary, if e.g. one party attacks and is defeated, then its payoff in case of defeat will strongly depend on the type of the antagonist.

For later considerations we define

$$R_X := \left[\frac{a_{vo}+b_{vo}}{a_{oo}+b_{oo}} - 1 \right] \cdot W_{vo}^X + \frac{b_{oo}-b_{vo}}{a_{oo}+b_{oo}} \tag{2-6a}$$

$$R_Y := \left[\frac{c_{ov}+d_{ov}}{c_{oo}+d_{oo}} - 1 \right] \cdot W_{ov}^Y + \frac{d_{oo}-d_{ov}}{c_{oo}+d_{oo}} \ . \tag{2-6b}$$

Then we have

Proposition 1

If both, X and Y, are f-types, then we have $R_X > 0$ and $R_Y > 0$.

Proof (for X only)

$R_X > 0$ is equivalent to

$$W_{vo}^X \cdot (a_{vo} + b_{vo} - a_{oo} - b_{oo}) > b_{vo} - b_{oo} \ .$$

Since the right-hand side of this inequality is smaller than zero, it is true for $a_{vo}+b_{vo}-a_{oo}-b_{oo} > 0$. If, however, $a_{vo}+b_{vo}-a_{oo}-b_{oo} < 0$, the assumption $R_X > 0$ is equivalent to

$$W^X_{vo} < \frac{b_{oo}-b_{vo}}{a_{oo}+b_{oo}-a_{vo}-b_{vo}} .$$

From (2-5h) follows that the right-hand side is greater one. Thus, this inequality is always true.

In the following, we consider four different games that correspond to the four different type constellations $(X_f,Y_f),\ldots,(X_i,Y_i)$ of players X and Y.

2.2 Solution of the Game (f,f)

A strategy B_i is called dominant if it yields a higher payoff for player i than any of his other strategies regardless of which strategies are chosen by his opponents. If a player has a dominant strategy he obviously does not have to enter into complicated strategy considerations since his best reply is independent on what the opponents do. Observe that every equilibrium point must prescribe the choice of dominant strategies if they exist. The choice of dominant strategies is obviously the most non-controversial rationality requirement since it is implied by payoff maximization. Unfortunately, in most games players do not have dominant strategies.

In the game considered here, the Defense Efficiency Hypothesis implies dominant strategies.

Proposition 2

Given the game (f,f). Then

(1) (v,v) is a strict equilibrium point.

(2) (o,o) is an equilibrium point $\Rightarrow$ DEH does not hold.

(3) DEH $\Rightarrow$ (v,v) is vector of dominant strategies.

Proof

Since the payoffs to X and Y are negative except for the strategy vector (v,v), the first assertion is fulfilled.

With (2-6), (o,o) is an equilibrium point iff

$$W^X_{oo} - W^X_{vo} > R_X, \quad W^Y_{oo} - W^Y_{ov} > R_Y .$$

Therefore, with Proposition 1 we are led to the second assertion.

(v,v) is vector of dominant strategies iff

$$(a_{vo}+b_{vo}) \cdot W^X_{vo} - b_{vo} - (a_{oo}+b_{oo}) \cdot W^X_{oo} + b_{oo} > 0$$

$$(c_{ov}+d_{ov}) \cdot W^{Y}_{ov} - d_{ov} - (c_{oo}+d_{oo}) \cdot W^{Y}_{oo} + d_{oo} > 0$$

which is equivalent to

$$R_X > W^X_{oo} - W^X_{vo}$$

$$R_Y > W^Y_{oo} - W^Y_{ov}$$

therefore, with Proposition 1 we are led to the third assertion.

2.3 <u>Solution of Game (f,i)</u>

From Assumptions (2-5a) and (2-5b) it follows immediately

$$(o,v) \quad \text{is no equilibrium point.}$$

Furthermore, (v,v) is an equilibrium point iff

$$P^Y_{vo} < \frac{d_{vo}}{c_{vo}+d_{vo}} \quad ,$$

or, with (2-2) iff

$$W^X_{vo} > \frac{1}{1 + \dfrac{d_{vo}}{c_{vo}}} \quad .$$

(o,o) is an equilibrium point iff

$$(a_{oo}+b_{oo}) \cdot P^X_{oo} - b_{oo} > (a_{vo}+b_{vo}) \cdot W^X_{vo} - b_{vo}$$

$$(c_{oo}+d_{oo}) \cdot P^Y_{oo} - d_{oo} > (c_{ov}+d_{ov}) \cdot W^Y_{ov} - d_{ov} \; ,$$

or, iff, with (2-6)

$$W^X_{oo} - W^X_{ov} > R_X$$

$$W^Y_{oo} - W^Y_{vo} > R_Y.$$

Thus, because of Proposition 1, we get

$$R_Y > 0 \Rightarrow (\text{DEH holds} \Rightarrow (o,o) \text{ is no equilibrium point}).$$

(v,o) is an equilibrium point iff

$$(a_{oo}+b_{oo}) \cdot W^X_{oo}-b_{oo} < (a_{vo}+b_{vo}) \cdot W^X_{vo}-b_{vo}$$

$$(c_{vo}+d_{vo}) \cdot P^Y_{vo}-d_{vo} > 0 ,$$

or, with (2-2), iff

$$W^X_{oo} < \frac{a_{vo}+b_{vo}}{a_{oo}+b_{oo}} \cdot W^X_{vo} + \frac{b_{oo}-b_{vo}}{a_{oo}+b_{oo}}$$

$$W^X_{vo} < \frac{1}{1+\frac{d_{vo}}{c_{vo}}} .$$

Because of (2-6a) the first inequality is equivalent to

$$W^X_{oo} - W^X_{vo} < R_X$$

which, because of Proposition 1, is fulfilled, if the DEH holds. Thus we get

<u>Proposition 3</u>

The DEH implies that the only equilibrium point of the game (f,i) is

$$\left.\begin{matrix} (v,o) \\ (v,v) \end{matrix}\right\} \text{ iff } W^X_{vo} \quad \left\{\begin{matrix} < \\ > \end{matrix}\right\} \frac{1}{1+\frac{d_{vo}}{c_{vo}}} .$$

In the latter case, the strategy v of player X is dominant.

<u>Remark</u>

From the two conditions that must be satisfied for (v,o) to be an equilibrium point we get

$$W^X_{oo} < \frac{a_{vo}+b_{vo}}{a_{oo}+b_{oo}} \cdot \frac{1}{1+\frac{d_{vo}}{c_{vo}}} + \frac{b_{oo}-b_{vo}}{a_{vo}+b_{oo}} ,$$

which is only non-trivial ($W^X_{oo} < 1$) if

$$\frac{a_{vo}+b_{vo}}{a_{oo}+b_{vo}} < 1 + \frac{d_{vo}}{a_{ov}} .$$

Obviously, for the game (i,f) the analogous results can be derived (see the summary given in Section 2.5).

2.4 <u>The Game (i,i)</u>

First, let us assume (v,v) to be an equilibrium point. This is equivalent to

$$W^X_{ov} > \frac{1}{1 + \frac{b_{ov}}{a_{ov}}}$$

$$W^Y_{vo} > \frac{1}{1 + \frac{d_{vo}}{c_{vo}}} .$$

Since we assumed $b_{ov} > a_{ov}$, $d_{vo} > c_{vo}$, this implies

$$(W^X_{vo}, W^Y_{ov}) > (\tfrac{1}{2}, \tfrac{1}{2})$$

thus, for high probabilities for successful defense, even imperialistic States choose defense-oriented military structures.

Let us assume next (o,o) to be an equilibrium point. This is, with (2-6), equivalent to

$$R_X < W^X_{oo} - W^X_{vo}$$

$$R_Y < W^Y_{oo} - W^Y_{ov} .$$

Thus, we see

$$(R_X, R_Y) > (o,o) \Rightarrow ((o,o) \text{ equilibrium point} \Rightarrow \text{DEH not true}),$$

or, more important for applications,

$$(R_X, R_Y) > (0,0) \Rightarrow (\text{DEH true} \Rightarrow (o,o) \text{ no equilibrium point}).$$

<u>1. Remark</u>

From $R_X > 0$ and DEH true we get

$$W_{vo}^{X} < \frac{1}{1+\frac{a_{oo}+a_{vo}}{b_{oo}+b_{vo}}},$$

which means that (o,o) and (v,v) are equilibrium points if

$$\frac{1}{1+\frac{d_{vo}}{c_{vo}}} < W_{vo}^{X} < \frac{1}{1+\frac{a_{oo}-a_{vo}}{b_{oo}-b_{vo}}}$$

which implies

$$\frac{a_{oo}-b_{vo}}{b_{oo}-b_{vo}} > \frac{d_{vo}}{c_{vo}}.$$

Furthermore, we get

$$W_{oo}^{X} > \frac{a_{vo}+b_{vo}}{a_{oo}+b_{oo}} \cdot \frac{1}{1+\frac{d_{vo}}{c_{vo}}} + \frac{b_{oo}-b_{vo}}{a_{oo}-b_{oo}};$$

similar conditions hold for Y.

2. Remark

(o,o) is for both players preferable to (v,v) if

$$W_{oo}^{X} > \frac{b_{oo}}{a_{oo}+b_{oo}}, \quad W_{oo}^{Y} > \frac{d_{oo}}{c_{oo}+d_{oo}},$$

which implies

$$W_{oo}^{X} > \max\left[\frac{b_{oo}}{a_{oo}+b_{oo}}, \frac{a_{vo}+b_{vo}}{a_{oo}+b_{oo}} \cdot \frac{1}{1+\frac{d_{vo}}{c_{vo}}} + \frac{b_{oo}-b_{vo}}{a_{oo}-b_{oo}}\right];$$

similar conditions hold for Y.

Finally, we assume (o,v) to be in equilibrium. This is equivalent to

$$R_Y > W_{oo}^{Y} - W_{ov}^{Y}$$

$$W^Y_{ov} < \frac{1}{1 + \frac{b_{ov}}{a_{ov}}} \quad .$$

Analogous conditions must be satisfied for (v,o) to be an equilibrium point. Thus, with a view to the implications of the DEH, we may formulate

<u>Proposition 4</u>

Given the game (i,i). In case the DEH holds we have

$$\left[R_X > 0, \quad W^Y_{ov} < \frac{1}{1 + \frac{b_{ov}}{a_{ov}}} \right] \Rightarrow \text{(o,v) equilibrium point}$$

$$\left[R_Y > o, \quad W^X_{vo} < \frac{1}{1 + \frac{d_{vo}}{c_{vo}}} \right] \Rightarrow \text{(v,o) equilibrium point.}$$

<u>3. Remark</u>

As already mentioned, $R_X > 0$ and DEH true implies

$$W^X_{vo} < \frac{1}{1 + \frac{a_{oo}-a_{vo}}{b_{oo}-b_{vv}}}$$

which means that the assumptions for (o,v) to be an equilibrium point imply that

$$W^X_{vo} < \min\left[\frac{1}{1 + \frac{a_{oo}-a_{vo}}{b_{oo}-b_{vo}}} \, , \, \frac{1}{1 + \frac{d_{vo}}{c_{vo}}} \right]$$

or, equivalently,

$$W^X_{vo} < \frac{1}{1 + \max\left[\frac{a_{oo}-a_{vo}}{b_{oo}-b_{vo}}, \frac{d_{vo}}{c_{vo}}\right]}$$

which leads to

$$W^X_{oo} < \frac{a_{vo}+b_{vo}}{a_{oo}+b_{oo}} \cdot \frac{1}{1 + \max\left[\frac{a_{oo}-a_{vo}}{b_{oo}-b_{vo}}, \frac{d_{vo}}{c_{vo}}\right]} \quad .$$

Similar conditions hold for Y.

X\Y	f	i
f	1) (v,v) is EP ($\Rightarrow$ (o,v) & (v,o) no EP's) 2) DEH $\Rightarrow$ (o,o) no EP 3) DEH $\Rightarrow$ (v,v) is vector of dominant strategies	1) (o,v) no EP 2) $W^X_{vo} > \frac{1}{1+\frac{d_{vo}}{c_{vo}}}$ $\Rightarrow$ (v,v) is EP 3) $W^X_{vo} < \frac{1}{1+\frac{d_{vo}}{c_{vo}}}$, DEH for X $\Rightarrow$ (v,o) is EP 4) $W^X_{vo} > \frac{1}{1+\frac{d_{vo}}{c_{vo}}}$, DEH for X $\Rightarrow$ v is dominant for X
i	1) (v,o) no EP 2) $W^Y_{ov} > \frac{1}{1+\frac{b_{ov}}{a_{ov}}}$ $\Rightarrow$ (v,v) is EP 3) $W^Y_{ov} < \frac{1}{1+\frac{b_{ov}}{a_{ov}}}$, DEH for Y $\Rightarrow$ (o,v) is EP 4) $W^Y_{ov} > \frac{1}{1+\frac{b_{ov}}{a_{ov}}}$, DEH for Y $\Rightarrow$ v is dominant for Y	1) $W^X_{vo} > \frac{1}{1+\frac{d_{vo}}{c_{vo}}}$, $W^Y_{ov} > \frac{1}{1+\frac{b_{vo}}{a_{ov}}}$ $\Rightarrow$ (v,v) is EP 2) $W^X_{vo} < \frac{1}{1+\frac{a_{oo}-a_{vo}}{b_{oo}-b_{vo}}}$, $W^Y_{ov} < \frac{1}{1+\frac{c_{oo}-c_{ov}}{d_{oo}-c_{ov}}}$ DEH $\Rightarrow$ (o,o) no EP 3) $W^X_{vo} < \frac{1}{1+\max\left[\frac{d_{vo}}{c_{vo}},\frac{a_{oo}-a_{vo}}{b_{oo}-b_{vo}}\right]}$, $W^Y_{ov} < \frac{1}{1+\max\left[\frac{b_{ov}}{a_{ov}},\frac{c_{oo}-c_{ov}}{d_{oo}-d_{ov}}\right]}$, DEH ,v) $\Rightarrow$ (o,v) & (v,o) EP's

Table 4: Summary of results for the games with complete information, v(o): Defensive (offensive) force structure, f(i): peaceful (imperialistic) player; DEH: Defense Efficiency Hypothesis; EP: Equilibrium point.

4. Remark

A non-degenerate non-cooperative two-person game has an odd number of equilibrium points. Thus, in addition to (o,v) and (v,o), there exists another equilibrium under the assumptions given, albeit one in terms of mixed strategies. However, a mixed strategy equilibrium point may be a solution only in case of a perfectly symmetrical game, i.e., in a highly special situation without any practical relevance. Thus, we are interested only in the pure strategy equilibrium points. Furthermore, we restrict our attention to generic situations where all pure strategy equilibrium points are strict.

2.5 Summary of Results

In Table 4 the important results obtained for the games with complete information are summarized. We observe that the validity of the Defense Efficiency Hypothesis has much stronger implications for f-type than for i-type players.

Both in Propositions 2 and 3 there are statements on the dominance of the strategy v. We summarize them as

Proposition 5

Given DEH holds, then v is a dominant strategy for an f-type player either if the antagonist is also an f-type player, or if the antagonist is an i-type player and if for the f-type player holds

$$W^{X}_{vo} > \frac{1}{1+\frac{d_{vo}}{c_{vo}}} \quad \text{resp.} \quad W^{Y}_{ov} > \frac{1}{1+\frac{b_{ov}}{a_{ov}}} .$$

2.6. Equilibrium Selection in the Game (i,i)

As pointed out in Section 2.4, under the assumptions

$$W^{Y}_{oo} < W^{X}_{ov} < \frac{1}{1+\max\left[\frac{d_{vo}}{c_{vo}},\frac{a_{oo}-a_{vo}}{c_{vo}-b_{vo}}\right]} \tag{2-7a}$$

$$W^{Y}_{oo} < W^{Y}_{ov} < \frac{1}{1+\max\left[\frac{b_{ov}}{a_{ov}},\frac{c_{oo}-c_{ov}}{d_{oo}-d_{ov}}\right]} \tag{2-7b}$$

the game (i,i) has the two equilibria (v,o) and (o,v).

$X_i \backslash Y_i$	o	v
o	$H^X_{ii}(o,o)$ $H^Y_{ii}(o,o)$	$H^X_{ii}(o,v)$ $H^Y_{ii}(o,v)$
v	$H^X_{ii}(v,o)$ $H^Y_{ii}(v,o)$	$H^X_{ii}(v,v)$ $H^Y_{ii}(v,v)$

Table 5a: Game I in normal form. The payoffs are explicitly given by Table 3. The arrows indicate the incentive directions.

If for one of these two equilibria the payoffs to both players are higher than those of the other, then the first one will be considered the solution of the game (according to the principle of 'payoff dominance'). Thus, e.g., (v,o) will be selected if

$$(a_{vo}+b_{vo}) \cdot W^X_{vo} - b_{vo} > a_{ov} - (a_{ov}+b_{ov}) \cdot W^Y_{ov}$$

$$c_{vo} - (c_{vo}+d_{vo}) \cdot W^X_{vo} > (c_{ov}+d_{ov}) \cdot W^Y_{ov} - d_{ov}$$

that is if

$$\frac{a_{ov}+b_{vo}}{a_{vo}+b_{vo}} - \frac{a_{ov}+b_{ov}}{a_{vo}+b_{vo}} \cdot W^Y_{ov} < W^X_{vo} < \frac{c_{vo}+d_{ov}}{c_{vo}+d_{vo}} - \frac{c_{ov}+d_{ov}}{c_{vo}+d_{vo}} \cdot W^Y_{ov}$$

which, in turn, requires that

$$\frac{a_{ov}+b_{vo}}{a_{vo}+b_{vo}} - \frac{c_{vo}+d_{ov}}{c_{vo}+d_{vo}} < \left[\frac{a_{ov}+b_{ov}}{a_{vo}+b_{vo}} - \frac{c_{ov}+d_{ov}}{c_{vo}+d_{vo}}\right] \cdot W^Y_{ov} .$$

Unless these conditions are fulfilled, there is no 'traditional' solution to the problem. In the following we will apply the equilibrium selection theory developed by HARSANYI AND SELTEN (1988).

For this purpose, we write the game under consideration in normal form as shown in Table 5a; we call it Game I.

Even though the indices (i,i) could be omitted in Table 5a, we have retained them for the purpose of clarification in the subsequent chapters. According to our assumptions, we have

$$H^X_{ii}(v,o) > H^X_{ii}(o,o); \quad H^X_{ii}(o,v) > H^X_{ii}(v,v)$$

$$H^Y_{ii}(o,v) > H^Y_{ii}(o,o); \quad H^Y_{ii}(v,o) > H^Y_{ii}(v,v). \tag{2-8}$$

Now, let us consider the non-cooperative two-person Game II as given in normal form by Table 5b.

0 0	$H^X_{ii}(o,v)-H^X_{ii}(v,v)$ $H^Y_{ii}(o,v)-H^Y_{ii}(o,o)$
$H^X_{ii}(v,o)-H^X_{ii}(o,o)$ $H^Y_{ii}(v,o)-H^Y_{ii}(v,v)$	0 0

Table 5b: Game II in normal form.

Game II is obtained from Game I by transforming Game I in a way which preserves the stability sets of both pure strategies of both players. This implies that all equilibria are preserved. 'Best reply invariance' means that such a transformation will not change the solution, i.e. a strategy vector is the solution of Game I iff it is the solution of Game II.

Finally, consider the non-cooperative two-person Game III as given in normal form by Table 5c.

Game III is obtained from Game II by applying positive linear transformations of individual cardinal utilities. A solution concept is called 'invariant with respect to positive affine transformations' iff the solution of two such games is the same.

0 / 0	1 / $\frac{H^Y_{ii}(o,v)-H^Y_{ii}(o,o)}{H^Y_{ii}(v,o)-H^Y_{ii}(v,v)} =: W$
$\frac{H^X_{ii}(v,o)-H^X_{ii}(o,o)}{H^X_{ii}(o,v)-H^X_{ii}(v,v)} =: U$ / 1	0 / 0

Table 5c: Game III in normal form.

The game at hand is called symmetric iff there exists a non-trivial way of renaming players such that the game is transformed into itself. A solution concept is called 'symmetry invariant' iff the solution is not affected by such a transformation. For $U = W$, Game III is obviously symmetric. Assume for $U = W$ that (o,v) would be the solution of the game. By changing the name of player 1 into player 2 and correspondingly for player 2, we do not change Game III, but the solution (o,v) of the original game to (v,o). This shows that selecting a pure equilibrium point in case of $U = W$ contradicts the requirement of symmetry invariance. Thus, the solution of Game III with $U = W$ must be the symmetry invariant mixed strategy equilibrium point.

Consider two games G and G' which differ from each other only in the sense that there exists a player i and a strategy vector s such that the payoffs $H_i(s)$ and $H_i'(s)$ in G and G' respectively differ by a positive constant in the sense that $H_i'(s)-H_i(s) > 0$. We say that G' results from G by *strengthening* s. A dominance relation between pure strategy equilibria is called (payoff-)monotonic iff strengthening a pure strategy equilibrium s will make s dominant if it was undominated before.

Consider again Game III with $U = W$ and assume that the dominance relation between the two pure strategy equilibria is symmetry invariant. This implies that neither (o,v) nor (v,o) is dominated. Any Game III with $U > W$ can be derived from such a symmetric game with $U = W$ by strengthening (o,v). This shows that, whenever $U > V$, (v,o) must be determined as the solution of the game by any solution concept based on (payoff-)monotonicity which selects, whenever this is compatible with symmetry invariance, one of the two pure strategy equilibria as the solution. Similarly, the solution of Game III is (o,v) if $U < W$.

As a result, we define, e.g., (v,o) as the solution of one game, iff

$$\frac{H^X_{ii}(v,o)-H^X_{ii}(o,o)}{H^X_{ii}(o,v)-H^X_{ii}(v,v)} > \frac{H^Y_{ii}(o,v)-H^Y_{ii}(o,o)}{H^Y_{ii}(v,o)-H^Y_{ii}(v,v)} .$$

Note that because of (2-8) both nominators and denominators are greater zero. Let us call these ratios relative defense incentives. Then we can formulate our results as

<u>Proposition 6</u>
Given the game (i,i). Then (v,o) is its solution if the relative defense incentive of player X is higher than that of Y and vice versa.

Explicitly, the inequality given above reads

$$\frac{(a_{vo}+b_{vo})\cdot W^X_{vo}-b_{vo}-(a_{oo}+b_{oo})\cdot W^X_{oo}+b_{oo}}{a_{ov}-(a_{vo}+b_{vo})\cdot W^X_{ov}} >$$

$$\frac{(c_{ov}+d_{ov})\cdot W^Y_{ov}-d_{ov}-(c_{oo}+d_{oo})\cdot W^Y_{oo}+d_{oo}}{c_{vo}-(c_{vo}+d_{vo})\cdot W^X_{vo}} .$$

Without special assumptions it does not seem easy to recognize simple necessary or sufficient conditions in terms of the payoff-parameters or defense probabilities for this inequality to hold.

3. Incompletely Informed Players

In this chapter we assume that either one, or both players do not know the type of the antagonist, however we assume that in case of incomplete information there exist subjective probabilities about the antagonist's type, and that these probabilities are common knowledge.

We assume furthermore that DEH holds, and

$$W^X_{vo} < \frac{1}{1+\max\left[\frac{d_{vo}}{c_{vo}}, \frac{a_{oo}-a_{vo}}{b_{oo}-b_{vo}}\right]} ,$$

$$W^Y_{ov} < \frac{1}{1+\max\left[\frac{b_{ov}}{a_{ov}}, \frac{c_{oo}-c_{ov}}{d_{oo}-d_{ov}}\right]} . \qquad (3\text{-}1)$$

According to the second chapter we consider the four games in normal form as given in Table 6 to be possible.

In the following we will use only the abbreviated payoffs as given in Table 6, together with their mutual relations which result from the incentive directions also given in this table.

Assumption 3

If one player plays v,. then the payoff to the antagonist is independent of the former's type, i.e. with the assumptions given in Table 6 we have

$$H^X_{ii}(\cdot,v) = H^X_{if}(\cdot,v) =: H^X_i(\cdot,v)$$
$$H^Y_{ii}(v,\cdot) = H^Y_{fi}(v,\cdot) =: H^Y_i(v,\cdot). \tag{3-2}$$

3.1 Player X_f Incompletely Informed

Let X be an f-type player and let X know with probability w resp. 1-w that Y is an f- resp. i-type player. Assume furthermore that all this is common knowledge. The equilibrium point of this game is given by

Proposition 7

Under the assumptions given the game with player X being of type f and asymmetric incomplete information about Y has the equilibrium point (v,o,v) for players (X_f, Y_i,Y_f).

Proof

According to our assumptions v is a dominant strategy both for players X_f and Y_f. Therefore, according to Tables 4 or 6, o is equilibrium strategy for player Y_i. ■

The result holds irrespective of the value of w ; in fact, we would not have needed to introduce subjective beliefs in order to derive this result. The situation is different in case player X_i is incompletely informed.

3.2 Player X_i Incompletely Informed

Now, let X be an i-type player and let X know with probability w resp. 1-w that Y is an f- resp. i-type player. Assume, again, that all this is common knowledge. Equilibrium points of this game are given by

Proposition 8

Under the assumptions given, the game with player X being of type i and asymmetric incomplete information about Y has the equilibrium point (o,v,v) for (X_i,Y_i,Y_f) and furthermore, the additional equilibrium point (v,o,v) iff

$$H^X_{ii}(v,o) - H^X_{ii}(o,o) > \frac{w}{1-w} \cdot H^X_i(o,v). \tag{3-3}$$

$X_f \backslash Y_f$	o	v
o	$H^X_{ff}(o,o)$ $H^Y_{ff}(o,o)$	$H^X_{ff}(o,v)$ $H^Y_{ff}(o,v)$
v	$H^X_{ff}(v,o)$ $H^Y_{ff}(v,o)$	⊗ $H^X_{ff}(v,v)=0$ $H^Y_{ff}(v,v)=0$

$X_f \backslash Y_i$	o	v
o	$H^X_{fi}(o,o)$ $H^Y_{fi}(o,o)$	$H^X_{fi}(o,v)$ $H^Y_{fi}(o,v)$
v	⊗ $H^X_{fi}(v,o)$ $H^Y_{fi}(v,o)$	$H^X_{fi}(v,v)=0$ $H^Y_{fi}(v,v)=0$

$X_f \backslash Y_f$	o	v
o	$H^X_{if}(o,o)$ $H^Y_{if}(o,o)$	$H^X_{if}(o,v)$ $H^Y_{if}(o,v)$
v	$H^X_{if}(v,o)$ $H^Y_{if}(v,o)$	⊗ $H^X_{if}(v,v)=0$ $H^Y_{if}(v,o)=0$

$X_f \backslash Y_i$	o	v
o	$H^X_{ii}(o,o)$ $H^Y_{ii}(o,o)$	$H^X_{ii}(o,v)$ $H^Y_{ii}(o,v)$
v	⊗ $H^X_{ii}(v,o)$ $H^Y_{ii}(v,o)$	$H^X_{ii}(v,v)=0$ $H^Y_{ii}(v,v)=0$

Table 6: The possible games expected by players X and Y. The arrows indicate incentive directions; equilibrium points are characterized by ⊗. Expected payoffs are abbreviations of those given in Table 3.

Proof

Since v is a dominant strategy for Y_f, this player actually does not take part in the game, and we can represent the game as a 2x2-game between players X_i and Y_i, the normal form of which is given in Table 7.

The incentive directions indicated in this table are taken from Table 6. Thus, we see immediately that (o,v) is an equilibrium point for (X_i, Y_i). Furthermore, with $H^X_{if}(v,v) = o$, the second assertion is also obvious.

■

It should be mentioned that because of

$$H^X_{ii}(v,o) - H^X_{ii}(o,v) > 0 \quad \text{and} \quad H^X_{if}(o,o) > 0,$$

$X_i \backslash Y_i$	o	v
o	$(1-w)\cdot H^X_{ii}(o,o)+w\cdot H^X_i(o,v)$ $H^Y_{ii}(o,o)$	$H^X_i(o,v)$ ⊗ $H^Y_{ii}(v,o)$
v	$(1-w)\cdot H^X_{ii}(v,o)+w\cdot H^X_i(v,v)$ $H^Y_i(v,0)$	$H^X_i(v,v)$ $H^Y_i(v,v)$

Table 7: Game with incomplete information in normal form. The arrows indicate the incentive directions.

depending on the value of w, there are always both possibilities for (o,v,v) to be an equilibrium point.

In case (3-3) holds, we may apply the equilibrium selection theory by Harsanyi and Selten (1988) in exactly the same way as in Section 2.6. Comparing Tables 5a and 7, we only need to replace

$$H^X_{ii}(o,o) \rightarrow (1-w) \cdot H^X_{ii} + w \cdot H^X_i(o,v)$$

$$H^X_{ii}(v,o) \rightarrow (1-w) \cdot H^X_{ii} + w \cdot H^X_i(v,v)$$

and obtain the result that (v,o,v) is the solution of our game iff

$$\frac{(1-w)\cdot[H^X_{ii}(v,o)-H^X_{ii}(o,o)]+w\cdot[H^X_{if}(v,v)-H^X_{if}(o,v)]}{H^X_{ii}(o,v)-H^X_{ii}(v,v)} >$$

$$> \frac{H^Y_{ii}(o,v)-H^Y_{ii}(o,o)}{H^Y_{ii}(v,o)-H^Y_{ii}(v,v)} \quad . \tag{3-4}$$

3.3 <u>Both Players Incompletely Informed</u>

Now, we assume that both players X and Y do not know the type of the antagonist, however, they have subjective probabilities about their antagonist's type which are common knowledge.

In general, we speak of *consistent incomplete information* if the subjective beliefs of all players about the opponent's types can be derived from the same probability distribution. In such a case we represent a game in extensive form as given in Figure 1. At the beginning there is a chance move which selects with probabilities p_{ii}, p_{if}, p_{fi} and p_{ff} the combinations of types (ii), (if), (fi) and (ff) for players X and Y. From these probabilities we deduce the following probabilities which are of interest here:

1) With probability $p_{ii}+p_{fi}$ player Y_i is 'chosen', with probability $p_{if}+p_{ff}$ player Y_f. Therefore, with probabilities

$$\frac{p_{ii}}{p_{ii}+p_{fi}} =: q_i \quad \text{and} \quad \frac{p_{fi}}{p_{ii}+p_{fi}} = 1-q_i \tag{3-5}$$

X_i and X_f are chosen under the condition that Y_i is chosen. With probabilities

$$\frac{p_{if}}{p_{if}+p_{ff}} =: q_f \quad \text{and} \quad \frac{p_{ff}}{p_{if}+p_{ff}} = 1-q_f \tag{3-6}$$

players X_i and X_f are chosen under the condition that Y_f is chosen.

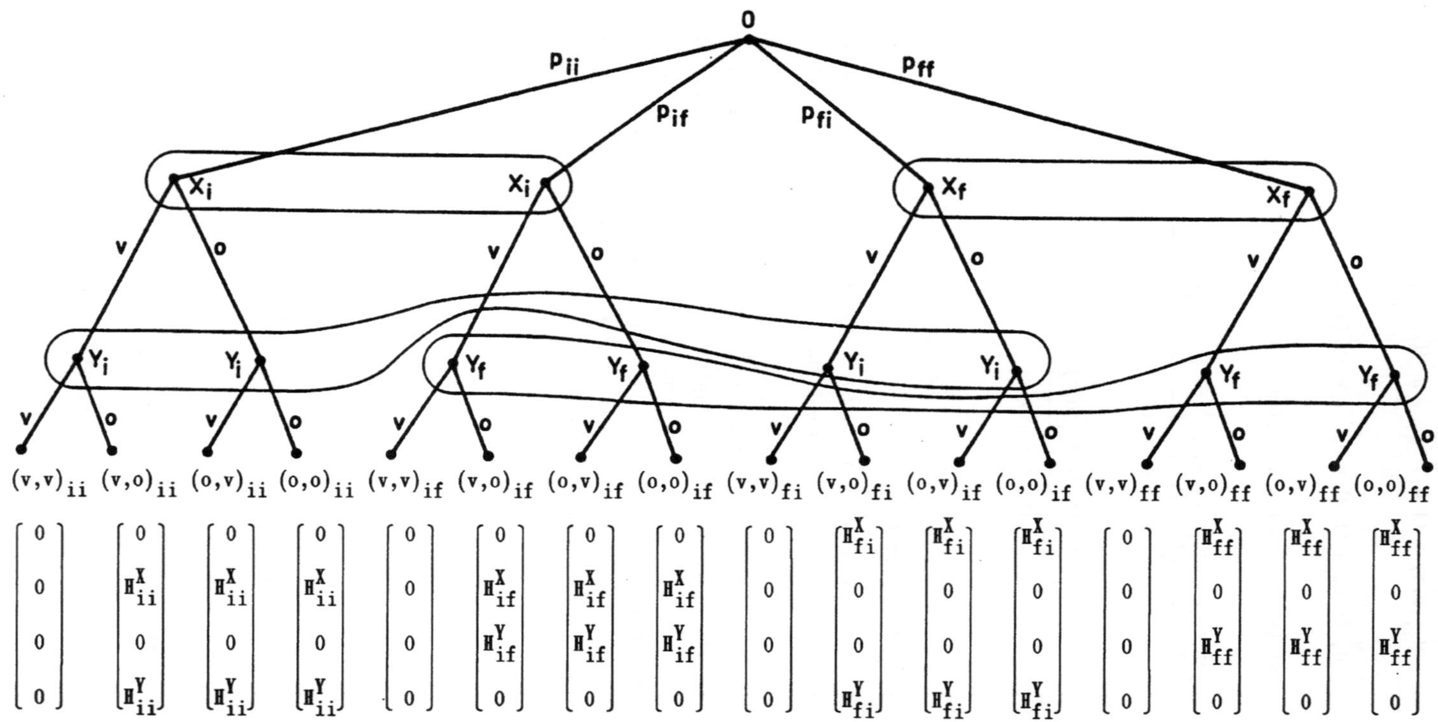

Figure 1: The game with mutually incomplete consistent information. The game starts with the chance move of player o which determines the type combinations of X and Y: With probabilities $p_{ii} \dots p_{ff}$ the combinations $(X_i, Y_i) \dots (X_f, Y_f)$ are chosen, where $p_{ii} + p_{if} + p_{fi} + p_{ff} = 1$. Information sets are graphically characterized such that all decision modes which that player who has to choose cannot distinguish are encircled. One play is a sequence of moves from the origin to the endpoint, i.e., that node at which the game ends. The evaluation of a play is determined by the payoff vector attached to the respective endpoint. At each endpoint payoffs are given in the order X_f, X_i, Y_f, Y_i, where we use the convention that the non-present type evaluates all plays with zero.

2) With probability $p_{ii}+p_{if}$ player X_i is chosen, with probability $p_{fi}+p_{ff}$ player X_f. Therefore, with probabilities

$$\frac{p_{ii}}{p_{ii}+p_{if}} =: p_i \quad \text{and} \quad \frac{p_{if}}{p_{ii}+p_{if}} = 1-p_i \tag{3-7}$$

Y_i and Y_f are chosen under the condition that X_i is chosen. With probabilities

$$\frac{p_{fi}}{p_{fi}+p_{ff}} =: p_f \quad \text{and} \quad \frac{p_{ff}}{p_{fi}+p_{ff}} = 1-p_f \tag{3-8}$$

players Y_i and Y_f are chosen under the condition that X_f is chosen.

As a result, p and q are the probabilities with which the imperialistic antagonist is expected; the index characterizes the own one's type. Now, we see that we have four probabilities q_i, q_f, p_i and p_f derived from the four probabilities $p_{ii}\ldots p_{ff}$ which have to sum up to one. Thus, we have only three free parameters.

In case the four probabilities q_i, q_f, p_i and p_f do not meet the required consistency, we speak of *inconsistent incomplete information*. In this case, the initial fictitious chance move which transforms the situation with incomplete information into a game with complete information has to be supplemented with one probability distribution for each player.

Let $p_{ii}\ldots p_{ff}$ be the distribution underlying the subjective beliefs of player Y, and $\tilde{p}_{ii}\ldots\tilde{p}_{ff}$ that of player X. Then, as before,

$$\frac{p_{ii}}{p_{ii}+p_{fi}} = q_i \,, \quad \frac{p_{if}}{p_{if}+p_{ff}} = q_f$$

are the subjective beliefs of Y. Similarly

$$\frac{\tilde{p}_{ii}}{\tilde{p}_{ii}+\tilde{p}_{if}} = \tilde{p}_i \,, \quad \frac{\tilde{p}_{fi}}{\tilde{p}_{fi}+\tilde{p}_{ff}} = \tilde{p}_f$$

are those of X. Thus, since we derive only two probabilities from each distribution, they need not meet any condition which means that now we have four free parameters instead of only three as before.

Now, we will see that the equilibrium conditions for the game in consideration depend on the subjective beliefs in a way such that it does not matter whether or not there is consistent or inconsistent incomplete information as long as one keeps in mind that in the latter case two underlying probability distributions have to be defined.

Equilibrium points of this game are given by

Proposition 9

Under the Assumption (3-1) the equilibrium points of the game given by Figure 2 are given as follows: Let p_{ic} and q_{ic} be given by

$$p_{ic} = \frac{H_i^X(o,v)}{H_i^X(o,v)+H_{ii}^X(v,o)-H_{ii}^X(o,o)} \tag{3-9}$$

$$q_{ic} = \frac{H_i^Y(v,o)}{H_i^Y(v,o)+H_{ii}^Y(o,v)-H_{ii}^Y(o,o)} \quad . \tag{3-10}$$

Let the strategies of players X_f, X_i, Y_f, Y_i be given in this order in any strategy vector. Then

1) for $p_i > p_{ic}$, (vvvo) is an equilibrium point,

2) for $q_i > q_{ic}$, (vovv) is an equilibrium point,

3) for $p_i < p_{ic}$ and $q_i < q_{ic}$, (vovo) is an equilibrium point.

Proof

1) $p_i > p_{ic}$.

For X_i the equilibrium condition is, if Y_f and Y_i choose v and o,

$$p_i H_{ii}^X(v,o)+(1-p_i)\cdot H_i^X(v,v)>p_i\cdot H_{ii}^X(o,o)+(1-p_i)\cdot H_i^X(o,v), \tag{3-11}$$

or, with $H(v,v) = 0$,

$$p_i\cdot(H_{ii}^X(o,o)-H_i^X(o,v)-H_{ii}^X(v,o)) < -H_i^X(o,v),$$

which is fulfilled because of (3-9).

For X_f :

$$p_f\cdot H_{fi}^X(v,o)+(1-p_f)\cdot H_{ff}^X(v,v)>p_f\cdot H_{fi}^X(o,o)+(1-p_f)\cdot H_{ff}^X(o,v) \tag{3-12}$$

which is fulfilled because of the relations indicated in Table 7.

For Y_i :

$$(1-q_i)\cdot H_i^Y(v,o)+q_i\cdot H_i^Y(v,o)>(1-q_i)\cdot H_i^Y(v,v)+q_i\cdot H_i^Y(v,v) \qquad (3\text{-}13)$$

which is again fulfilled because of Table 6.

For Y_f :

$$q_f\cdot H_{if}^Y(v,v)+(1-q_f)\cdot H_{ff}^Y(v,v)>q_f\cdot H_{if}^Y(v,o)+(1-q_f)\cdot H_{ff}^Y(v,o) \qquad (3\text{-}14)$$

which is again fulfilled because of Table 6.

ii) $q_i > q_{ic}$

Here, the considerations are exactly the same as given above. Therefore, they are not repeated here.

iii) $p_i < p_{ic}$ & $q_i < q_{ic}$

For X_i :

$$p_i\cdot H_{ii}^X(o,o)+(1-p_i)\cdot H_i^X(o,v)>p_i\cdot H_i^X(o,v)+(1-p_i)\cdot H_i^X(o,v) \qquad (3\text{-}15)$$

which is fulfilled because of (3-9).

For X_f :

$$p_f\cdot H_{fi}^X(v,o)+(1-p_f)\cdot H_{ff}^X(v,v)>p_f\cdot H_{fi}^X(o,o)+(1-p_f)\cdot H_{ff}^X(o,v) \qquad (3\text{-}16)$$

which is fulfilled because of Table 6.

For Y_i :

$$(1-q_i)\cdot H_i^Y(v,o)+q_i\cdot H_{ii}^Y(o,o)>(1-q_i)\cdot H_i^Y(v,v)+q_i\cdot H_{ii}^Y(o,v);$$

which is equivalent to

$$q_i\cdot(H_{ii}^Y(o,v) - H_{ii}^Y(o,o) + H_i^Y(v,o)) < H_i^Y(v,o);$$

this is fulfilled because of (3-10).

For Y_f :

$$q_f \cdot H^Y_{if}(o,v)+(1-q_f)\cdot H^Y_{ff}(v,v)>q_f \cdot H^Y_{if}(o,o)+(1-q_f)\cdot H^Y_{ff}(o,v) \qquad (3\text{-}17)$$

which is fulfilled because of Table 6.

1. Remark

According to Table 7 we have

$$H^X_f(o,v) > 0, \quad H^X_{ii}(v,o) > H^X_{ii}(o,o)$$

therefore $0 < p_{ic} < 1$. In the same way, naturally, $o < q_{ic} < 1$.

2. Remark

In Figure 2, the areas for the equilibrium points given in Proposition 2 in the (p_i,q_i)-plane are illustrated.

Qualitatively these results can be interpreted as follows: If an imperialistic player expects with high probability an imperialistic antagonist, then he chooses a defense oriented force structure, otherwise an offensive one.

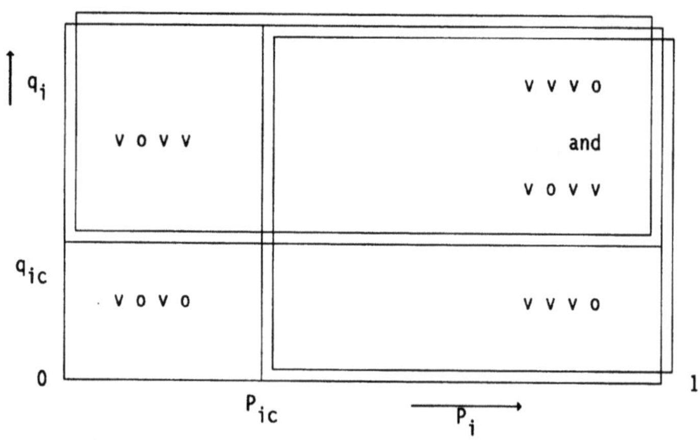

Figure 2: Areas in the (p_i,q_i)-plane for the equilibria according to Proposition 9.

3. Remark

The payoffs implied by the equilibrium are given by the left-hand sides of inequalities (4-12) ff. and will not be repeated here.

4. Remark

The game is *partially revealing*: Since the f-type player's equilibrium choice is always v, it is clear that the player who chooses o must be an i-type player.

5. Remark

Since for an f-type the strategy v is dominant, his decision behaviour does not depend at all on his subjective beliefs concerning the type of his opponent. This explains why all conditions in Proposition 9 refer to beliefs of i-types; only thebeliefs of X_i and Y_i, expressed by p_i and q_i, respectively, have to meet certain conditions in order to assure the equilibrium property for a given strategy combination. This shows that there is no need here to require consistency of incomplete information or to argue whether consistency is acceptable or not.

3.4 Equilibrium Selection

In order to select in the area $\{p_i>p_{ic}\ \&\ q_i>q_{ic}\}$ a unique solution, we need only consider the game between X_i and Y_i, since in both equilibria the strategies of X_f and Y_f are v. Taking into account that both X_i and Y_i do not know of which type their respective antagonists are, but that in case the latter ones are of the f-type, they will choose v, we have to consider a two-person game the normal form of which is given in Table 8a.

$X_i \backslash Y_i$	o	v
o	$p_i \cdot H^X_{ii}((o,o)+(1-p_i)\cdot H^X_i(o,v)$ $q_i \cdot H^Y_{ii}(o,o)+(1-q_i)\cdot H^Y_i(v,o)$	$H^X_i(o,v)$ $p_i \cdot H^Y_{ii}(o,v)+(1-q_i)\cdot H^Y_i(v,v)$
v	$p_i \cdot H^X_{ii}(v,o)+(1-p_i)\cdot H^X_i(v,v)$ $H^Y_i(v,o)$	$H^X_i(v,v)$ $H^Y_i(v,v)$

Table 8a: The reduced game with mutual incomplete information. Payoffs H^X_{ii} and H^Y_{ii} are given in Table 6. H^X_i and H^Y_i are defined by (3-2). The arrows indicate the incentive directions; (ov) and (vo) are equilibrium points in pure strategies.

In order to select one solution, we proceed as described in Section 2.6. For this purpose, we define the *incentives* to attack

$$J_i^X := H_i^X(o,v) - H_i^X(v,v)$$

$$J_i^Y := H_i^Y(v,o) - H_i^Y(v,v), \qquad (3\text{-}18)$$

the *incentives for efficient defense*,

$$E_{ii}^X := H_{ii}^X(v,o) - H_{ii}^X(o,o)$$

$$E_{ii}^Y := H_{ii}^Y(o,v) - H_{ii}^Y(o,o), \qquad (3\text{-}19)$$

and, finally, the relative defense incentives

$$D_i^X := \frac{E_{ii}^X}{J_i^X}, \quad D_i^Y := \frac{E_{ii}^Y}{J_i^Y}. \qquad (3\text{-}20)$$

Then, by going through the same procedure as described in Section 2.6, we arrive at a two-person Game III whose normal form is given in Table 8b. Thus, e.g., (o,v) is selected as the solution of the game iff

$$\frac{1}{p_i\cdot(D_i^X+1) - 1} > \frac{1}{q_i\cdot(D_i^Y+1) - 1}$$

or, equivalently, if

$$D_i^Y + 1 > \frac{p_i}{q_i} \cdot (D_i^X + 1). \qquad (3\text{-}21)$$

The result we formulate as

Proposition 10

Given the game with mutually incomplete information as defined by Figure 1. Then for $p_i > p_{ic}$, $q_i > q_{ic}$ where p_{ic} and q_{ic} are defined by (3-9) and (3-10), (v,o,v,v) resp. (v,v,v,o) is the solution of the game iff the ratio of the relative defense incentives of players Y_i and X_i is larger resp. smaller than the ratio p_i/q_i of their probabilities with which they expect their antagonists to be imperialistic.

	0, 0	1, $\frac{1}{q_i \cdot (D_i^Y+1)-1}$
	$\frac{1}{p_i \cdot (D_i^X+1)-1}$, 1	0, 0

Table 8b: Game III in normal form. D_i^X and D_i^Y are given by (3-20).

Graphically, this result is depicted in Figure 3. If we compare (3-21) to the result of Section 2.6, then we see that we get the same result for $q_i = p_i$: If both players have the same subjective beliefs about their antagonists type, then in case those probabilities are larger than $\max(p_{ic}, q_{ic})$, the same equilibrium point will be selected for the i-types as in case both i-types have complete knowledge about their antagonist's types.

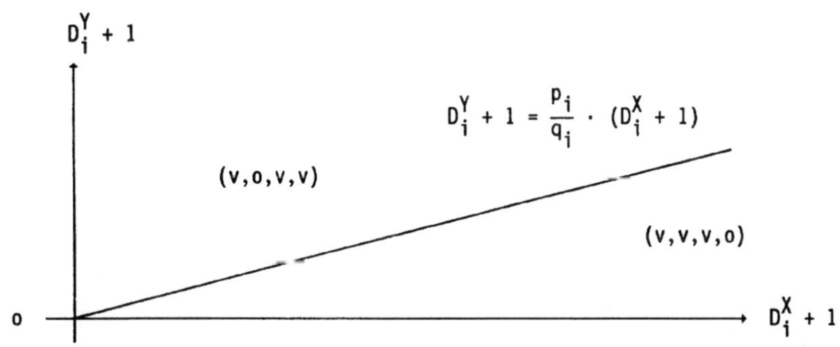

Figure 3: Areas of selected equilibrium points in the region $(p_i > p_{ic},\ q_i > q_{ic})$ according to Proposition 10.

References

Clausewitz, G. von (1984). On War. Edited and Translated by M. Howard and P. Paret. Princeton, N.J.: Princeton Univ. Press.

Harsanyi, J. (1967/68). Games with Incomplete Information Played by 'Bayesian' Players, Parts I-III. Manag. Sci. **14**: 159-182, 320-334 and 486-502,

Harsanyi, J. and R. Selten (1988). A General Theory of Equilibrium Selection in Games. Cambridge, Mass.: The MIT Press.

Hofmann, H.W., R.K. Huber, and K. Steiger (1986) On Reactive Defense Options - A Comparative System Analysis of Alternatives for the Initial Defense Against the First Strategic Echelon of the Warsaw Pact in Central Europe. In: Huber, R.K. (ed.), Modeling and Analysis of Conventional Defense in Europe. New York: Plenum Press, pp. 97-140.

Huber, R.K., and H.W. Hofmann (1984). Gradual Defensivity - An Approach to Stable Conventional Force Equilibrium in Europe? In: Bran, J.P. (ed.), Operational Research '84. Amsterdam-New York-Oxford, (Verlag?) pp. 197-211.

Huber, R.K. (1987). Über strukturelle Voraussetzungen für Krisenstabilität in Europa ohne Kernwaffen: Eine systemanalytische Betrachtung. Oper. Res. Spektrum Nr. **1**: 1-12.

Quester, G.M. (1977). Offense and Defense in the International System. New York: John Wiley and Sons.

IMPLICATIONS OF THE DEFENSE EFFICIENCY HYPOTHESIS FOR THE CHOICE OF MILITARY FORCE STRUCTURES. PART II: A SEQUENTIAL GAME INCLUDING THE POSSIBILITY OF RESTRUCTURING FORCES[1]

by

Rudolf Avenhaus, Werner Güth, and Reiner K. Huber

Abstract: The so-called Defense Efficiency Hypothesis (DEH) implies that conventional forces designed specifically for defensive operations and, thus, incapable of any major offensive operations can exploit the intrinsic defense advantage more efficiently than forces designed to operate in all combat modes including offensive operations. If this were true, conventional stability between two antagonistic parties could be brought about by a mutual adoption of defensive force structures.

However, with a view to the uncertainty about the true strategic intentions of the opponent, the soundness of such a force conversion policy is frequently called into question. For this reason, an attempt has been made in the first part of this study to shed some light on the issue by studying a simple game-theoretic model of defense policy interactions between two antagonistic parties. In particular, the equilibria are determined in a series of games in which the antagonistic parties are free to choose, simultaneously, between offensive and defensive force structures given the DEH were true.

In this second part, a model is considered in which both antagonists, after having chosen offensive force structures independently, may consider to restructure their forces from offensive to defensive ones. Contrary to the original choice modelled in the first part of the paper, the restructuring choice is assumed to be sequential and asymmetric, i.e., one party initiates the restructuring process and the other may or may not follow suit.

1. Defense Efficiency Hypothesis

At present, heavy armor and its associated combat support equipment are considered to be the quintessential tools of modern conventional forces. However, since armor is the backbone of offensive operations as well, it follows that a so-called "worst case" military assessment would very likely advise the leadership of two antagonistic parties to consider the opponent's military plans as offensive, irrespective of what his true strategic objectives and political intentions may be. In a crisis, such perceptions could become a powerful motor of instability because either side might be tempted to preempt an expected attack by the other. Short of more or less complete disarmament, mutual arms reductions may even exacerbate this instability problem.

---

[1]The two parts deal with related, but structurally different game models and can be read independently.

This is because, if surprised, a defense thinned by the negotiated force reductions would face considerably more unfavorable local force ratios at the points where the attacker pursues his main thrusts. In addition, the reductions would deprive the attacker of some of his operational reserves. Thus, he must be expected to rely even more on a quick initial success by means of surprise.

Thus, with regard to crisis stability there is clearly a dilemma that cannot be solved by the classical approach to conventional arms control as, for example, practiced in the Mutual Balanced Force Reductions (MBFR)-talks between NATO and the Warsaw Pact (WP). They have been aimed at symmetrical reductions of troop strengths that were to leave the basic force structures essentially intact. For this reason, several experts have argued that, before sizeable reductions are implemented, the structures of the conventional forces need to be changed in a manner that they may not be perceived by either opponent as being capable of major offensive operations. Most of the respective proposals recommend to limit force mobility and replace armor by anti-armor systems of some kind. Indeed, if both sides were incapable of offensive operations, the military situation between them would obviously be highly stable.

However, based on a conceptual transarmament model HUBER (1987) has shown that there is no possibility for a transition to that ultra-stable situation without a temporary deterioration of crisis stability unless the Defense Efficiency Hypothesis (DEH) proposed by HUBER and HOFMANN (1984) is true. That hypothesis says that, in a conventional war, "reactive" military forces designed specifically for defensive operations and incapable of major offensive actions can exploit the intrinsic defense advantage[2] more efficiently, i.e., at higher cost-effectiveness, than the traditional "active" forces that are designed to operate in all combat modes.

Thus, if the Defense Efficiency Hypothesis were true, it is obvious that some degree of replacement of active by reactive units would improve the defense potential of a force while simultaneously reducing its offensive capabilities. For a given operational depth available for defensive operations, the optimum degree of replacement that maximizes the defense potential would depend, among other things, primarily on the configuration of the reactive force elements and on the degree to which the opponent has replaced active by reactive units in his forces.

Whether and under which circumstances the Defense Efficiency Hypothesis might hold has not yet been convincingly demonstrated.[3] But assuming it were true, there remains the question whether the adoption of a largely reactive, i.e., defensive

---

[2]The fact that a well organized defense has an inherent advantage over the attacker has been substantiated by many a historical battle. It has led VON CLAUSEWITZ to state *"that the defensive form of warfare is intrinsically stronger than the offensive"* (On War, Book Six, Chapter 2). See also QUESTER (1977).

[3]A major part of the research at the institute of the first and third named authors is devoted to testing the DEH by means of combat simulation experiments (see HUBER (1987) and HOFMANN, HUBER, STEIGER (1986)).

force structure would be a rational policy, especially with a view to the uncertainties about the opponent's true political and strategic intentions. In order to somewhat illuminate that issue, an attempt is made to determine equilibrium[4] points in a series of games between two antagonistic parties which are free to choose between offensive and defensive structures and the strategic intentions of which may be either defensive or offensive.

In the first part of the analysis, we have looked at situations of independent choice, i.e., both parties are assumed to select their force structures simultaneously without regard to any previous choices of their own or their opponent. In this second part, the games are extended into a two-stage decision process such that both sides may consider to restructure their forces in case they both had previously chosen offensive forces.

Contrary to the original choice modelled in Part I, the initial decisions on restructuring the military forces are not assumed to represent a situation of simultaneous choice, but rather a sequential one in which one side initiates the conversion to defensive forces and the other may or may not follow suit when learning about its opponents decision to restructure.

As in the games modelled in Part I, the sequential games allow for incomplete information about the opponent's true intentions. Thus, parties might try to reveal their intentions by signaling their type. If the costs for restructuring the armed forces are considerable, the respective games yield unique solutions by means of the equilibrium selection theory. If restructuring involves only minor costs, the more complex games are analyzed using the method of equilibrium scenarios.

The results of these two-stage games show that despite knowing that it will have to restructure its military forces at some later time, a party may initially choose offensive force structures in an attempt to induce the opponent in adopting defensive force structures as well. Of course, the incentives for using such a force structuring strategy are higher the lower the costs for restructuring and the stronger the DEH is supported. Unfortunately, the implications of the DEH are not strong enough to rule out the "dreadful world" in which both types of both parties choose offensive force structures and only permit the "peaceful world" characterized by both types of both parties relying on defense oriented force structures.

Part II is organised as follows: Section 2 introduces our basic game model consisting of two successive decision phases after an initial fictitious chance move that determines the types of players. The class of games with considerable costs for restructuring armed forces is analysed in Section 3 by applying the theory of

[4]It should be noted that, in the game-theoretic context of this paper, the term equilibrium does not necessarily imply a numerical force balance. Rather, it refers to a situation that represents a rational choice for both sides in the sense that any other individual decision would result in less favorable outcomes for the deviator.

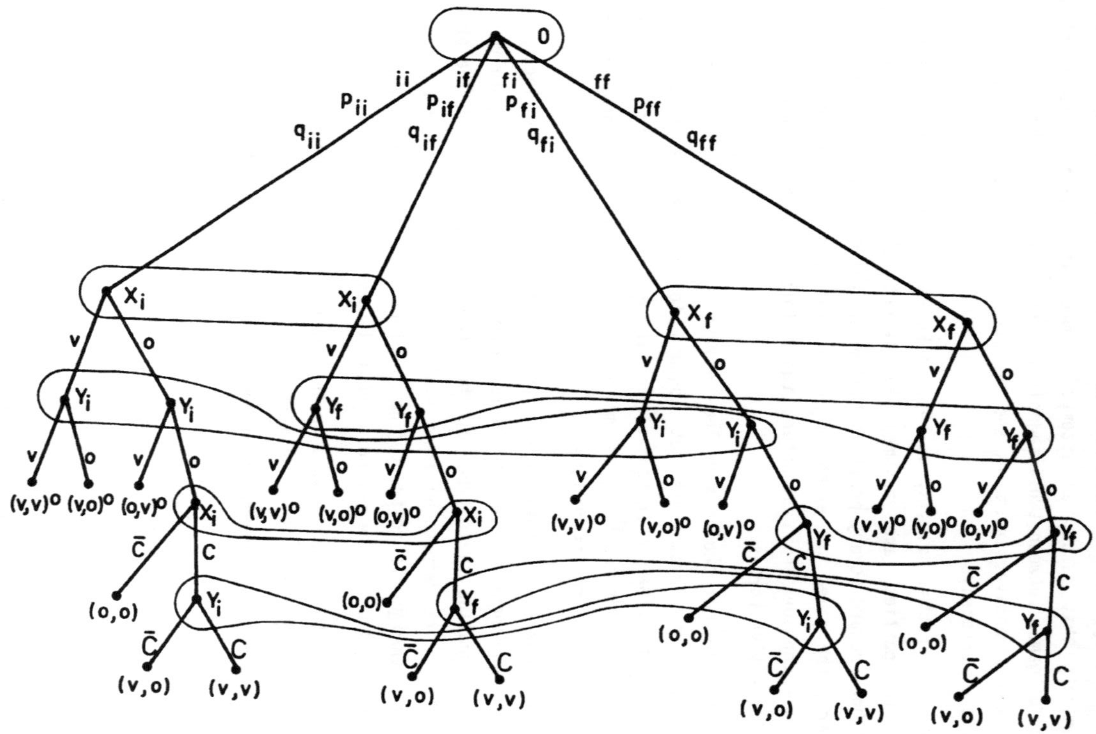

Figure 1: The basic game model: the game starts with the chance move of player 0 which determines the type combinations of X and Y. Then X chooses force structure v or o, thereafter Y. In case X and Y have chosen o, X has the possibility of restructuring from o to v (C) or not ($\bar{C}$). If X restructures its forces, Y may do likewise. The information sets are encircled. They include all decision nodes which the choosing player cannot distinguish.

equilibrium selection (HARSANYI and SELTEN, 1988). In Section 4, we analyse the games with minor costs for restructuring armed forces by using the method of equilibrium scenarios (GÜTH, 1984). We conclude by summarizing and comparing the results of Parts I and II and by indicating some possible perspectives for future research.

2. Game Model and Solution Requirements

In Part I, we have analysed the normal form game with incomplete information in which two antagonistic parties X and Y, which are either peaceful (f-type) or imperialistic (i-type), can choose once and for all between an offensive (o-choice) or a defensive oriented (v-choice) force structure. In this part, we extend the analysis to include a second choice in case both antagonists had chosen offensive force structures before. Also, we assume that both antagonists are only incompletely informed about the other's type (f or i) and that subjective beliefs about the opponent's type may be inconsistent, i.e., they need not be derived as conditional probabilities of the same probability distribution.

The model is illustrated by the complete game tree in Figure 1. The game starts with the fictitious chance move (of the chance player 0) which determines the type constellation (t_X,t_Y) of party X and Y respectively where

$$t_X,t_Y \in \{f,i\} . \qquad (2\text{-}1)$$

The probability distribution p expresses probabilistic beliefs of X concerning all possible type constellations; for instance, p_{ii} is the probability with which X expects the type constellation $(t_X,t_Y) = (i,i)$. The probabilistic beliefs of Y are given by the probability distribution q.

As in Part I, both players, knowing their own type but not the one of their opponent, must independently decide between an offensive (o) or a defensive force structure (v). Formally, this has been represented by letting first choose X between o and v and then Y where Y is not informed about X's previous move. Information conditions are graphically represented by information sets. An information set consists of decision nodes of a given player which this player cannot distinguish when deciding about his move. The game tree, where first Y and then X would decide between o and v, together with an appropriate information condition for X, would therefore describe the same conflict situation, i.e., the two games are strategically equivalent.

In case X and Y, respectively their actual types, both have chosen o, player X has the chance to initiate a process of restructuring the armed forces from offensive to defensive ones. Here, the choice C means that X implements a defensive structure on his part whereas $\bar{C}$ indicates the decision not to change the force structure. With the choice of C by X the game is over. Otherwise, player Y can either follow the move C of X by also choosing C or not ($\bar{C}$). Whereas the positions of X and Y concerning their choice between o and v are symmetric, the positions in the restructuring process following the initial (o,o)-choice are completely different. It has been assumed that one party is responsible for initiating the process by making the first decisive step with the risk that the antagonist may not follow its example.

The main reason for modelling the restructuring process in an asymmetric way is that, in most of the controversies between two antagonistic parties, it is only one of the parties that is dissatisfied with the prevailing territorial situation or which is fundamentally more afraid of being invaded by the other. Our model tries to capture such basic asymmetries by assuming that the strategic positions of both parties in the restructuring of armed forces are quite different. Whereas party X is the initiator facing the risk associated with the first (unilateral) step, party Y merely reacts to X's previous move C.

In the following, we do not determine expected payoffs by assuming probabilities of winning and losing war and cardinal utilities for all possible results as done in the first section of Part I. Instead, it is assumed that each type of both players has a complete preference order on the set of all possible plays of the game in Figure 1. A play always starts with an initial chance move and ends with a node (endpoint) of the game tree at which no player has to move anymore. Thus, different endpoints correspond to different plays. The military constellation implied by a play is indicated by denoting the vector of military force structures of the parties X and Y at the corresponding endpoint. So (v,v), for example, is the military constellation resulting from all plays where, after the initial chance move, both parties first choose o and where party X restructures its forces from offensive to defensive ones being followed suit by party Y.

Of course, it is possible to reach some military constellation either directly or via (o,o) by restructuring the armed forces. For example, the military constellation (v,o), X being defense oriented and Y having offensive armed forces, may result directly from the choice of (v,o) and indirectly via the choice of (o,o) followed by X choosing C and Y refraining from the choice of C. Clearly, a direct achievement of a given military constellation is more efficient than the indirect one requiring a restructuring effort with all its waste of scarce resources. To distinguish both military constellations, an upper index 'o' attached to the endpoint indicates that a military constellation has been reached directly. Without this index, the military constellation has been reached via (o,o). Preference relations over the set of military constellation will be denoted by $>^J_t$ with $J \in \{X,Y\}$ and $t \in \{f,i\}$. So,

$$(v,o)^0 \;>_t^X\; (v,o) \quad \text{for} \quad t \in \{f,i\} \tag{2-2}$$

means, for instance, that both types of State X rather prefer to reach (v,o) directly than by choosing o first and then restructuring its armed forces. Given the sequential decision process, our main assumptions are restrictions for the set of possible preference orderings $>_t^J$ for the different types of both players.

Assumption 1 ("type distinction"):

1.1 An i-type prefers o if his opponent chooses v .

1.2 An f-type prefers v if his opponent chooses v .

■

Observe that both assumptions can apply to several pairs of military constellations. For instance, (1.1) implies that $(v,o)^0 >_i^Y (v,v)^0$ as well as $(v,o) >_i^Y (v,v)$. Clearly, the preference relations of the i-and the f-type have to be different if the type distinction is to be useful. Loosely speaking, an i-type is willing to take advantage of the possibility of coercive threats when faced with an opponent who cannot threaten to attack. An f-type always prefers the peaceful world where no one can threaten another.

Since cardinal utilities are only uniquely defined up to positive affine transformations, the following situation implies no loss of generality:

Assumption 2 ("payoff normalization"):
The resource loss implied by first choosing o and then C instead of choosing v immediately is normalized to 1.

■

Our main assumptions are related to the Defense Efficiency Hypothesis (which has been discussed in more detail in Part I; AVENHAUS et al., 1988) and the preference for an opponent which cannot attack:

Assumption 3 ("Defense Efficiency Hypothesis" (DEH)):
An f-type prefers v if his opponent chooses o.

■

As shown in Part I, it is always dominant for an f-type to choose v instead of o regardless of whether the opponent uses v or o. Thus, Assumption 3 is strongly supported by our results of Part I. However, we do not necessarily claim that the DEH will always be satisfied. Rather, we attempt to explore the possible implications of this hypothesis to eventually provide a rationale for developing defensive weapon systems for which the DEH would be true.

Assumption 4 ("preference for defense oriented opponents"):
Each type of both players prefers an opponent which uses v instead of o.

■

This assumption does not appear to require lengthy justifications. However, in view of the DEH, Assumption 4 could imply a serious restriction: If an i-type player is firmly determined to attack his opponent, he might prefer an opponent with an offensive rather than a defensive force structure. Such an opponent could attack him too, but due to the DEH he could also be less effective in defending his own territory.

Assumption 5 ("case distinction"):
The preference of both f-type players using v for an opponent which chooses v instead of 0, is outweighed by the resource loss, i.e.,

$$(o,v)^0 >_f^Y (v,v) \text{ and } (v,o)^0 >_f^X (v,v) . \qquad (2\text{-}3)$$

■

The reason for assuming (2-3) is that games with (2-3) and games where the opposite preference relation is true are structurally very different. We first will solve the game satisfying Assumption 5 and later on consider the games where the opposite preference relation holds.

Given the type distinction as specified by Assumption 1, only Assumptions 3 and 4 entail essential restrictions. Assumption 2 is a pure convention and Assumption 5 is only used to distinguish two different classes of games.

Of course, Assumptions 1 to 5 are far from being sufficient to uniquely derive the preference relations of both types of both players. This shows that we are actually solving classes of games. Or, in other words: Not all political evaluations of possible results will matter for the solution of the game. Thus, only after a thorough analysis of a given conflict situation one knows which political evaluations are truly relevant and necessary to be answered by the responsible political actors.

In the following, we will invariably rely on Assumptions 1 to 4. In the next Section, we will also use Assumption 5. The class of games, which results if (2-3) is reversed, will be studied thereafter. Verbally, we will refer to games satisfying (2-3) as games with *considerable resource costs for restructuring armed forces*. Games with (2-3) reversed will be called games with *minor resource costs for restructuring armed forces* from offensive to defensive ones.

Before starting to analyse the games, we would like to indicate briefly the solution concept for the games at hand. If one strategy is sometimes better than another strategy and never worse, we say that the first one dominates the latter one. Clearly, no player will want to use a strategy which is dominated. However, elimination of dominated strategies may yield new dominated strategies. Thus, a strategy vector is said to be *in undominated strategies* if repeated elimination of dominated strategies will not remove any of its components. Essentially, we require the solution to be a *(subgame)perfect equilibrium point* in undominated strategies.

In an *equilibrium point* no player can gain by unilateral deviations. A subtree of a given game tree is said to describe a subgame if no information set, containing decision nodes of the subtree, contains decision nodes which do not belong to the subtree. In other words: A subgame is an informationally closed subtree of the game tree. An equilibrium point is called *(subgame)perfect* if it induces equilibrium points for all the subgames of the given game. If a game has more than one (subgame)perfect equilibrium point in undominated strategies, a unique solution will be selected by means of the dominance relation among equilibria proposed by HARSANYI AND SELTEN (1988) as part of their equilibrium selection theory. The dominance relation among equilibria can be characterized axiomatically by three intuitively convincing rationality requirements for the class of games studied here.

Sometimes, one may not be interested so much in knowing all of the equilibria of a given game or class of games but only in whether certain political constellations, e.g., the actually observed policies of the interacting parties, are consistent with basic game theoretic rationality requirements such as the equilibrium concept. Often, certain moves that appear to be irrational at a first glance may be well explained after a more thorough investigation of an appropriate game model. It is in this sense, the *method of equilibrium scenarios*, which restricts itself to testing the equilibrium conditions for certain interesting policy constellations, can help to resolve cognitive dissonance about political developments like, for instance, nuclear arms races or different policies of different parties or countries.

Especially when a game model is rather complex and therefore not easily analyzable, the method of equilibrium scenarios could be a first step in investigating such a game model without engaging into lengthy derivations and complicated case distinctions. We will partly solve games uniquely as described before and partly restrict ourselves to using the method of equilibrium scenarios. By this, we hope that our study does not only explore an interesting class of game models but also propagates some not yet commonly known tools of modern game theory.

3. The Class of Games with Considerable Resource Costs of Restructuring Armed Forces

Here we are considering the class of games satisfying Assumption (2-3), i.e., the restructuring of armed forces involves considerable costs. It should be noticed that this assumption concerns only the f-types of players X and Y. Assumption (2-3) is important because of the following reason: If both players choose C in the restructuring phase and, e.g., Y_f expects his opponent to use o, then Y_f can induce $(o,v)^0$ by choosing v and (v,v) by choosing o.

<u>Proposition 1</u>

Under Assumptions 1 to 5, peaceful types choose v immediately and would choose C in case they are asked to decide between C and $\bar{C}$, an imperialistic player (Y_i) never responds to a structural change of his opponent.

Proof

The result follows from repeated elimination of dominated strategies: Y_f chooses C because of Assumption (1.2). Similarly, Y_i uses $\bar{C}$ since he prefers (v,o) to (v,v) due to (1.1). Anticipating this, X_f chooses C since he prefers (v,o) to (o,o) because of Assumption 3 and (v,v) to (o,o) since $(v,v) >_f^X (v,o)$ due to Assumption 4. Now, Y_f will not use o since he prefers $(v,v)^0$ to $(v,o)^0$ due to (1.2) and $(o,v)^0$ to (o,o) as well as (v,v) because of Assumptions 3 and (2-3), respectively. Similarly, one can conclude that X_f chooses v: $(v,v)^0 >_f^X (o,v)^0$ due to (1.2); $(v,v)^0 >_f^X (v,o)$ due to (2-2), and $(v,o)^0 >_f^X (v,v)$ because of (2-3).

■

Due to Proposition 1, the behaviour of the f-types of both countries can be predicted unambiguously. Thus, strategic uncertainty is restricted to decisions of the i-type opponent. The strategic interaction of the two i-type players can be graphically illustrated if one removes all dominated moves, i.e., the choices of $\bar{C}$ and o by the two f-types and the choice of C by Y_i from the game tree in Figure 1. The reduced game tree is presented in Figure 2. It illustrates that the restructuring phase will be reached only in case of the type constellation $(t_X,t_Y) = (i,i)$. Thus, whenever the restructuring phase results because both countries have chosen o, each country knows that its opponent is an i-type player.

Corollary 1

Under Assumptions 1 to 5 the initial choices are type-revealing in the sense that players know that their opponent is an i-type whenever the play yields the initial choice of offensive force structures.

■

Of course, if the restructuring phase is not reached, a player might still be uncertain about his opponent's type. So, for instance, player X_i does not know the type of player Y in case of a play yielding the military constellation $(v,v)^0$.

The game in Figure 2 describes the strategic interaction of the agents controling the decision behaviour in the three information sets of players X_i and Y_i. Now, the optimal choice of player X_i in case of the play ((i,i), (o,o)) is independent of what the others do. Formally, this can be expressed by the fact that the subtree following this play describes a subgame of the game in Figure 2. The solution of this subgame is obviously the choice of $\bar{C}$ by X_i whenever

$$(o,o) \;>_i^X\; (v,o)^0 \quad \text{and} \quad (o,o) \;>_i^Y\; (o,v)^0 \tag{3-1}$$

whereas X_i will choose C if

$$(v,o) \;>_i^X\; (o,o) . \tag{3-2}$$

The contradicting conditions (3-1) and (3-2) are closely related to our Assumption 3 describing the DEH. In case of (3-2), not only the f-type players but also X_i prefers the choice of v if the opponent uses o. This is why we refer to Assumption 3 in conjunction with (3-2) as to the *Strong Defense Efficiency Hypothesis (SDEH)*. The other constellation, i.e., Assumption 3 together with (3-1), is called the ***Weak Defense Efficiency Hypothesis (WDEH)*** since here the comparative advantage of defensive force structures is not sufficient to guarantee that X_i prefers (v,o) to (o,o) and that Y_i prefers (o,v) to (o,o). It should be noted that Assumptions (3-1) and (3-2) are not exhaustive. First of all, preferences could be different for parties X and Y. Secondly, (o,o) could be viewed by X_i as intermediate to the other situations appearing in (3-1) and (3-2).

From Figure 2, it is obvious that players X_i and Y_i have to entertain different expectations for the event of (o,o) in case of SDEH and WDEH. Thus, in the following we distinguish between the games satisfying (3-2) and those satisfying (3-1).

Proposition 2

Under Assumptions 1 to 5 as well as (3-1) the solution of the game is given by the choice of $(o,\bar{C})$ by the i-type players and the choice of (v,C) by the f-type players.

Proof

The strategic interaction of players X_i and Y_i in Figure 2, where both anticipate the choice of $\bar{C}$ by X_i, can be described by the 2×2-normal form game in Table 1. There, X_i and Y_i can choose between o and v, respectively. The payoff of player X_i appears in the upper right-hand, that of Y_i in the lower right-hand corner of the matrix component. $H_i^X(\cdot)$ and $H_i^Y(\cdot)$ are utility functions representing the preference relations $>_i^X$ and $>_i^Y$, respectively. The arrows next to the matrix indicate the deviation incentives. The upward pointing arrow to the left of the matrix in Table 1 means, for instance, that X_i prefers the situation (o,o) to the choice of v by himself and o by player Y_i. Due to the deviation incentives of Table 1, the game has a unique equilibrium point. Thus, to prove Proposition 2, one only has to prove that the deviation incentives are as claimed by the arrows attached to the matrix of Table 1. That this is true, follows from Assumptions 1.1 and (3-1): $(v,o)^0 \;>_i^Y\; (v,v)^0$ and $(o,v)^0 \;>_i^X\; (v,v)^0$ due to

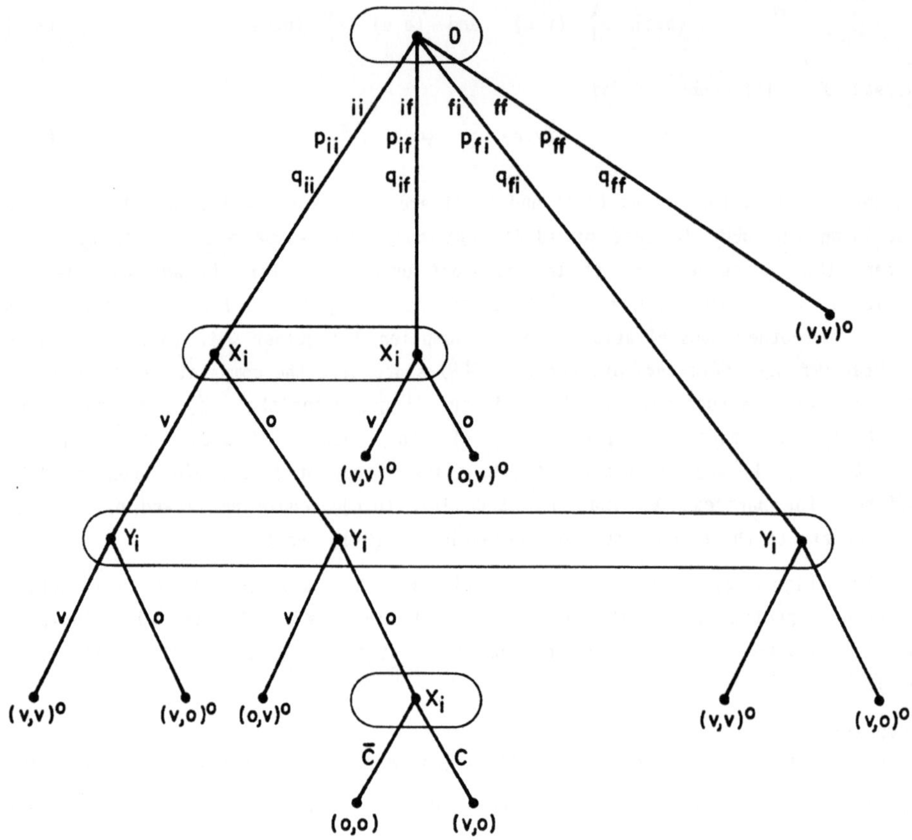

Figure 2: The reduced game model as implied by Proposition 1.

$X_i \backslash Y_i$	o	v
o	$p_{ii}\cdot H_i^X(o,o)+p_{if}\cdot H_i^X((o,v)^0)$ $q_{ii}\cdot H_i^Y(o,o)+q_{fi}\cdot H_i^Y((v,o)^0)$	$p_{ii}\cdot H_i^X((o,v)^0)+p_{if}\cdot H_i^X((o,v)^0)$ $q_{ii}\cdot H_i^Y((o,v)^0)+q_{fi}\cdot H_i^Y((v,v)^0)$
v	$p_{ii}\cdot H_i^X((v,o)^0)+p_{if}\cdot H_i^X((v,v)^0)$ $q_{ii}\cdot H_i^Y((v,o)^0)+q_{fi}\cdot H_i^Y((v,o)^0)$	$p_{ii}\cdot H_i^X((v,v)^0)+p_{if}\cdot H_i^X((v,v)^0)$ $q_{ii}\cdot H_i^Y((v,v)^0)+q_{fi}\cdot H_i^Y((v,v)^0)$

<u>Table 1</u>: Matrix representation of the strategic interaction between X_i and Y_i when X_i chooses $\bar{C}$ in Figure 2 (arrows indicate the directions of deviation incentives).

Assumption 1.1 and $(o,o) >_i^Y (o,v)^0$ and $(o,o) >_i^X (v,o)^0$ are the conditions postulated by (3-1).

■

Proposition 2 shows that the WDEH-hypothesis implies rather strong results. The i-type of both parties chooses o and would not change his force structure in case the (o,o)-situation is reached whereas the f-type does not choose o and would change from offensive to defensive force structures in case of (o,o). It is in this sense, the Weak Defense Efficiency Hypothesis implies a *complete type signaling behavior*. If a single move of the opponent becomes known, one knows which type he is.

In case of (3-2), player X_i will choose C so that both players X_i amd Y_i can anticipate that their initial choice of o leads to (v,o). As a consequence, the strategic interaction of X_i and Y_i can be described by the 2×2-normal form game of Table 2. We want to prove that the deviation incentives are as postulated by the arrows contained in Table 2. Since Tables 1 and 2 differ only in the (o,o)-component, the arrows below the matrix and at its right-hand side follow from the proof of Proposition 2. For direction of the arrow above the matrix, one has to show that $(v,o) >_i^Y (o,v)^0$: Because of Assumption 1.1, one has $(v,o) >_i^Y (v,v)^0$. Furthermore, Assumption 4 implies $(v,v)^0 >_i^Y (o,v)^0$. Thus, $(v,o) >_i^Y (o,v)^0$.

$X_i \backslash Y_i$	o	v
o	$p_{ii}\cdot H_i^X(v,o)+p_{if}\cdot H_i^X((o,v)^0)$ $q_{ii}\cdot H_i^Y(v,o)+q_{fi}\cdot H_i^Y((v,o)^0)$	$p_{ii}\cdot H_i^X((o,v)^0)+p_{if}\cdot H_i^X((o,v)^0)$ $q_{ii}\cdot H_i^Y((o,v)^0)+q_{fi}\cdot H_i^Y((v,v)^0)$
v	$p_{ii}\cdot H_i^X((v,o)^0)+p_{if}\cdot H_i^X((v,v)^0)$ $q_{ii}\cdot H_i^Y((v,o)^0)+q_{fi}\cdot H_i^Y((v,o)^0)$	$p_{ii}\cdot H_i^X((v,v)^0)+p_{if}\cdot H_i^X((v,v)^0)$ $q_{ii}\cdot H_i^Y((v,v)^0)+q_{fi}\cdot H_i^Y((v,v)^0)$

Table 2: Matrix representation of the strategic interaction between X_i and Y_i when X_i chooses C in Figure 2 (arrows indicate the direction of deviation incentives).

From the deviation incentives we can conclude

Proposition 3

Under Assumptions 1 to 5 as well as (3-2), the solution of the game is given by the strategy vector $((o,C),(v,C),(o,\bar{C}),(v,C))$, where strategies are listed in the order X_i, X_f, Y_i, Y_f whenever

$$u_i^X((o,v)^0) - u_i^X((v,v)^0) > p_{ii}/p_{if}, \tag{3-3}$$

whereas the solution is the strategy vector $((v,C),(v,C),(o,\bar{C}),(v,C))$ if condition (3-3) is reversed.

Proof

Given the deviation incentives in Table 2, it follows that only the two strategy vectors mentioned by Proposition 3 are eligible for an equilibrium of the game in Table 2. Considering that the payoff difference $u_i^X((v,o)^0) - u_i^X((v,o))$ has been normalized to 1 by our Assumption 2, one can see that Condition (3-3) would imply an upward pointing deviation incentive at the left of Table 2, whereas this arrow would have to point downward if (3-3) is reversed.

■

The left-hand side of (3-3) can be interpreted as X_i's incentive for being able to threaten country Y. Since country Y uses v, it cannot attack X. Our Assumption 1.1 concerning i-type players implies that this incentive for being able to threaten is positive for both i-type players. The ratio on the right-hand side of (3-3) can be described as X_i's relative probability for facing an i-type opponent. Proposition 3 implies that both i-types will choose o when X_i's incentive for being able to threaten Y is large and/or his relative probability for facing an i-type opponent is low whereas all players except Y_i will choose v when X_i's incentive to threaten Y is low and/or his subjective probability to meet Y_i is large.

In Table 3, we have described the solution plays for all possible situations covered by the Assumptions 1 to 5 as well as the conditions (3-1) and (3-2). Accordingly, the restructuring phase can be reached only by the type constellation $(t_X,t_Y) = (i,i)$. In case of (3-1), this type constellation does not use the possibility to restructure the military forces, whereas in cases of (3-2) and (3-3), only X_i restructures his forces from offensive to defensive ones. For (3-2) and (3-3) reversed, not even the (i,i)-type constellation reaches the restructuring phase. In this case, both types of State X use the same strategy so that State Y cannot learn anything about X's true type by observing State X's behavior. Thus, for (3-2) and (3-3) reversed, the solution is only partially type revealing due to the basic asymmetry of the role of parties X and Y in the restructuring phase.

It has been mentioned before that conditions (3-1) and (3-2) do not exhaust all possible subclasses of games satisfying Assumption (2-3). If we exclude the politically insignificant cases of indifference, the situations not covered by (3-1)

Assumption \ Type Constellation	(i,i)	(i,f)	(f,i)	(f,f)
(3-1)	$(o,o,\bar{C})$	(o,v)	(v,o)	(v,v)
(3-3) (3-2)	$(o,o,C,\bar{C})$	(o,v)	(v,o)	(v,v)
(3-3) reversed	(v,o)	(v,v)	(v,o)	(v,v)

<u>Table 3</u>: The solution plays for the four possible type constellations (t_X,t_Y) and situations covered by Assumptions (3-1) and (3-2) according to Propositions 2 and 3.

and (3-2) do either satisfy

$$(o,o) \;\succ_i^X\; (v,o)^0 \quad\text{and}\quad (o,v)^0 \;\succ_i^Y\; (o,o) \qquad (3\text{-}4)$$

or

$$(v,o)^0 \;\succ_i^X\; (o,o) \;\succ_i^X\; (v,o)\;. \qquad (3\text{-}5)$$

Condition (3-4) describes situations with a fundamental discrepancy of the i-types of both parties: Whereas X_i prefers o to v if his opponent uses o, the opposite is true for Y_i. This situation could be described as one where the Weak Defense Efficiency Hypothesis (WDEH) applies for party X and its strong version (SDEH) for party Y.

Situations satisfying (3-5) result if for X_i neither the condition in (3-1) nor the one in (3-2) hold. We can refer to these situations as the *games with an intermediate evaluation of mutual offensive force structures*. In the following, we will solve the classes of games satisfying (3-4) and (3-5) in addition to Assumptions 1 to 5 specified in Section 2.

Proposition 4

$$q_{ii}\cdot u_i^Y((o,o)) + q_{fi}\cdot u_i^Y((v,o)^0) > q_{ii}\cdot u_i^Y((o,v)^0) + q_{fi}\cdot u_i^Y((v,v)^0) \qquad (3\text{-}6)$$

is satisfied, and $(o,\bar{C}),(v,C),(v,\bar{C})(v,C))$ in case (3-6) is reversed.

Proof

In case of (3-4), the direction of the arrow above the matrix of Table 1 is ambiguous since $(v,o)^0 \succ_i^Y (v,v)^0$ due to Assumption 1.1 and $(o,v)^0 \succ_i^Y (o,o)$ because of (3-4). Condition (3-6) implies that the arrow points to the left, whereas it has to point to the right when (3-6) is reversed. Since all other arrows in Table 1 are unchanged, this proves the proposition.

■

Observe that according to Proposition 4 the solution depends crucially on the beliefs of Y_i about the type of his opponent. If q_{ii} is large, (3-6) cannot be satisfied implying that X_i chooses o and Y_i adopts v. If q_{ii} is relatively small, both i-types use o which implies the same plays as in the top row of Table 3. The subjective beliefs of X_i do not matter at all. This asymmetry is, of course, no surprise. First of all, our model is basically asymmetric due to the different roles of X and Y in the restructuring phase. Furthermore, condition (3-4) imposes another basic asymmetry as explained above.

We now turn our attention to the remaining case (3-5). Because of (3-5), the strategic interaction of X_i and Y_i is as described by Table 1. Compared to (3-1), the directions of arrows below and to the right of the matrix have not been changed. But the direction of the two remaining arrows is ambiguous if condition (3-5) is true. Due to $(v,o)^0 >_i^X (o,o)$, the arrow to the left of Table 1 may point downward as well as upward. Similarly, if $(o,v)^0 >_i^Y (o,o)$, as in condition (3-4), the arrow above the matrix may point to the right as well as to the left.

Proposition 5

Under Assumptions 1 to 5 as well as (3-5) the solution of the game is $((o,\bar{C}),(v,C),(o,\bar{C}),(v,C))$, if

$$p_{ii}\cdot u_i^X((o,o))+p_{if}\cdot u_i^X((o,v)^0) > p_{ii}\cdot u_i^X((v,o)^0)+p_{if}\cdot u_i^X((v,v)^0) \tag{3-7}$$

and (3-6) are satisfied; in case of (3-6) and the reversed inequality of (3-7) the solution is $((v,\bar{C}),(v,C),(o,\bar{C}),(v,C))$; for (3-7), but not (3-6), the solution is $((o,\bar{C}),(v,C),(v,\bar{C}),(v,C))$; if both inequalities (3-6) and (3-7) are reversed, both strategy vectors are (subgame)perfect equilibria in undominated strategies.

Proof

From what has been said about the arrows describing the deviation incentives, it follows that $((o,\bar{C}),(v,C),(o,\bar{C}),(v,C))$ is the solution in case of (3-6) if the arrow to the left of Table 1 points upward. The condition for this is expressed by (3-7). The results follows immediately if one remembers that (3-6) is the condition that the arrow above the matrix points to the left.

■

Unlike in case of (3-4), the solutions of the games satisfying condition (3-5) depend on the subjective beliefs of both i-type players about the true type of their opponent. Since (3-7) is nothing else but the condition of X_i corresponding to condition (3-6) for Y_i, one can explain (3-7) in the same way as (3-6). As pointed out before, (3-6) could be described as the condition for having a dominant incentive of Y_i to threaten X in the sense that the effect of $(v,o)^0 >_i^Y (v,v)^0$ is more decisive for the deviation incentive than the one of $(o,v)^0 >_i^Y (o,o)$. Thus, situations where both conditions (3-6) and (3-7) are valid can be described as *games with strong imperialistic incentives of the i-type players*. If only one of the two conditions is valid, one could speak of a situation with *asymmetric imperialistic incentives*. If neither (3-6) nor (3-7) is satisfied, we face a situation with *weak imperialistic incentives of the i-type players*. To derive a unique solution for all such games, we apply the risk dominance relation of the Harsanyi and Selten theory of equilibrium selection (Harsanyi and Selten, 1988), which has a nice axiomatic

	o	v
o	0 0	$(p_{ii}+p_{if})\cdot[u_i^X((o,v)^0)-u_i^X((v,v)^0)]$ $q_{ii}\cdot[u_i^Y(o,v)^0)-u_i^Y((o,v))]+$ $q_{fi}\cdot[u_i^Y((v,v)^0)-u_i^Y((v,o)^0)]$
v	$p_{ii}\cdot[u_i^X((v,o)^0)-u_i^X((o,o))]+$ $p_{if}\cdot[u_i^X((v,v)^0)-u_i^X((o,v)^0)]$ $(q_{ii}+q_{fi})\cdot[u_i^Y((v,o)^0)-u_i^Y((v,v)^0)]$	0 0

	o	v
o	0 0	δ_i^X 1
v	1 δ_i^Y	0 0

Table 4: Two transformed games of Table 1 which are strategically equivalent to the game of Table 1 according to the Harsanyi and Selten theory (the left matrix is derived from Table 1 by applying best reply invariance, the right one from the left one by using invariance with respect to isomorphisms, see Harsanyi and Selten, 1988). The variables δ_i^X and δ_i^Y are defined by

$$\delta_i^X = \frac{(p_{ii}+p_{if})\cdot[u_i^X((o,v)^0)-u_i^X((v,v)^0)]}{p_{ii}\cdot[u_i^X((v,o)^0)-u_i^X((o,o))]+p_{if}\cdot[u_i^X((v,v)^0)-u_i^X((o,v)^0)]} \tag{3-8}$$

and

$$\delta_i^Y = \frac{(q_{ii}+q_{fi})\cdot[u_i^Y((v,o)^0)-u_i^Y((v,v)^0)]}{q_{ii}\cdot[u_i^Y((o,v)^0)-u_i^Y((o,o))]+q_{fi}\cdot[u_i^Y((v,v)^0)-u_i^Y((v,o)^0)]} \tag{3-9}$$

characterization for a 2×2-normal form game with two strict equilibria. Here, an equilibrium is called *strict* if both players lose by unilaterally deviating from the equilibrium. It follows from our Assumptions 1 to 5 as well as from (3-5) and the reversed inequalities (3-6) and (3-7) that both equilibria mentioned in Proposition 5 are strict. The remaining mixed strategy equilibrium point is non-strict and therefore no (inital) solution candidate.

In Table 4, one finds two normal form games which can be derived from the game in Table 1 by certain transformations. The left matrix is derived from Table 1 by subtracting, in each column, player X_i's non-equilibrium payoff from his equilibrium payoff in that column and by a similar operation, in each row, for Y_i's payoff. Such a transformation does not affect the stability sets, i.e., the sets of all strategy vectors to which a pure strategy is a best reply and, therefore, not the best reply structure of the game. A solution concept is *best reply invariant* if it depends only on the best reply structure of the game.

The right matrix in Table 4 is derived from the left one by renormalization of utility functions, namely by dividing by one of the equilibrium payoffs. Harsanyi and Selten say that two games, which can be transformed into each other by such an operation, are isomorphic and that a solution concept is *invariant with respect to isomorphisms* if the isomorphism maps the solution of one game into the one of the other game.

Another aspect of invariance with respect to isomorphisms is symmetry invariance. Consider the case where the two parameters δ_i^X and δ_i^Y, defined by (3-8) and (3-9), of the right matrix in Table 4 are equal. Obviously, renaming players X_i and Y_i will not change the game since it is completely symmetric with respect to both players. As a consequence, symmetric games are also considered to be isomorphic. Now, in case of $\delta_i^X = \delta_i^Y$, the only solution of the right game in Table 4 that satisfies invariance with respect to isomorphisms is the mixed strategy equilibrium point. This is because the choice of one of the two strict equilibria has to be arbitrary because of the basic symmetry of the game. Here, we do not want to focus our attention on the special case $\delta_i^X = \delta_i^Y$ which has no political relevance. The main reason for considering this highly special case at all is that it provides an excellent starting point for applying the third and last axiom.

The third requirement, in addition to best-reply invariance and invariance with respect to isomorphisms, is monotonicity. A dominance relation among equilibria is said to be *monotonic* if an equilibrium, which is not dominated by other equilibria, becomes dominant whenever its payoff is increased for at least one player.

The concept for solving games with two strict equilibria according to Proposition 5 is based on the (unique) dominance relation between strict equilibria which satisfies best reply invariance, invariance with respect to isomorphisms, and monotonicity.

Proposition 6

Under Assumptions 1 to 5 as well as (3-5) and the reversed inequalities of (3-6) and (3-7), the solution determined by best reply invariance, invariance with respect to isomorphisms, and monotonicity is

$$((o,\bar{C}),(v,C),(v,\bar{C}),(v,C)) \tag{3-10}$$

for $\delta_i^X > \delta_i^Y$ and

$$((v,\bar{C}),(v,C),(o,\bar{C}),(v,C)) \tag{3-11}$$

for $\delta_i^Y > \delta_i^X$.

Proof

For $\delta_i^X = \delta_i^Y$ neither (3-10) can dominate (3-11), nor vice versa, if the dominance relation is invariant with respect to isomorphisms. Thus, for $\delta_i^X > \delta_i^Y$ dominance of (3-10) over (3-11) follows from monotonicity. Similarly, $\delta_i^Y > \delta_i^X$ implies that (3-11) dominates (3-10). Since the right game of Table 4 can be obtained from Table 1 by transformations which do not change the best reply structure or transform the game into an isomorphic one, the results for the right game of Table 4 translate into the same results as for the game of Table 1.

∎

With Propositions 2 to 6 we have obtained a complete view of the solutions of all generic games satisfying our initial Assumption (2-3) of considerable costs for restructuring armed forces (only special cases of indifference and the special games with $\delta_i^X = \delta_i^Y$ have been neglected). In Table 5, we list all the plays implied for the possible type constellations by extending our previous Table 3. The first three rows are already contained in Table 4. The play of the first row is also implied by (3-4) and (3-6) as well as (3-5), (3-6), and (3-7).

Similarly, the play implied by (3-2) and (3-3) reversed is also induced by (3-5), (3-6), and (3-7) reversed as well as by (3-5) and (3-6) and (3-7) both reversed with $\delta_i^Y > \delta_i^X$.

The play implied by (3-4) and (3-6) reversed and also by (3-5), (3-6) reversed and (3-7) or (3-7) reversed with $\delta_i^X > \delta_i^Y$ is one where the roles of X_i and Y_i are interchanged when compared with the play implied by (3-2) and (3-3) reversed.

Assumption \ Type Constellation		(i,i)	(i,f)	(f,i)	(f,f)
	(3-1)	(o,o,$\tilde{C}$)	(o,v)	(v,o)	(v,v)
(3-2)	(3-3)[1]	(o,o,C,$\tilde{C}$)	(o,v)	(v,o)	(v,v)
	(3-3) reversed	(v,o)	(v,v)	(v,o)	(v,v)
(3-4)	(3-6)	(o,o,$\tilde{C}$)	(o,v)	(v,o)	(v,v)
	(3-6) reversed	(o,v)	(o,v)	(v,v)	(v,v)
	(3-6); (3-7)	(o,o,$\tilde{C}$)	(o,v)	(v,o)	(v,v)
(3-5)	(3-6); (3-7) reversed	(v,o)	(v,v)	(v,o)	(v,v)
	(3-6) rev. (3-7)	(o,v)	(o,v)	(v,v)	(v,v)
	(3-6) rev. (3-7) rev. $\delta_i^X>\delta_i^Y$ 2)	(o,v)	(o,v)	(v,v)	(v,v)
	(3-6) rev. (3-7) rev. $\delta_i^Y>\delta_i^X$ 2)	(v,o)	(v,v)	(v,o)	(v,v)

Table 5: The solution plays for all four possible type constellations as implied by Propositions 2 to 6.

1) Only active use of restructuring.

2) Multiple equilibria.

Thus, only in case of (3-2) and (3-3) and the (i-i)-type constellation the restructuring phase will be used in an essential way. In all other cases, players will either not want to use it (this is always true for Player Y_i) or never reach it in order to use it (as in case of both f-type players). Even in case of (3-2) and(3-3) and $(t_X,t_Y) = (i,i)$ only player X_i will change his force structures from offensive to defensive ones.

The rare use of restructuring armed forces in Table 5 should be of no surprise. After all, we are considering the case (2-3) with high costs of restructuring armed forces which provides a low incentive for using the possibility to change force structures. Furthermore, the Defense Efficiency Hypothesis (DEH) implies that the immediate choice of v is rather attractive.

4. Games with Minor Resource Costs of Restructuring Armed Forces

In the following, we will consider games which satisfy Assumptions 1 to 4 but not Assumption 5, i.e., we will rely on

$$(v,v) \succ_f^Y (o,v)^0 \text{ and/or } (v,v) \succ_f^X (v,o)^0 . \tag{4-1}$$

According to (4-1), it is impossible that the costs for restructuring armed forces are high for both players X_f and Y_f, i.e., at least one of the two must have minor costs for restructuring his armed forces.

<u>Proposition 7</u>
Under Assumptions 1 to 4 as well as (4-1) both peaceful types choose C and Y_i adopts move $\bar{C}$.

<u>Proof</u>
The proof is the same as for Proposition 1 when one skips all arguments relying on Assumption (2-3).

■

It is easy to demonstrate with the help of examples which satisfy the qualifications of Proposition 7 that no further strategies can be removed by repeated elimination of dominated strategies. As in Section 3, the effect of repeated elimination of dominant strategies is graphically illustrated by the reduced game tree of Figure 3. There are still five strategically interacting agents ruling the decision behavior of X and Y in one of their information sets, i.e., we face a far more complex game situation than in Figure 2 with only three strategically interacting agents. Another major difference of the games in Figures 2 and 3 is that, contrary to the latter, the game of Figure 2 has a proper subgame. Thus, by definition, any equilibrium point of the game in Figure 3 is (subgame)perfect.

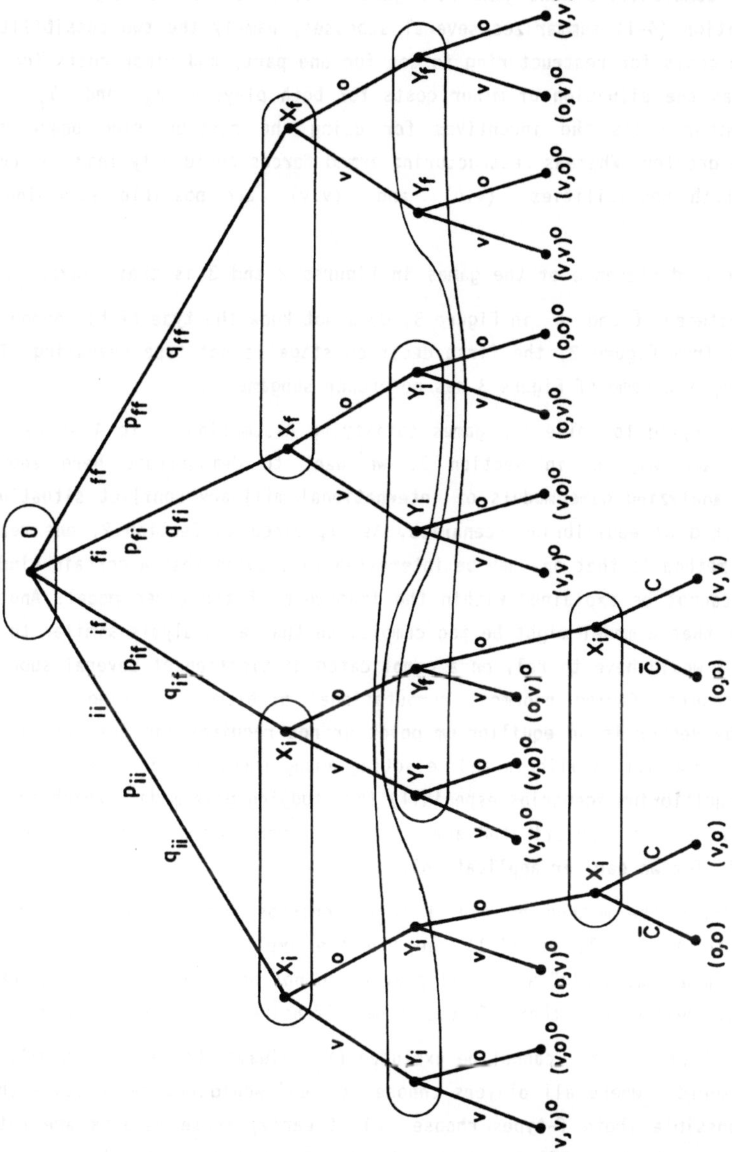

Figure 3: The reduced game model as implied by Theorem 7.

The higher complexity of the game in Figure 3 is, of course, partly due to the fact that condition (4-1) summarizes several subcases, namely the two possibilities of considerable costs for restructuring forces for one party and minor costs for the other as well as the situation of minor costs for both players X_f and Y_f. In addition, at minor costs the incentives for using the restructuring phase more essentially are greater. Whereas restructuring armed forces could only lead to (v,o) in Figure 2, both possibilities (v,o) and (v,v) are possible according to Figure 3.

Another major difference of the games in Figures 2 and 3 is that player X_i, when choosing between $\bar{C}$ and C in Figure 3, does not know the type of his opponent, i.e., different from Figure 2, the first decision stage is not type revealing. This is the reason why the game of Figure 3 has no proper subgame.

Instead of trying to solve the games satisfying Assumptions 1 to 4 as well as (4-1) in the same way as in Section 3, we want to demonstrate here another possibility of analyzing game models of international military conflict situations, namely, the method of equilibrium scenarios. As explained in Section 2, one reason for using this method is that one may be interested in showing that a certain kind of policy can or cannot be explained within the framework of the given model. Another reason might be that a model might be too complex so that an analysis similar to the one of Section 3 would have to rely on a complicated distinction of several subcases with varying results facing enormous computational problems. Checking whether a certain strategy vector is an equilibrium point or not requires far less analytical effort than deriving all equilibria and selecting among them. We therefore recommend the method of equilibrium scenarios especially for studying game models which are too complex to allow a more sophisticated analytical treatment within a reasonable time (see Güth, 1984, for an earlier application).

To demonstrate the method of equilibrium scenarios for the games satisfying Assumptions 1 to 4 as well as (4-1), we use the implication of Proposition 7. Obviously, the "peaceful world" where all players do not only choose v, but where they also would restructure their forces from offensive to defensive ones, must remain a dream if player Y_i cannot be excluded (Y_i always chooses $\bar{C}$). Similarly, the "dreadful world", where all players choose o and would not restructure their forces, is impossible (both f-types choose C). Clearly, these results are rather straightforward implications of Assumption 1. But is a less restrictive peaceful world possible in which all players use v and therefore avoid the need to restructure their forces, given the Assumptions 1 to 4 as well as (4-1)? Similarly, one can ask whether there exists an equilibrium which prescribes the choice of o by all players.

Another possibility is to find out whether certain plays are at all possible. To demonstrate this, consider the game of Figure 3. In Section 3, it has been shown that a play leading to (v,o) is possible and one might well expect that this is also possible for condition (4-1) instead of (2-3). But is there an equilibrium play yielding (v,v)? In order to enable the play yielding (v,v) for the type constellation $(t_X,t_Y) = (i,f)$, X_i must choose (o,C) and Y_f has to use his move o. But this leaves open whether Y_i and X_f use o or v. Since Y_f chooses o, it seems natural to require that also Y_i plays o. Thus, there remain only two strategy vectors to bring about the play (v,v), namely

$$s^1 = ((o,C),,(o,C),(o,\bar{C}),(o,C)) \tag{4-2}$$

and

$$s^2 = ((o,C),(v,C),(o,\bar{C}),(o,C)) \;. \tag{4-3}$$

Proposition 8

Under Assumptions 1 to 4 as well as (4-1) both strategy vectors s^1 and s^2 are equilibrium scenarios, i.e., there exist non-empty parameter regions such that s^1 and s^2 are equilibrium points of the corresponding classes of games.

Proof

Assume that s^1 is played. Our assumptions are consistent with $(v,v) >_i^X (o,o) >_i^X (v,o)$. Thus, the choice of C and o is X_i's best reply against s^1 if p_{if} is relatively large and X_i's costs for restructuring his forces are relatively low. In case $(v,o) >_i^Y (o,v)^0$, the condition for Y_i choosing o can be guaranteed by low costs of restructuring armed forces since $(v,o) >_i^Y (o,o) >_i^Y (o,v)$ due to Assumptions 4 and 1.1. If both conditions in (4-1) hold, it follows that o is optimal for Y_f and X_f if p_{ff} is sufficiently large. Clearly, all these requirements are mutually independent (p_{if} is an expectation parameter of X_i, p_{ff} of X_f).

If s^2 is played, the same conditions can be applied for X_i since only X_f's strategy is different from the one in s^1. Assume that the condition for Y_f in (4-1) is satisfied and the one for X_f not. This implies that v is X_f's best reply against s^2 and that o is Y_f's best reply if q_{if} is sufficiently large. From $(v,o)^0 >_i^Y (v,v)^0$ and q_{fi} sufficiently large, it follows that o is Y_i's best reply against s^2. Again these requirements are mutually independent showing that there exists a parameter region so that s^2 is an equilibrium point of the corresponding games.

■

Our initial intuition that minor costs for restructuring armed forces in the sense of (4-1) can induce a more essential use of the restructuring phase is confirmed by Proposition 8. If X is of type i and Y of type f, both strategy vectors s^1 and s^2 predict that both countries will first build up offensive force structures and then restructure them.

A more naive analysis might have come to the conclusion that player X_i's behaviour according to s^1 and s^2 is irrational: since both types of Y choose o according to s^1 and s^2, player X_i knows, because of Proposition 8, from the very beginning that he has to restructure his forces if he chooses o. This shows that X_i's choice of o is not an attempt to threaten Y but a measure to prevent the threat by Y (if X_i meets Y_f he can avoid the situation $(v,o)^0$ permitting Y to threaten him). An important conclusion of Proposition 8 is therefore that offensive force structures are justifiable by the argument that they will make the antagonist more inclined to agree to restructuring forces from offensive to defensive ones. Exactly this argument has been used to justify the increase of nuclear missiles in Western Europe after 1982. By now, there is also a lot of evidence that this has been the true intention of that measure and, furthermore, that the measure has been rather successful.

Now, s^1 and s^2 imply the same play whenever X_f is not present. This illustrates that the more essential use of the restructuring phase in the sense of plays leading to (v,v) is consistent with more than just one strategy combination and that, usually, one will not be able to identify the strategy combination from observing the play. In general, many different strategy vectors can imply the same plays. Therefore, game-theoretic studies of international policy problems face the fundamental problem that consistency of its predictions with observed policies provides only necessary but not sufficient evidence for their adequacy. Since one rarely will be able to check directly the adequacy of the game model by identifying the actual decision process, political evaluations, expectation parameters, and information conditions of all strategically interacting actors, this indicates that one often must be satisfied with having found one model consistent with observed policies and can hardly hope to identify the true model of a political conflict situation at hand.

Another implication of Proposition 8 is that the "dreadful world", where both types of both countries use o, is far from being unlikely as shown by the equilibrium scenario s^1. But as discussed above, the motivation for choosing o is very different for the different players. Whereas X_i and X_f only want to prevent the situation $(v,o)^0$ by choosing o, player Y_i chooses o since this makes him the only possible agressor (he always obtains (v,o)). The reason of Y_f to choose o is similar to the one of X_i and X_f.

Knowing that the "dreadful world" with all players choosing o is possible, it is interesting to ask whether the "peaceful world", where all players use v, is also possible or not. In view of Proposition 7, there are only two strategy vectors which could induce the peaceful world, namely

$$s^3 = ((v,\bar{C}),(v,C),(v,\bar{C}),(v,C)) \tag{4-4}$$

and

$$s^4 = ((v,C),(v,C),(v,\bar{C}),(v,C)) . \tag{4-5}$$

Unfortunately, because of Assumption 1.1, the choice of v cannot be optimal for an i-type player when both types of his opponent use v.

<u>Proposition 9</u>

Under Assumptions 1 to 4 as well as (4-1) neither s^3 nor s^4 are equilibrium scenarios, i.e., there are no parameters such that s^3 and s^4 are equilibrium points of the corresponding games.

■

This pessimistic result can easily be explained by the often heard argument that unilateral disarmament, here unterstood as unilaterally giving up the possibility to attack the other, will induce an imperialistic opponent to take advantage of this move. But since this argument is only a justification of our Assumption 1.1 describing the i-type players, the impossibility of the "peaceful world" is more an assumption than a result of our analysis.

5. Final Remarks

A major difference of our analyses in Part I and Part II is that choosing o in Part I means to rely definitely on armed forces which are capable of a major attack. In Part II, the choice o of offensive force structures is more ambiguous: it can also be seen as an attempt to induce the opponent, who has chosen o, to restructure his forces from offensive to defensive ones. Our analysis in Section 4 has shown that both interpretations for the choice of o can apply. A State might choose o although it knows that it has to restructure its forces later on simply because it wants to induce its opponent to change his forces from offensive to defense oriented ones.

Whereas in Part I the Defense Efficiency Hypothesis (DEH) has first been formulated in probabilities of loosing or winning a major combat, our DEH-Assumption 3 has been expressed by a fundamental preference relation of both f-type players: given that his opponent has an offensive force structure, an f-type player prefers to rely on the intrinsic comparative advantage of defense oriented forces. The two formulations of the DEH are closely related because of our results in Part I. There, it has been shown that v is the only undominated strategy of an f-type player. Thus, v is not only the best reply against the opponent's choice of o, as required by our Assumption 3, but also against the opponent's choice of v. In this sense, Assumption 3 is rather weak if compared with the results of Part I. But one should keep in mind that in Part II we have a more complex situation. Thus, similar assumptions for winning or losing probabilities as in Part I might imply different conclusions for the behaviour of f-type players. For instance, an f-type player might react to o by choosing o himself since he wants to induce his opponent to restructure his forces from offensive to defensive ones.

In our view, the game model of Figure 1 is probably the simplest way of analysing a strategic conflict where both countries may first build up offensive force structures and later on restructure them if this leads to the highly unstable constellation of mutual fears of attack with all of its preemption incentives. Actually, the model has been designed in a way such that the play, where both countries first choose offensive weapons and then restructure them, is possible. Our analysis in Section 4 shows that this play can be implied by an equilibrium point, i.e., the possibility of restructuring forces from offensive to defensive ones is used in an essential way.

As shown in Section 3, the use of the restructuring phase depends crucially on the costs for restructuring armed forces. If these costs are high for both f-type players in the sense of our condition (2-3), it never happens in equilibrium that one reaches (v,v), i.e., whenever both States choose o, the result (v,v) with defensive forces on both sides can never be achieved. The most one can obtain for considerable costs of restructuring armed forces is that one State, the initiator X, restructures its forces. As a consequence, military innovations enabling cheaper changes of offensive into defensive forces could improve the prospects of reaching (v,v) by avoiding condition (2-3).

Whereas in Part I the role of both parties X and Y was the same, here this is only true for the first phase of the decision process. In the restructuring phase, the roles of both parties are fundamentally different: Party X has to start the restructuring process whereas Party Y only considers to restructure its forces when Party X has done so before. One justification for this fundamental asymmetry has been that usually one of the two parties has strong unsatisfied claims and that it is this party which has to signal that it will not try to achieve its ends by military means. In such situations, a more symmetric type of model appears to be less realistic.

Of course, a party will not be able to change immediately all its forces from offensive to defensive ones. Furthermore, some weapons might be indispensable for both, the attack and an effective defense. Thus, the choice of C should be interpreted as the first step in the process of possible changes in the force structure. In our view, further studies of the implications of the DEH should focus on models with a multistage restructuring phase which reduce the risk of the initiator's first move C and thereby make the asymmetry of both parties more acceptable. Of course, one may also distinguish more than just an imperialistic and a peaceful type.

The DEH is an assumption and not a result of our analysis. We wanted to elaborate the implications of the DEH since it obviously has some relevance at least for special weapon systems as, for instance, those used in conventional wars. Knowing the implications of the DEH makes it possible to analyse also the incentives for military innovations inducing the intrinsic defense advantage underlying the DEH. In Section 3, it has been shown that the consequences of the DEH can be different for the weak and the strong version of the Defense Efficiency Hypothesis. Since the strong version requires that even an imperialistic X_i prefers v to o if Y chooses o, innovations favouring the DEH could bring about conditions that surmount the imperialistic incentives of i-type opponents. Furthermore, our analysis has clearly illustrated how crucial the costs for restructuring armed forces are, an aspect that is neglected in most discussions of the DEH.

Another basic motivation of this study is to propagate the methodological tools of modern game theory for analysing international conflict situations such as the modeling of incomplete information and sequential decision processes as well as the solution methods of equilibrium selection and equilibrium scenarios. Whereas equilibrium selection allows us to derive unique policy conclusions as in Section 3, the method of equilibrium scenarios restricts itself to asking whether certain policies can be in equilibrium or not. Thus, if unique policy conclusions are what one wants to achieve, there is an - although sometimes rather cumbersome - way to derive such unambiguous results. But if one is more interested in checking whether, for a given and often very complex game model, a certain political constellation is consistent with basic rationality requirements, there is a more direct and usually more comfortable way to derive such specific results. Not every game theoretic analysis of international policy problems needs to use the same level of sophistication. Usually, more simple models can be solved uniquely whereas more complex game models can be more easily analysed with the method of equilibrium scenarios. A more realistic, and thereby more complex, model will usually require a less ambitious game theoretic analysis. It will depend on the specific problem whether it is more important to derive unambiguous results or to study a more realistic game model.

References

Avenhaus, R., Güth, W., and Huber, R.K.: Implications of the Defense Efficiency Hypothesis for the Choice of Military Force Structures. Part I: Games with and without Complete Information about the Antagonist's Intentions. To appear in: Lecture Notes in Economics and Mathematical Systems, Springer, Heidelberg.

Clausewitz, G. von (1984). On War. Edited and Translated by M. Howard and P. Paret. Princeton: Princeton University Press.

Güth, W. (1984). Egoismus und Altruismus - Eine spieltheoretische und experimentelle Analyse. In: Todt, H. (ed.) Normengeleitetes Verhalten in den Sozialwissenschaften: 35-58, Berlin: Duncker und Humblot.

Harsanyi, J. and Selten, R. (1988). A General Theory of Equilibrium Seletion in Games, Cambridge, Mass.: The MIT Press.

Hofmann, H.W., Huber, R.K., and Steiger, K. (19).On Reactive Defense Options - A Comparative System Analysis of Alternatives for the Initial Defense Against the First Strategic Echelon of the Warsaw Pact in Central Europe. In: Huber, R.K. (ed.) Modeling and Analysis of Conventional Defense in Europe: 97-140, Plenum Press.

Huber, R.K., and Hofmann, H.W. 1984). Gradual Defensivity - An Approach to Stable Conventional Force Equilibrium in Europe? Oper. Res. '84: 197-211 (J.P. Bran, Ed.), Amsterdam-New York-Oxford: Elsevier Science Publ., North-Holland.

Huber, R.K. (1987). Über strukturelle Voraussetzungen für Krisenstabilität in Europa ohne Kernwaffen: Eine systemanalytische Betrachtung. Oper. Res. Spectr. Nr. 1: 1-12.

Quester, G.H. (1977). Offense and Defense in the International System. New York: John Wiley and Sons.

INSPECTOR LEADERSHIP WITH INCOMPLETE INFORMATION

by

Rudolf Avenhaus, Akira Okada and Shmuel Zamir

Abstract

Inspection problems arise when an inspector has to decide whether an inspectee has behaved legally, i.e., according to a formal agreement. For such a decision the inspector uses a well-defined procedure which is based on observations of random variables and which may or may not be announced to the inspectee.

Whereas in the latter case (no announcement) in equilibrium the inspectee will behave illegally with positive probability, in the former case he will act legally with certainty: the inspector deters the inspectee from illegal behavior by means of his "inspector leadership procedure".

In the case of complete information on both sides the leadership solution is instable. It is shown that in case the inspector has only incomplete information about the inspectee's payoff for undetected illegal action this instability disappears.

For the purpose of illustration the results are applied to material accountancy and data verification problems, and it is shown that the inspector's equilibrium strategies are the statistical tests commonly used in these contexts.

1. Introduction

Inspection problems arise when a person or organization (an *inspector*) has to decide whether a person, a group of persons, an organization or a State (*inspectee* in the following) has behaved legally, i.e., according to a law or a formal agreement. The inspector responsible for the surveillance of the behavior has two possibilities: He can declare the inspectee's behavior to be legal, or he can question it and induce a *second action level* in order to settle his suspicion.

Typically, the decision on these two alternatives has to b made with only uncertain information. Then stochastic methods have to be used. In terms of our theory, this means that the inspector is observing random variables whose distributions depend on the inspectee's behavior, and he uses these observations for deciding b tw n two alternative hypotheses. In order to find the "best" decision procedure the inspector needs information about the utilities of the inspect e and about his possible actions. This means that inspection problems have to be modelled game theoretically in contrast to probl ms like process or quality control, where the adversary is inanimate natur or a technical process.

Apparently, the first analyses of this kind are due to DRESHER (1962) and KUHN (1963) who developed their ideas in the context of arms control and disarmament. MASCHLER (1967/68) makes decisive progress[1]: He abandons the zero-sum assumption and introduces the *price-leadership principle* into the field which he takes from STACKELBERG's work (1934) on market equilibria.

Without knowledge of this work, BIERLEIN (1969) analyzed zero-sum game theoretic models for nuclear material safeguards in the fram work of the Non-Proliferation Treaty. In later papers (e.g., 1970,1983) he kept the zero-sum assumption but avoided the explicit use of payoffs to the inspector: BIERLEIN interpreted the zero-sum assumption such that ideally the inspector should consider the gain resp. loss of the inspectee as his loss resp. gain, and he was primarily interested in

[1] For the inspectee he uses the term *violator*. We do not use it, sinc it is not certain a priori (and, in fact, the inspector seeks to prevent this) that the inspectee violates the law or agr m nt. In the nuclear material safeguards application to be discuss d later, the term *operator* of a nuclear plant under safeguards is sometimes used. We prefer the neutral term *inspectee* in order to indicate the wide range of applications of the theory.

reliable inspection strat gies. HÖPFINGER (1974) also determin d reliable inspection strategies for a special inspection problem, where the total inspection effort is determined stochastically.

Except for a few attempts (MASCHLER (1967), BRAMS (1985)) the papers mentioned so far, and many more, are purely game theoretical in their scope; they do not deal with the evaluation of measurements and their errors which are typical features of many inspection problems. One of the first approaches in this direction was made by AVENHAUS and FRICK (1977) who emphasized that, in general, false alarms cannot be avoided and that, therefore, the interests of both the inspector and th inspectee are only partially opposing: Both parties want to avoid false alarms which cause losses to both. Inspection problems of this kind have to be modelled by non-zero-sum games.

The solution of concrete problems of this kind may become very complicated even in rather simple applications (see, e.g., FRICK (1976), BATTENBERG (1982)). Therefore, it turned out to be v ry convenient that the solution of games of this kind can be performed in two steps. In the first, the equilibrium false alarm probability is determined. In the second (and, for applications more important) step, a zero-sum game is solved. The solution is found with the help of the Lemma of Neyman and Pearson, well-known in statistical theory. In this way, a series of applied problems was solved (see, e.g., AVENHAUS (1986)).

A disadvantage of models where both players act simultaneously is that in equilibrium the inspectee behaves illegally with positive probability, whereas the primary objective of any inspection is the total deterrence of such behavior. FICHTNER (1985) was the first to show that the inspector leadership method (as mentioned, already introduced by MASCHLER) solves this problem. It has to be emphasiz d, however, that this method means that the inspector commits himself in a credible way to stay to the announced strategy. If one doubts that this is always possible for private or even national "inspectors", it can be safely assumed for the case of international inspection problems because of the attention paid to these procedures by the Stat s involved; thus, here this method has a satisfactory justification.

This paper shows how the leadership game differs from the "simultaneous" game considered earli r and how in equilibrium the inspect behaves legally. (FICHTNER, who in lin with MASCHLER and STACKELBERG used the term leadership *solution*, did not formulate this explicitly.) Nevertheless, the result is not yet fully satisfying, since, if the inspector plays the equilibrium strategy, the inspectee's best reply is not exclusively legal behavior and, furthermore, because of what is called the "knife-edge" property of this equilibrium: a slight mistake by the inspector in determining his equilibrium strategy can lead to a major loss. It turns out, however, that the treatment of problems of this kind with incomplete information models solves this problem effectively. This approach is the more justified if one admits that the knowledge of the insp ctor about the gain of the inspectee in case of undetected ill gal action is indeed incomplete.

Clearly, the above outlined problems are very closely relat d to a variety of economic models known as *principal agent problems* extensively studied in the economic literature (see, e.g., KANODIA (1985), DYE (1986)). The model and the applications in this pap r are not primarily economic but rather what one may call "political". The main issues here are: violation, deterrence, detection, false alarms, etc. The traditional approach to such problems has been basically that of statistical analysis. The statistical tests were occasionally incorporated as strategies in a game theoretical model.

This paper which is based on two earlier versions (AVENHAUS and OKADA, and AVENHAUS and ZAMIR (1988)) is organized as follows:

In the next section, the nuclear material safeguards example, with its basic tools material accountancy and data verification, is introduced since it represents an important application of the models to be discussed subsequently; in fact, it stimulated the development of these models. In the third section, quite a general game theoretical model is proposed which generalizes the statistical procedure such that by appropriate *calibration* of the model - i.e., restrictions on payoffs and strategies - the well-known Neyman-Pearson test emerges as the unique Nash equilibrium of the game. The model is sequential with the potential of capturing the importance of the time in situations in which *time is valuable*. This is not done in existing statistical and game theoretical models which are basically static.

In the fourth section, the differences between games without and with a leadership of the inspector are analyzed with the help of simplified models which lead, as outlined above, to the formulation and treatment of inspector leadership games with incomplete information; this is achieved in the fifth chapter. In the concluding section, the results are applied to the problems outlined in the second one. Among others, it is shown that practitioners' questions like those for best decision procedures can indeed be answered without precise knowledge of payoff parameter values which especially in the application considered here are hard to estimate.

Since in this paper statistical and game theoretical methods are matched, sometimes there arises the problem of having to use symbols which are common in the one, but not in the other field, or even worse which are used in both with different meanings. Thus, compromises have to be found which occasionally are not too statisfying both for the statistician and the game theorist.

2. The Nuclear Material Safeguards Example

There are many important applications of the general framework which will be developed subsequently (see, e.g., AVENHAUS (1986)). The case of nuclear material safeguards, however, is of primary importance; in fact, this special application stimulated the development of the formalism presented here as already mentioned before, therefore, it will be discussed in some detail in the following.

In partial fulfillment of the *Treaty for the Non-Proliferation of Nuclear Weapons* (NPT), the International Atomic Energy Agency (IAEA) in Vienna controls the nuclear material of the peaceful sectors[2] of those States' nuclear fuel cycles who have ratified this Treaty (by now more than 100 States) in order to achieve the objective of safeguards which is according to the model agreement (IAEA 1971)

[2] There are five States who officially have non-peaceful nuclear fuel cycles; three of them signed and ratified the NPT. According to the NPT provisions, these States are not obliged to have IAEA safeguards applied to their peaceful nuclear fuel cycles; they accept it, however, for good will reasons.

> "...the timely detection of diversion of significant quantities of nuclear material from peaceful nuclear activities to the manufacture of nuclear weapons or of other nuclear explosive devices or for purposed unknown and deterrence of such diversion by the risk of early detection."

The basic principle of IAEA safeguards is *material accountancy*, i.e., the comparison of book with physical inventory at the end of a given inventory period. The procedures for the performance of safeguards were also laid down in the model agreement: The operator of a nuclear plant generates all source data for the establishment of a material balance and reports these data to a national or regional authority; the international authority *verifies these reported data with the help of independent measurements* made on a random sampling basis. If there are no significant differences between the operator's reported data and the international authority's own findings, then all the operator's data are accepted and the material balance is closed on the basis of the operator's data alone.

According to this procedure, the operator of a nuclear plant, who intends to divert material, has two basic possibilities: Either he can simply divert material without data falsification, hoping that the measurement uncertainty of the whole balance covers this diversion, or he can falsify data such that the balance appears to be correct, hoping that the measurement and sampling uncertainty of the data verification procedure covers this falsification. Naturally, he can use both 'strategies'; we will come back to this point.

There are model agreement statements about the *maximum routine inspection effort* which provides the basis for all random sampling procedures. In order to make meaningful statements about this effort, the importance of the nuclear material in the sense of the NPT processed in the different plants of the nuclear fuel cycle had to be defined (the "critical" mass of about 8 kg plutonium or highly enriched uranium is necessary for the construction of one nuclear device). Thus, the concept of *effective kilogram* was introduced. Accordingly 1 kg of plutonium corresponds to one effective kilogram, whereas 1 kg of uranium with an enrichment of 0.01 and above corresponds to a quantity in effective kilograms that is obtained by taking the square of enrichment. The maximum routine inspection effort, given in inspection man-hours spent in a nuclear plant, is determined on the basis of the annual throughput or inventory of nuclear material expressed in effective kilograms.

Now, let us look at material accountancy more closely: Consider a so-called material balance area (a nuclear plant or a part of it). At time t_{i-1} the beginning inventory I_{i-1} is measured; during the inventory period $[t_{i-1}, t_i]$ receipts R_i and shipments S_i are measured which together with the beginning inventory add up to the book inventory B_i at t_i. This is compared to the ending inventory I_i of this period such that the difference between B_i and I_i is used (which for traditional reasons is called Material Unaccounted For, or shortly MUF),

$$MUF_i \equiv X_i := B_i - I_i = I_{i-1} + R_i - S_i - I_i, \; i = 1,\ldots,n. \tag{2-1}$$

Since in observing these quantities measurement errors cannot be avoided, the variances of which are known to all parties involved, the vector $X' = (X_1,\ldots,X_n)$ is a random vector with covariance matrix Σ. If there are no systematic errors, then the only correlation between the differences (2-1) for two subsequent inventory periods is given by the variance of the common intermediate inventory, i.e., the elements of the covariance matrix Σ are given by

$$(\Sigma)_{ij} = \text{cov}(X_i, X_j) = \begin{cases} \text{var}(X_i) & \text{for} \quad i = j \\ -\text{var}(I_i) & \text{for} \quad i \pm j \\ 0 & \text{otherwise.} \end{cases} \tag{2-2}$$

In case the operator of the plant, being the inspectee, does not divert any material, the expectation of the random vector X is zero due to the conservation of matter. If, according to general experience, we assume the measurement errors to be normally distributed, then the random vector X is multivariate normally distributed with the expectation zero and covariance matrix Σ,

$$X \sim N(0,\Sigma) \quad \text{for legal behavior.} \tag{2-3a}$$

In case the operator diverts the amount μ_i of material in the i-th period, the expectation vector of X is $\mu' = (\mu_1,\ldots,\mu_n)$, and we get

$$X \sim N(\mu,\Sigma) \quad \text{for illegal behavior.} \tag{2-3b}$$

Let us assume that the illegal behavior of the operator consists in choosing a vector μ such that

$$\sum_{i=1}^{n} \mu_i = e'\cdot\mu = M, \; e' = (1,\ldots,1). \tag{2-4}$$

Then the problem of the inspector is to decide with the help of an observation of the random vector X if the operator behaved legally, or on the contrary, he diverted the amount M of material. The value of M will be discussed later.

Let us consider now a simplified data verification problem: A nuclear plant is shut down for some time, and there is an inventory of nuclear material which is collected in n items. The operator of the plant presents via his national authority his estimates Y_i, $i = 1,\dots,n$ either correctly, or he may falsify the ith-estimate by the amount μ_i, $i = 1,\dots,n$; if he does this and is not detected, he can divert the corresponding amount of material and the material balance still appears to be correct. The inspector estimates k data independently; his estimates are Z_i, $i = 1,\dots,k$ without loss of generality. Since he is only interested in deviations, he uses the differences

$$X_i := Y_i - Z_i, \quad i = 1,\dots,k \tag{2-5}$$

of reported estimates and independently generated ones in order to solve his decision problem which is very similar to the one described above: In case the inspector takes the maximal sample size (k=n), X is multivariate normally distributed with expectation zero and some known covariance matrix Σ. In case the operator falsifies the data Y_i by the amounts μ_i, $i = 1,\dots,n$, the expectation of X is $\mu' = (\mu_1,\dots,\mu_n)$. If, again, it is assumed that the total falsification is M, then we have the same problem as before, only the covariance matrix being different.

We assume the total amount of material or, equivalently, the total falsification M to be in the order of magnitude of the critical mass described above. Naturally, one may assume that the total diversion consists of severall small diversions using different *paths* (inventories at different points of time, diversion without data falsification by using the measurement uncertainty of the material balance and others). There are, however, both practical and theoretical reasons to assume that in case of diversion the operator will use only one path if he really wants to divert material.

3. Simple Inspection Games

There are two players, namely the inspector and the inspectee. $M_1, M_2, \dots, M_n$ are the (pure) action sets of the inspectee in stages

1,2,...,n, respectively. The action set of the inspector is the same for all stages and consists of two elements A (alarm) and A (clear or continue).

$E_1,\dots,E_n$ are the signals (or observations) sets in stages 1,2,... . E_t is typically a finite dimensional Eucledian space or a subset of it. $f_1,\dots,f_n$ are transition probabilities:

$$f_t : M_1 \otimes E_1 \otimes \dots \otimes M_{t-1} \otimes E_{t-1} \otimes M_t \rightarrow E_t,$$

here, $M_0 = E_0 = \emptyset$ by convention. That is, for any actions $\mu = (\mu_1,\dots,\mu_t)$, $\mu_i \in M_i$ of the inspectee in stages 1,...,t and observations $x_1,\dots,x_{t-1}$, $x_i \in E_i$ at previous stages 1,...,t-1, $f_t(\mu_1,x_1,\dots,\mu_{t-1},x_{t-1},\mu_t)$ is the probability density of the random variable X_t observed at stage t.

For t = 1,...,n, I_t and O_t are real functions on M_1 x...x M_t. These are the payoff functions to the inspector and to the inspectee, respectively, if the game is stopped by alarm after stage t < n. Finally, I and O are two real functions on M_1 x...x M_n, the payoff functions when the game ends after n stages with no alarm called.

The game is played as follows:

- At stage 1, the inspectee chooses an action $\mu_1 \in M_1$ (not observable by the inspector). An observation x_1 is drawn from a random variable X_1 according to the density function $f_1(.|\mu_1)$. This observation becomes common knowledge to both players.

- The inspector chosses either $\bar{A}$ in which case the game continues to stage 2 or A in which case the game terminates with payoffs $(I_1(\mu_1),O_1(\mu_1))$.

- Inductively: At stage t, t = 1,...,n, the inspectee chooses $\mu_t \in M_t$, an observation x_t is made from a random variable,

  $$X_t \quad f_t(.|\mu_1,x_1,\dots,\mu_{t-1},x_{t-1},\mu_t).$$

  The inspector chooses either $\bar{A}$ in which case the game continues to stage t + 1 if t < n, or terminates with payoffs $(I(\mu_1,\dots,\mu_n), O(\mu_1,\dots,\mu_n))$ if t = n, or A in which case the game ends with payoffs $(I_t(\mu_1,\dots,\mu_t),O_t(\mu_1,\dots,\mu_t))$. Note that by this notation we assume that the payoffs are determined only by the actions of the inspectee and not by the observations. This is obvious if no alarm is called. In case of alarm this requires the implicit assumption that the actual diversions are checked and found out after the alarm.

The extensive form of this game is scetched in Figure 1.

The first step in analyzing our model is to consider a special case corresponding to situations in which *time is not important* - inspite of the sequentiality of the problem; the only issues of interest are whether or not there was illegal behavior of the operator, and whether or not an alarm was called.

In addition to being a natural simple case to start with, this first step may be considered as *calibration* of the model, namely relating it to the existing, mainly statistical analyses of the problem. To do this, we choose a special, simple payoff structure capturing the underlying assumptions of most existing analyses.

The formal reduction to the non-sequential case is obtained by the following set of assumptions.

Assumption 1

o The inspector decides on whether to call an alarm only once at the end of the n-th stage, i.e., after having observed the whole random vector $X' = (X_1, \ldots, X_n)$. Note that this does not exclude the possibility that the inspector decides on calling an alarm already after $t < n$ observations. However, since *time is not important,* he may as well postpone his announcement to the end of the game. In other words, this assumption still leaves room to some inherent sequentiality and intrinsic order of the test procedure.

o The inspectee's diversion strategy is completely determined at the beginning of the game, i.e., he decides either not to divert ($\mu_t = 0, t = 1, \ldots, n$) or to divert according to the plan $\mu' = (\mu_1, \ldots, \mu_n)$ where $\mu_t \geq 0$, $\Sigma\, \mu_t = M$. The constant $M > 0$ is the "critical diversion" as motivated in the previous section. ■

Under these assumptions the game in extensive form can now be described as given by Figure 2. We denote this game by Γ_0.

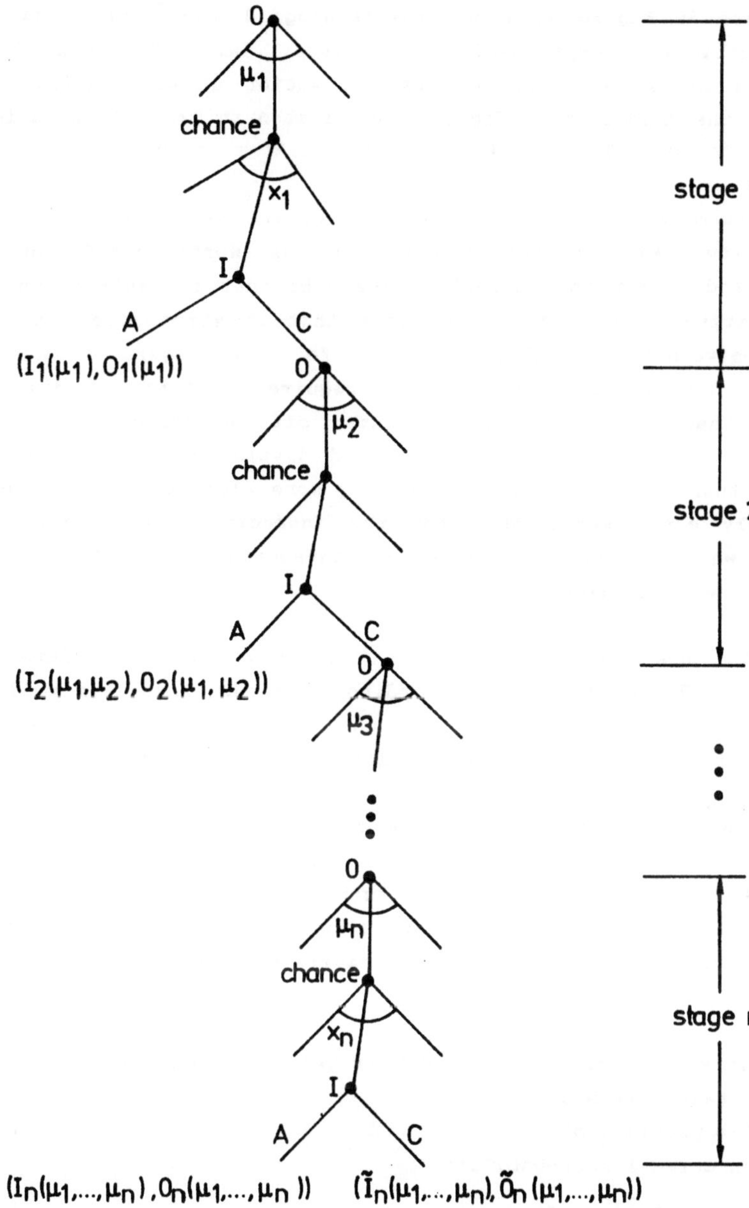

Figure 1: The sequential game in extensive form

The notation in Figure 2 is self-explaining: 1 and $\bar{1}$ are legal and illegal actions, respectively. $\mu' = (\mu_1, \dots, \mu_n) : \Sigma \mu_i = M$ is a diversion plan. $x' = (x_1, \dots, x_n)$ is the vector of observations. The payoffs at the endpoints of Figure 1 are listed in Table 1 which is to be read as follows: $I(1,\bar{A}) = 0(\bar{1},\bar{A}) = 0$, $I(1,A) = -e$, $0(\bar{1},A) = -b$ etc.

Here, we normalized the utilities so that the status quo (A,1) has zero utility, and undetected diversion is worth $d > 0$ for the inspectee and $-c$ for the inspector. The fact that all entries in line A are negative is to reflect the idea that an alarm (i.e., an open conflict between the two players) is bad for both players, compared to the status quo. $a < c$ is the obvious requirement that, if diversion occurred, then the inspector is better off detecting it than not detecting it. $h < b$ means that detected diversion is worse for the inspectee than a false alarm (reflecting some element of punishment), and finally $e < a$ means that for the inspector a detection of a diversion which means "a failure" of safeguards is worse than the inconvenience of a false alarm.

A general pure strategy of the inspector is an alarm set $s \subset E_1 \otimes \dots \otimes E_n$, that is: According to s the inspector calls an alarm at the end of period n if and only if $(x_1, \dots, x_n) \in s$. A mixed strategy σ is a probability distribution on pure strategies.

A pure strategy of the inspectee is a choice between 1 and $\bar{1}$ and a diversion plan $\mu' = (\mu_1, \dots, \mu_n)$ satisfying $\Sigma \mu_t > M$. Thus, a mixed strategy is a joint distribution on such pairs. This is equivalent to a behavioral strategy $\tau = (q,p)$ where q is the probability of $\bar{1}$, and p is the probability distribution over diversion plans μ if he chooses $\bar{1}$.

Any pair of strategies (σ,τ) determines two conditional probabilities, namely the false alarm probability

$$\alpha(\sigma,\tau) = \text{prob}(A|1,\sigma) \qquad (3\text{-}1)$$

and the non-detection probability

$$\beta(\sigma,\tau) = \text{prob}(\bar{A}|\bar{1},\sigma,\tau). \qquad (3\text{-}2)$$

Clearly, $\alpha(\sigma,\tau)$ does not depend on τ while $\beta(\sigma,\tau)$ depends on τ only through p, and it is linear in p; hence, we shall write $\alpha(\sigma)$ and $\beta(\sigma,p) = \int \beta(\sigma,\mu)\, dp(\mu)$.

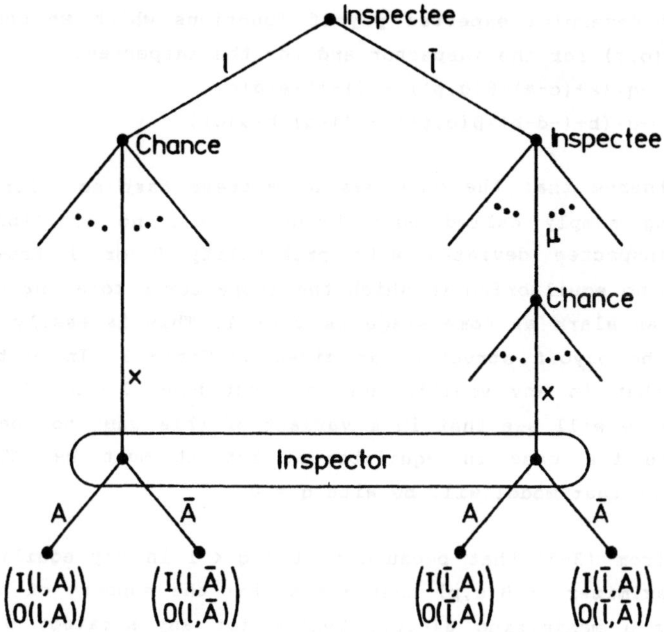

Figure 2: The "essentially" non-sequential game

Inspectee / Inspector	Legal behavior l	Illegal behavior $\bar{l}$
No alarm $\bar{A}$	0 / 0	d / -c
Alarm A	-h / -e	-b / -a

Table 1: The simplified payoff structure for the essentially non-sequential inspection game as given by Figure 1.
$0 < e < a < c$, $0 < h < b$, $d > 0$.

Any σ and τ determine expected payoff functions which we denote by $I(\sigma,\tau)$ and $O(\sigma,\tau)$ for the inspector and for the inspectee:

$$I(\sigma,\tau) = -q\cdot(a+(c-a)\cdot\beta(\sigma,p)) - (1-q)\cdot e\cdot\alpha(\sigma) \tag{3-3}$$

$$O(\sigma,\tau) = -q\cdot(b+(-d-b)\cdot\beta(\sigma,p)) - (1-q)\cdot h\cdot\alpha(\sigma). \tag{3-4}$$

First, we observe that the game has no extreme Nash equilibrium (in the following simply called equilibrium), i.e., no equilibrium in which the inspectee deviates with probability 0 or 1 from legal behavior and no equilibrium in which the inspector's total probability for calling an alarm at some stage is 0 or 1. This is easily seen by looking at the payoff structure as given by Table 1. In particular, this means that in any equilibrium one must have $0 < q < 1$. In the next section we will see that in a variant of this game not only that $q = 0$ can be the case in equilibrium, but it must be: The only equilibrium of that model will be with $q = 0$.

It follows from (3-3) that because of $0 < q < 1$ in any equilibrium p must be a maximizer of $\beta(\sigma,p)$ that means that the support of p has to contain only μ maximizing $\beta(\sigma,\mu)$. Denote for any strategy σ of the inspector

$$\beta(\sigma) := \sup_p \beta(\sigma,p) = \sup_\mu \beta(\sigma,\mu). \tag{3-5}$$

If (σ,μ) is an equilibrium, we can rewrite (3-3) and (3-4) as

$$I(\sigma,\tau) = -q\cdot(a+(c-a)\cdot\beta(\sigma)) - (1-q)\cdot e\cdot\alpha(\sigma) \tag{3-6}$$

$$O(\sigma,\tau) = -q\cdot(b+(-d-b)\cdot\beta(\sigma)) - (1-q)\cdot h\cdot\alpha(\sigma). \tag{3-7}$$

Note that we actually eliminated the diversion strategy plan μ from the equilibrium strategy of the inspectee, since it is automatically determined by (3-5). It will reappear in the auxiliary game to be discussed later.

Let us proceed to determine q and σ in equilibrium. The inspectee's payoff is $(d+b)\cdot\beta(\sigma) - b$ if he chooses $\bar{1}$ and $-h\cdot\alpha(\sigma)$ if he chooses 1. Therefore, in the (α,β)-space the inspectee's best reply is to choose $\bar{1}$ or 1 or both according to whether

$$\beta(\sigma) \gtreqless \frac{b}{d+b} - \frac{h}{d+b}\cdot\alpha(\sigma), \tag{3-8}$$

see Figure 3. In this figure, also the corresponding zones in the (α,β)-plane and the resulting payoff functions for the inspector are given.

Note that the inspector's payoff is discontinuous along the line L where it is also not defined since the inspectee, being indifferent between $\bar{1}$ and 1, can use any mixture of them as a best reply to th strategy σ.

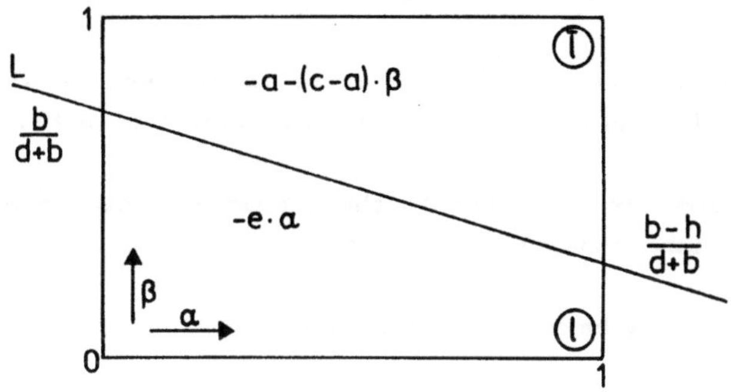

Figure 3: The payoff for the inspector when using the strategy σ, and the inspectee is doing his best reply

In order to maximize the inspector's payoff function, we first notice that its domain in the $(\alpha(\sigma),\beta(\sigma))$-space is not the whole unit square, but is restricted by the line $\beta = \beta(\alpha)$ of the relation between α and β for most powerful tests, that is

$$\beta(\alpha) = \inf_{\sigma}\{\beta(\sigma)\,|\,\alpha(\sigma)=\alpha\}$$

or, with (3-5),

$$\beta(\alpha) = \inf_{\sigma:\alpha(\sigma)=\alpha} \sup_{\mu} \beta(\sigma,\mu). \tag{3-9}$$

Obviously, the function $\beta(\alpha)$ is determined by parameters of the problem, in particular, by the distributions of the random vector $(x_1,\dots,x_n)$. We make the usual assumptions on $\beta(\alpha)$, satisfied, for instance, for the multinormal distribution:

Assumption 2

The line $\beta = \beta(\alpha)$ is well-defined for all $\alpha \in [0,1]$. It is differentiable, convex, strictly decreasing and satisfies $\beta(0) = 1$, $\beta(1) = 0$. ■

Consider now the following auxiliary game played by the inspector and the inspectee which we d note by G_α and call the ***axiliary game with parameter*** α:

- The strategy set of the inspector is $\Sigma_\alpha := \{\sigma | \alpha(\sigma) \leq \alpha\}$.

- The pure strategy set of the inspectee is the set of all diversion plans $M := \{\mu' = (\mu_1, \ldots, \mu_n) : \mu_i \geq 0, i=1,\ldots,n, \Sigma \mu_i \geq M\}$.

- The payoff from the inspectee to the inspector, when (σ,μ) are used, is the detection probability $1 - \beta(\sigma,\mu)$.

Assumption 2 can now be written as follows:

Assumption 2'

For any α, $0 < \alpha < 1$, the game G_α has a value $v(\alpha) = 1 - \beta(\alpha)$, and both players have optimal strategies (i.e., sup and inf may be replaced by max and min, respectively). Furthermore, $v(\alpha)$ is differ ntiable, concave, strictly increasing and satisfies $v(0) = 0$, $v(1) = 1$. ■

Note that G_α depends on α, but not on the parameters of the original game. Of course, the value of α, for which Γ_α plays a role, will depend on the data of the original game. Later on, we show that Assumption 2' is satisfied under the usual assumptions of multinormally distributed signals.

Now, we add the line $\beta = \beta(\alpha)$ to Figure 3 and maximize the inspector's payoff, see Figure 4. In region 1, the payoff decreases in α for any fixed β, whereas in region 1 it decreases in β for fixed α. It follows that the supremum in σ is reached on the line $\beta = \beta(\alpha)$.

On that part of the line, in region 1 the payoff decreases in σ, similarly on the part in region 1 it decreases in β. It follows that the supremum is at the intersection of the line $\beta = \beta(\alpha)$ with th straight line L. Thus, we conclude, if (σ^*,τ^*) is an equilibrium, that the corresponding (α^*,β^*) is the unique solution of

$$\beta(\alpha^*) = \frac{b}{b+d} - \frac{h}{b+d} \cdot \alpha^* . \tag{3-10}$$

Sinc (α^*,β^*) lies on L, by its definition σ^* makes the inspect e indifferent between legal and illegal behavior, so, an illegal behavior with any probability q will be a best reply against σ^*. However, the payoff to the inspector at (α^*,β^*) is a function of q,

$$I(\sigma^*,q) = q\cdot(-a-(c-a)\cdot\beta(\alpha^*)) - (1-q)\cdot e\cdot\alpha^* . \qquad (3\text{-}11)$$

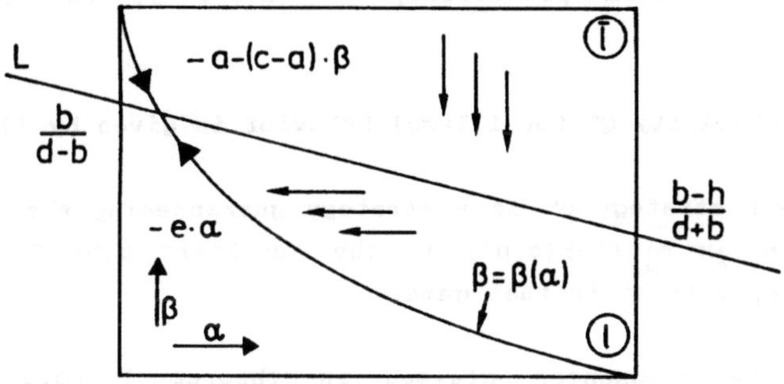

Figure 4: The Nash equilibrium values (α^*,β^*)

The value q^* of q now has to be determined so as to satisfy the second condition for an equilibrium which is that α^* is a best reply to q, that is, it maximizes

$$q^*\cdot(-a-(c-a)\cdot\beta(\alpha)) - (1-q^*)\cdot e\cdot\alpha$$

as a function of α. By the assumption of differentiability of $\beta(\alpha)$, this function has one local maximum at α^* satisfying the first derivate condition which is

$$\frac{e}{q^*} = e - (e-a)\cdot-\frac{d}{d\alpha}\beta(\alpha)\Big|_{\alpha^*} . \qquad (3\text{-}12)$$

Thus, q^* is determined since α^* is fixed by (3-10).

Note that $0 < q^* < 1$ as expected from our assumptions. By straightforward verification it can be shown that this local maximum α^* is also the global maximum, i.e., it is the unique best reply to q^*.

As soon as (α^*,q^*) are uniquely determined in equilibrium, the rest of the strategies are determined as follows: the inspector chooses σ^* as a maxmin strategy in the game Γ_{α^*}, and the inspectee uses p^* which is any probability distribution over best replies μ^* to σ^* in that game. We summarize this discussion in

Theorem 1

The game Γ_0 as given by Figure 2 has a unique equilibrium $(\sigma^*,(q^*,\mu^*))$ in which

(i) the false alarm probability $\alpha^* = \alpha(\sigma^*)$ is the solution of (3-10),

(ii) the probability q^* for illegal behavior is given by (3-12),

(iii) the test strategy σ^* is a strategy guaranteeing the lower value $v(\alpha^*) = \max_{\sigma} \min_{\mu}(1-\beta(\sigma,\mu))$ in the auxiliary game G_{α^*}; μ^* is a best reply to σ^* in that game. ■

Note that the uniqueness claimed in Theorem 1 holds only for (α^*,β^*,q^*). There may well be more than one σ^*, or μ^*, although as we will see later, may also be unique, e.g., in the multinormal case.

A special case worth noticing is when $\beta(\alpha) = 1 - \alpha$. In this case, equations (3-10) and (3-12) yield

$$\alpha^* = \frac{d}{d+b-h}, \quad q^* = \frac{e}{c+e-a},$$

which is the (unique) mixed equilibrium of the matrix game in Table 1. This describes a situation in which the inspector has no specific test procedure for detecting illegal behavior and, therefore, $(\alpha^*,1-\alpha^*)$ appears as mixed equilibrium in that game.

The equilibrium of the inspection game Γ_0 has two undesirable and related features: First, as was already mentioned, the unique equilibrium has a positive probability for illegal behavior ($q^* > 0$). Second, the inspector's payoff at the equilibrium point is drastically discontinuous. This is particularly disturbing in view of the fact that the inspectee is indifferent between legal and illegal behavior at that point. Therefore, by "doing as well as in the equilibrium point" he may inflict a severe loss to the inspector by behaving illegally with probability 1.

In the following sections, we address these two problems: First, we propose a variant of the model which takes into account the natural leadership role of the inspector. We show that in the leadership game there is a unique equilibrium in which the inspector uses the same strategy σ^* as in the equilibrium of the game Γ_0, but the inspectee uses $q^* = 0$, i.e., he behaves legally in equilibrium.

We then turn to the discontinuity problem and introduce some uncertainty of the inspector about the utility d of the inspectee in case of undetected illegal action. We show that the result of such uncertainty removes the discontinuity of the inspector's payoff at equilibrium.

4. Deterrence and Commitment

An important feature of safeguards games is the fundamental asymmetry in the players' roles. The inspector, generally representative of an official authority, can, and usually does, announce his inspection strategy while it is inconceivable for the inspectee to make public his deviation strategy.

Furthermore, the inspector has the *power of commitment*: not only can he make public his inspection strategy, but he can credibly commit himself to follow the announced strategy (e.g., to sign a contract implemented and enforced by court). In game theoretial terminology this means that the inspector can become a *Stackelberg leader* in the game if he chooses so.

As mentioned in the introduction, the idea of granting a leadership role to the inspector was already brought into the literature on safeguards and arms control (see, e.g., MASCHLER (1967/68), FICHTNER (1985)). However, to our best knowledge, the precise distinction between the equilibrium with and without inspector leadership was not explicitly formulated.

Does this power of commitment give an advantage to the inspector, and hence, does it induce a significant change in the analysis? Let us first look at a very simple example. Before, we recall the notion of Stackelberg leadership which we would like to emphasize as being a version of the game rather than a version of the solution.

Definition 1

Given any game Γ (in extensive or in strategic form) with players set $N = \{1,...,n\}$ and corresponding (mixed) strategy sets $\Sigma_1,...,\Sigma_n$. The version of the game with commitment power to player i is the game Γ_i played as follows: Player i chooses $\sigma_i \in \Sigma_i$ which is made common knowledge to all players. Then, the game Γ is played by the players $(N\backslash\{i\})$ where the moves of player i are mady by a "machine" executing the instructions of σ_i. ∎

At a first glance, one may suspect that such a modification should not be very important, since after all in Nash equilibrium reasoning each player assumes the strategies of the others to be given. In order to see this more carefully, consider the two-person game G in strat gic form as given by Table 2.

The game G has a unique equilibrium which is (1/2,1/2) for player 1 and (3/4,1/4) for player 2, yielding expected payoffs (3/4,1/2).

2 1	Left L	Right R
Top T	0 1	1 0
Bottom B	1 0	0 3

Table 2: Strategic or normal form game G

The game $\tilde{G}$ with commitment power to player 1 is a game the extensive form of which is sketched in Figure 5.

In equilibrium, at any information set U_x, player 2 will choose L if $x < 1/2$, R if $x > 1/2$ and be indifferent if $x = 1/2$. The only parameter needed to complete the specification of the second play r's strategy in equilibrium is $0 \leq q \leq 1$, the probability of playing L at

$U_{1/2}$. Denote the resulting strategy by τ_q. The first player's payoff against τ_q is given by, see Figure 6,

$$f_q(x) = \begin{cases} x & 0 \le x < \frac{1}{2} \\ q\cdot x + 3\cdot(1-q)\cdot(1-x) \quad \text{for} & x = \frac{1}{2} \\ 3\cdot(1-x) & \frac{1}{2} < x \le 1. \end{cases}$$

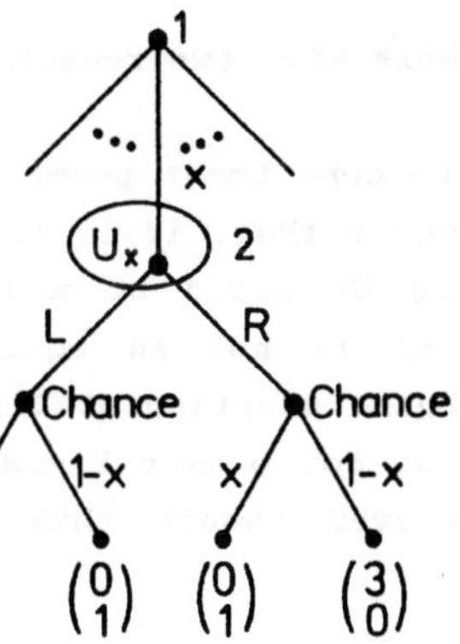

Figure 5: The game G with commitment power

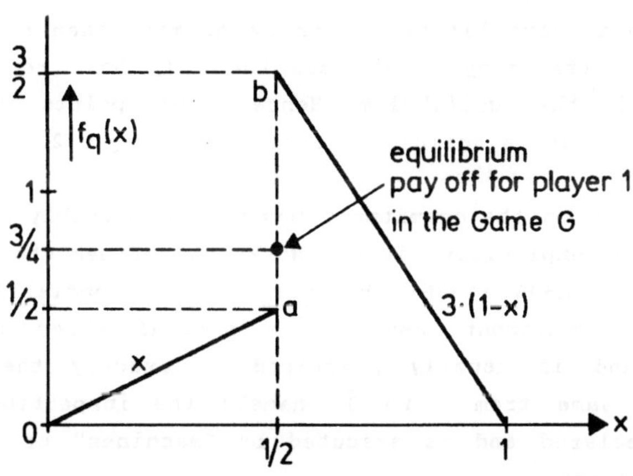

Figure 6: The function $f_q(x)$

Now, an equilibrium point is a pair (x^*, τ_{q^*}) such that τ_{q^*} is best reply to x^* which is satisfied automatically for all (x^*, q^*) by the construction of τ_{q^*}, and $f_{q^*}(x)$ has a global maximum over [0,1] at x^*. But $f_{q^*}(x)$ *has no maximum* unless $q^* = 0$, i.e., player 2 plays R. It follows that the game G has a unique equilibrium point at $x^* = 1/2$, $q^* = 0$ with expected outcome (3/2,1/2).

So, each of the two games G and $\tilde{G}$ has a unique equilibrium (which is also a perfect equilibrium), but *they are* not the same. While they yield the same payoff for player 2 the outcome is higher for player 1 in the equilibrium of G. The difference 3/2 - 3/4 may be viewed as the *value of his commitment power.*

Let us conclude this example with two remarks.

1) In order to see how the commitment power changes the equilibrium of the original game, observe that, if in the game tree of G the moves at the nodes following U_x would be made by player 1 instead of chance, then (x^*=1/2,R) is not an equilibrium, since it is not equilibrium in the subgame starting at $O_{1/2}$: The best reply to R is not $x_* = 1/2$, but x = 0. So, player 1 commits himself not to *change his mind* and play B even though this would be more profitable against R.

2) Equilibrium of $\tilde{G}$ can be viewed as a limit of points ($x=1/2+\varepsilon$,R) which are not equilibria, but from the point of view of player 1, all he can improve his payoff is by no more than ε, and in reward to this "inefficiency" R becomes the *only best reply*, unlike the situation in the equilibrium. Hence, these points are more stable than the equilibria against deviations of player 2.

Our observations on the commitment power were probably noticed, though not spelled out explicitly, in the works on refinements of equilibria, see, e.g., VAN DAMME (1986). However, it is especially relevant for our context: Commitment power is a natural thing for inspection authorities and is usually practiced in exactly the same way we modified our game from G to $\tilde{G}$, namely the inspection strategy is officially declared and is executed by "machines" or agents without decision authority.

To say this more precisely, let us first give a second representation of the game Γ_0 which is equivalent to that in Figure 2, see Figure 7. In this figure, we used the same notation as before: μ is the inspectee's diversion plan, σ is the test strategy of the inspector,

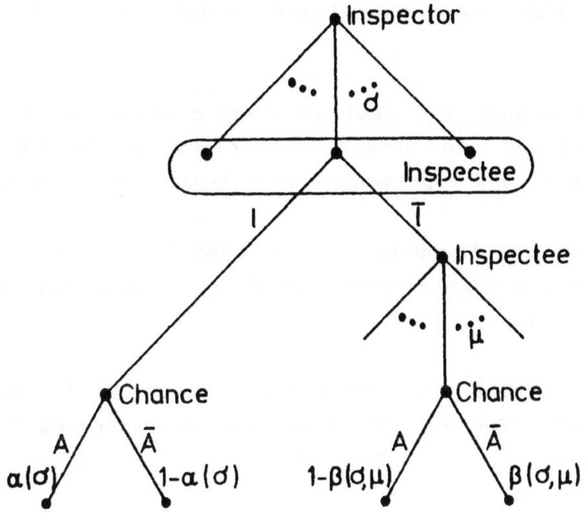

Figure 7: An equivalent representation of the game Γ_0

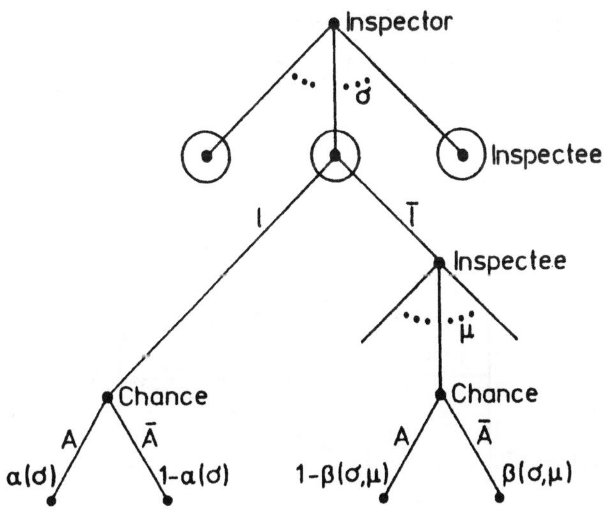

Figure 8: The inspector leadership game Γ_1

and $\alpha(\sigma)$ and $\beta(\sigma,\mu)$ are the false alarm and not detection probabilities.

If the inspector announces his test strategy σ which is then executed by a neutral mechanism, we obtain the following "leadership game" which is denoted by Γ_1 and represented in extensive form in Figure 8.

A "pure" strategy of the inspector in the game Γ_1 is a test procedure σ. For the inspectee, a pure strategy is a map τ which assigns to each σ a strategy $\tau(\sigma) = (q,\mu)$.

The analysis of the game Γ_1 is very similar to that of Γ_0 (modified as done in our example). For any fixed α, the corresponding μ and σ is given by the auxiliary game G_α. The values α^*, β^* and μ^* are therefore identical to those in Γ_0.

The inspectee's best reply as a function of α is (fixing $\beta = \beta^*$, see Figure 4)

$$\text{play} \quad \begin{matrix} \bar{1} \\ \\ 1 \end{matrix} \quad \text{if} \quad \begin{matrix} 0 \le \alpha < \alpha^* \\ \\ \alpha^* < \alpha \le 1, \end{matrix}$$

and if $\alpha = \alpha^*$, play any mixed strategy $(q,1-q)$. The corresponding payoff to the inspector is, see Figure 9,

$$f_q(\alpha) = \begin{cases} -a - (c-a)\cdot\beta(\alpha) & 0 \le \alpha < \alpha^* \\ q\cdot(-a-(c-a)\cdot\beta^*) - (1-q)\cdot e\cdot\alpha^* \quad \text{for} & \alpha = \alpha^* \\ -e\cdot\alpha & \alpha^* < \alpha \le 1. \end{cases}$$

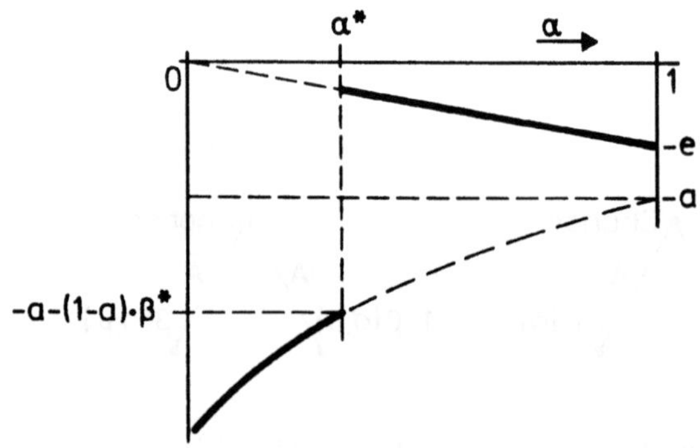

Figure 9: Payoff to the inspector in the leadership game Γ_1

Now, in equilibrium, $f_q(\alpha)$ has to have its maximum at α^*, but $f_q(\alpha)$ does not have a maximum unless $q = 0$. We thus get

Theorem 2

The leadership game Γ_1 has a unique Nash equilibrium point (hence also perfect) in which the inspector commits himself to the most powerful test with false alarm probability α^* determined by (3-10), and the inspectee surely does not deviate from legal behavior. ■

Note that, inspite of having $q = 0$ in equilibrium, the inspectee is still indifferent between $\bar{1}$ and 1. Nevertheless, this equilibrium may be called a *deterrence equilibrium* since it is the limit of points of the form $(\alpha^*+\varepsilon,1)$. These are not equilibria, but rather deterrence points in which the inspector gets an outcome ε-close to that in the equilibrium, and the inspectee strictly prefers legal to illegal behavior.

5. Insp ctor Leadership with Incomplete Information

In the last section, we assumed that the inspector and the inspectee know all payoff parameters in Table 1. There are, however, many real situations in which this assumption of complete information is too strong to be satisfied. In particular, it rarely happens that the inspector has complete knowledge of the inspectee's payoff. Therefore, in this section we will modify our rule of the inspector leadership game Γ_1 described in Figure 8 so that it can capture a situation in which the inspector is uncertain about the inspectee's payoff parameter. We will use HARSANYI's model (1967/68) of games with incomplete information and assume:

(i) The inspectee's payoff d in case the inspectee acts illegally ($\bar{1}$) and the inspector does not raise an alarm is a random variable on the interval $[0,\infty)$ with the (cumulative) distribution F(.). This distribution F(.) is known to both players. In what follows we will use the symbol t for the value of this random variable.

(ii) The inspectee knows the realized value t, but the inspector does not. All other payoff parameters in Table 1 are common knowledge. We may justify this in such a way that the losses of the inspectee in case of false alarms or illegal behavior are economic losses or well-defined sanctions which are publicly known, whereas his gain in case of undetected illegal action can only be guessed by the inspector.

Let $x = (x_1, \dots, x_n) \in \mathbb{R}^n$ denote a vector of observations. A general test procedure of the inspector is represented by a measurable function from $\mathbb{R}^n$ to $\{A, \bar{A}\}$, therefore, we denote by

$$\Sigma := \{\sigma \mid \sigma: \mathbb{R}^n \rightarrow \{A, \bar{A}\};\ \sigma \text{ is measurable}\} \tag{5-1}$$

the set of all test procedures.

Remark that we defined here the set of pure strategies. We could have defined the set of mixed strategies (distributions on Σ) or the set of behavioral strategies

$$\Sigma = \{\sigma \mid \sigma: \mathbb{R}^n \rightarrow [0,1];\ \sigma \text{ is measurable}\}.$$

However, for our equilibrium analysis, this larger set will not be needed, and pure strategies will suffice.

Recall that a behavioral strategy of the inspectee in Γ_0 is $\tau = (q, \mu)$ where q is the probability for illegal behavior and $\mu \in \Delta_M$ is the diversion plan in case of illegal behavior. For q = 1 or q = 0 we shall also write $\tau = (\bar{1}, \mu)$ or $\tau = 1$, respectively. Any pair (σ, τ) determines a false alarm probability $\alpha(\sigma)$ which depends on σ only and non-detection probability $\beta(\sigma, \mu)$ which depends on τ only through μ. The rules of our game with incomplete information are described as follows:

- First, a chance move selects a value of the inspectee's payoff t in case of 'no detection' according to the distribution F(.). The inspectee is informed of this value (his own type), and the inspector is not.

- The inspector selects a test procedure $\sigma \in \Sigma$ and announces it to the inspectee.

- With complete knowledge of (t, σ), the inspectee selects $\tau = (q, \mu) \in \mathcal{T}$.

o Finally, a chance move selects "alarm" (A) or "no alarm" ($\bar{A}$) with the probability distribution

$$(\alpha(\sigma), 1-\alpha(\sigma))$$

in case the inspectee selects l or else

$$(1-\beta(\sigma,\mu), \beta(\sigma,\mu))$$

in case the inspectee selects $\bar{l}$ with a diversion plan μ.

We call this game the inspector leadership game with incomplete information and denote it by Γ_2. The extensive form of this game is described in Figure 10.

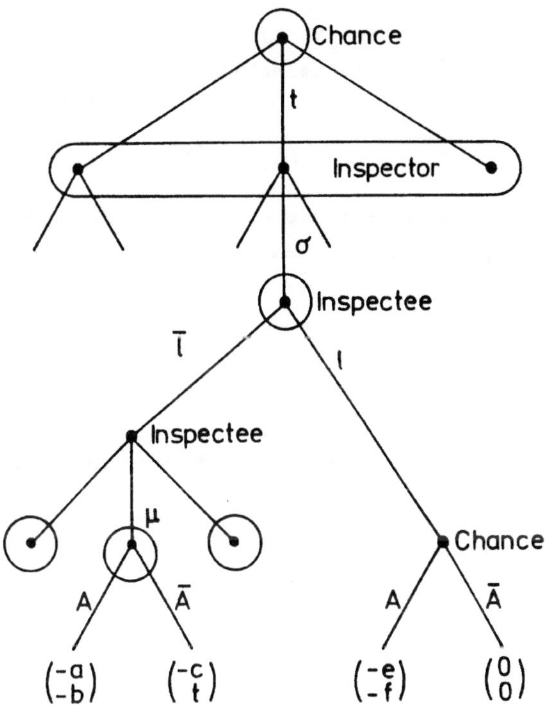

Figure 10: The extensive form of the inspector leadership game Γ_2 with incomplete information

Note that the set of pure strategies for the inspector is Σ as in Γ_0 or Γ_1 while for the inspectee the set of pure strategies is the set of maps

$$[0,\infty) \otimes \Sigma \rightarrow \mathcal{T} .$$

We shall denote this map also by τ and use for instance $\tau(t,\sigma)$ for the element (q,p) in chosen by the inspectee of type t when facing the test procedure $\sigma \in \Sigma$.

In the following analysis it turns out that, since the inspectee makes the last personal move, he may as well do without randomizing over 1 and $\bar{1}$ or over diversion plan μ, therefore, from now on we restrict the strategies τ to be pure, i.e.,

$$\tau : [0,\infty) \otimes \Sigma \to \mathcal{T}_0 := \{1\} \cup \{(\bar{1},\mu) \mid \mu \in \Delta_M\}. \tag{5-2}$$

Given a strategy combination (σ,τ), the conditional expected payoff for the inspector, given that t is selected, is

$$I_2(\sigma,\tau|t) = \begin{cases} -e\cdot\alpha(\sigma) & \text{for } \tau(t,\sigma) = 1 \\ -a\cdot(1-\beta(\sigma,\mu)) - c\cdot\beta(\sigma,\mu) & \phantom{\text{for }} \tau(t,\sigma) = (\bar{1},\mu), \end{cases} \tag{5-3}$$

and the conditional expected payoff for the inspectee, given that t is selected, is

$$O_2(\sigma,\tau|t) = \begin{cases} -f\cdot\alpha(\sigma) & \text{for } \tau(t,\sigma) = 1 \\ -b\cdot(1-\beta(\sigma,\mu)) + t\cdot\beta(\sigma,\mu) & \phantom{\text{for }} \tau(t,\sigma) = (\bar{1},\mu), \end{cases} \tag{5-4}$$

respectively. The expected payoff to the inspector is defined by

$$I_2(\sigma,\tau) = \int_0^\infty I_2(\sigma,\tau|t)\, dF(t),$$

whereas that to the inspectee is (5-5)

$$O_2(\sigma,\tau) = \int_0^\infty O_2(\sigma,\tau|t)\, dF(t).$$

An equilibrium point of the game Γ_2 is defined as follows:

Definition 2

A strategy combination (σ^*, τ^*) of Γ_2 is a (subgame perfect) equilibrium point of Γ_2 if and only if

(1) $EI_2(\sigma^*, \tau^*) \geq EI_2(\sigma, \tau^*)$ for all $\sigma \in \Sigma$,

(2) $O_2(\sigma^*, \tau^*(t, \sigma^*) | t) \geq O_2(\sigma^*, \tau | t)$ for all $\tau \in \mathcal{T}_0$ and for all $t \in [0, \infty)$. ■

In order to simplify the subsequent Theorem and its proof, in line with previous assumptions we use

Assumption 3

(1) The distribution $F(.)$ has the density function $f(.)$, i.e., it can be written as

$$F(t) = \int_0^t f(t')dt' \quad \text{for all} \quad t \in [0, \infty).$$

(2) For every $\alpha \in [0,1]$, the minmax problem

$$\min_{\sigma \in \Sigma_\alpha} \max_{\mu \in \Delta_M} \beta(\sigma, \mu)$$

has a solution. ■

We denote by $\beta(\sigma)$ the maximum and by $\beta(\alpha)$ the minmax value of $\beta(\sigma,\mu)$,

$$\beta(\sigma) = \max_{\mu \in \Delta_M} \beta(\sigma, \mu) \tag{5-6}$$

$$\beta(\alpha) = \min_{\sigma \in \Sigma_\alpha} \max_{\mu \in \Delta_M} \beta(\sigma, \mu) = \min_{\sigma \in \Sigma_\alpha} \beta(\sigma). \tag{5-7}$$

The next theorem characterizes a (subgame perfect) equilibrium point of the game Γ_2.

Theorem 3

Under Assumption 4.1, a pure strategy combination (σ^*,τ^*) of the inspector leadership game Γ_2 with incomplete information is an equilibrium point of Γ_2 if and only if it satisfies the following conditions:

(1) The false alarm probability $\alpha^* = \alpha(\sigma^*)$ is given by

$$\alpha^* = \arg \max_{\alpha \,\varepsilon\, [0,1]} \Big[(-e\cdot\alpha)\cdot F(K(\alpha)) + (-a\cdot(1-\beta(\alpha))-c\cdot\beta(\alpha))\cdot(1-F(K(\alpha))\Big] \tag{5-8}$$

where $K(\alpha)$ is given by

$$K(\alpha) = \frac{b-\alpha\cdot f}{\beta(\alpha)} - b \tag{5-9}$$

and where $\beta(\alpha)$ is given by (5-6).

(2) σ^* is an optimal strategy of the inspector in the auxiliary game G_{α^*}, i.e., it is given by

$$\sigma^* = \arg \max_{\sigma \,\varepsilon\, \Sigma_{\alpha^*}} \min_{\mu \,\varepsilon\, \Delta_M} (1-\beta(\sigma,\mu)).$$

(3) For every pair $(t,\sigma) \,\varepsilon\, [0,\infty) \otimes \Sigma$,

$$\tau^*(t,\sigma) = \begin{cases} 1 & h(t,\sigma) < 0 \\ 1 \text{ or } (\bar{1},\mu^*(t,\sigma)) \quad \text{if} & h(t,\sigma) = 0 \\ (\bar{1},\mu^*(t,\sigma)) & h(t,\sigma) > 0 \end{cases} \tag{5-10}$$

where

$$h(t,\sigma) = -b + (b+t)\cdot\beta(\sigma) + f\cdot\alpha(\sigma)$$

and

$$\mu^*(t,\sigma) = \arg \max_{\mu \,\varepsilon\, \Delta_M} \beta(\sigma,\mu).$$

Proof

We will prove this theorem by "backward induction": Let (σ^*,τ^*) be an equilibrium point of Γ_2. First, for every pair $(t,\sigma) \,\varepsilon\, [0,\infty) \otimes \Sigma$ and every $\tau = (\bar{1},\mu)$ we have from (5-4)

$$O_2(\sigma,\tau|t) = (b+t)\cdot\beta(\sigma,\mu) - b.$$

Since $b + t > 0$, the left hand side is maximized with respect to μ by maximizing $\beta(\sigma,\mu)$, and this maximum is attained by Assumption 4.2.

Second, given a pair $(t,\sigma) \varepsilon [0,\infty) \otimes \Sigma$, the inspectee's expected payoff is

$$\begin{array}{lll} -f\cdot\alpha(\sigma) & & 1 \\ & \text{if he selects} & \\ -b\cdot(1-\beta(\sigma)) + t\cdot\beta(\sigma) & & \bar{1}. \end{array}$$

Choice of the alternative which gives a higher payoff yields (5-9).

To verify condition (1) in Definition 2 we have to show that

$$\sigma^* = \arg \max_{\sigma \varepsilon \Sigma} \int_0^\infty I_2(\sigma,\tau^* | t) \, dF(t) \qquad (5\text{-}11)$$

where by (5-3) and (5-9) $I_2(\sigma,\tau^* | t)$ is

$$I_2(\sigma,\tau^* | t) = \begin{cases} -e\cdot\alpha & \text{if } \quad h(t,\sigma) < 0 \\ -a\cdot(1-\beta(\sigma)) - c\cdot\beta(\sigma) & \text{if } \quad h(t,\sigma) > 0. \end{cases}$$

To do this, we first prove the following

Proposition

If (σ,τ) is Nash Equilibrium and if we let

$A = \{t | h(t,\sigma) = 0, \; 0 < q(t,\sigma) < 1\}$,

then $\text{prob}(A) = 0$.

Proof

Let D be given by

$$D = \inf_{\alpha,\beta}(-e\cdot\alpha-[-a\cdot(1-\beta)-c\cdot\beta]) \geq a - e > 0;$$

in words: D is a lower bound for the inspector's payoff difference between 1 and $\bar{1}$ (keeping his own strategy unchanged). For $\varepsilon > 0$, denote by σ_ε the modification of σ such that $\alpha(\sigma_\varepsilon) = \alpha(\sigma) + \varepsilon$. The consequences of changing σ to σ_ε are the following:

- At any value of t in which the inspectee played 1 against σ, he will certainly play 1 against σ_ε, and the inspector's payoff difference will change by the order of ε.

- At any value of t in the event A the inspectee strictly prefers 1 over 1 against σ_ε, while he was indifferent against σ where he played $\bar{1}$ with probability $q(t,\sigma)$. The inspector's payoff thus increases by at least $q(t,\sigma)\cdot D$.

Summing up, the change in the inspector's payoff resulting from changing σ to σ_ε is at least of the order $prob(A)\cdot q(t,\sigma)\cdot D-k\cdot\varepsilon$ which is positive for ε small enough contradicting the equilibrium property of (σ,τ). ■

Returning to (5-11) and noting

$$\Sigma = \bigcup_{\alpha\,\varepsilon\,[0,1]} \Sigma_\alpha$$

we can rewrite the maximization problem posed by (5-11) as

$$\max_{\alpha\,\varepsilon\,[0,1]} \; \max_{\sigma\,\varepsilon\,\Sigma_\alpha} \int_0^\infty I_2(\sigma,\tau^*\,|t)\; dF(t). \tag{5-12}$$

We first consider the first maximization in (5-12):

$$\max_{\sigma\,\varepsilon\,\Sigma_\alpha} \int_0^\infty I_2(\sigma,\tau^*\,|t)\; dF(t). \tag{5-13}$$

For every $\alpha\;\varepsilon\;[0,1]$ and every $\sigma\;\varepsilon\;\Sigma_\alpha$ we have, using the Proposition,

$$\int_0^\infty I_2(\sigma,\tau^*\,|t)\; dF(t) = \int_{t:h(t,\sigma)\,<\,0} (-e\cdot\alpha)\cdot dF(t) \; + $$

$$+ \int_{t:h(t,\sigma)\,>\,0} (-a\cdot(1-\beta)-c\cdot\beta)\; dF(t) = \tag{5-14}$$

$$= \int_0^{\frac{b-\alpha\cdot f}{\beta}-b} (-e\cdot\alpha)\cdot dF(t) + \int_{\frac{b-\alpha\cdot f}{\beta}-b}^\infty (-a\cdot(1-\beta)-c\cdot\beta)\cdot dF(t)$$

where $\beta \equiv \beta(\sigma)$. Since the sum of the two integrals in (5-14) is motonically decreasing as a function of β, $0 \le \beta \le 1$, see AVENHAUS and OKADA (1988), the subproblem (5-13) has the same solution as

$$\min_{\sigma \in \Sigma_\alpha} \beta(\sigma). \tag{5-15}$$

According to (5-7) we have denoted this minimum by $\beta(\alpha)$. Then the maximization problem (5-10) is equivalent to the maximization problem

$$\max_{\alpha \in [0,1]} \int_0^{K(\alpha)} (-e\cdot\alpha)\cdot dF(t) + \int_{K(\alpha)}^{\infty} (-a\cdot(1-\beta(\alpha))-c\cdot\beta(\alpha))\cdot dF(t) =$$

$$= \max_{\alpha \in [0,1]} (-e\cdot\alpha)\cdot F(K(\alpha)) + (-a\cdot(1-\beta(\alpha))-c\cdot\beta(\alpha))\cdot(1-F(K(\alpha)))$$

where $K(\alpha)$ is given by (5-9). Therefore, we can prove (1) and (2) in the theorem. Conversely, we can prove without much difficulty that a pure strategy combination (σ^*,τ^*) satisfying (1) to (3) in the Theorem is an equilibrium point of Γ_2. ∎

Theorem 3 gives us the following decisions for the inspector and the inspectee at an equilibrium point of the inspector leadership game Γ_2 with incomplete information.

For any $\sigma \in \Sigma$, the inspectee

(1) selects the illegal strategy μ which maximizes the non-detection probability $\beta(\sigma,\mu)$, and

(2) behaves legally if his payoff t in case of no detection is smaller than the critical level

$$K(\sigma) = \frac{b-f\cdot\alpha(\sigma)}{b(\sigma)} - b$$

and behaves illegally if t is greater than $K(\sigma)$.

The inspector

(3) selects the false alarm probability α^* which maximizes his expected payoff given in (5-8), and

(4) selects the test procedure $\sigma \in \Sigma_{\alpha^*}$ which minimizes the non-detection probability $\beta(\sigma,\mu)$ under the assumption that the inspectee maximizes $\beta(\sigma,\mu)$.

When the inspector employs a test procedure $\sigma(\alpha)$ which is a solution of

$$\min_{\sigma \in \Sigma_\alpha} \max_{\mu \in \Delta_M} \beta(\sigma,\mu), \quad 0 \leq \alpha \leq 1,$$

the inspectee behaves legally if $t < K(\alpha)$ and illegally if $t > K(\alpha)$ where $K(\alpha)$ is given by (5-9). This means that, from the inspector's point of view, the inspectee behaves legally with probability $F(K(\alpha))$.

For any false alarm probability α, $K(\alpha)$ represents the critical value of the inspectee's payoff in case of no detection which determines his behavior. We can show that $K(\alpha)$ is monotonically increasing in α (see AVENHAUS and OKADA (1988)). This implies that the probability $F(F(\alpha))$ for legal behavior, when the inspector selects a false alarm probability α, is also monotonically increasing in α.

We now assume that the distribution $F(.)$ has a bounded support $[d_0, d_1]$ where $0 < d_0 \leq d_1$. Figure 11 illustrates the critical value $K(\alpha)$ in (5-9) between legal behavior l and illegal behavior l, the probability $F(K(\alpha))$ for legal behavior l, and the inspector's expected payoff function in both cases of complete information ($d=d_0=d_1$) and incomplete information ($d_0 < d_1$). With the help of Figure 11, we can see that the discontinuity of the inspector's payoff with respect to the false alarm probability in case of complete information is removed by the introduction of incomplete information.

We have two possibilities about the equilibrium false alarm probability α^* in case of incomplete information, i.e., (1) an interior solution in (α_0, α_1) where $d_0 = K(\alpha_0)$ and $d_1 = K(\alpha_1)$, and (2) the boundary solution $\alpha^* = \alpha_1$. From the viewpoint of the inspector, the boundary solution $\alpha^* = \alpha_1$ is important because the probability for illegal behavior $\bar{l}$ is zero if he selects α^*. We call an equilibrium point with the boundary solution $\alpha^* = \alpha_1$ the legal behavior equilibrium point. In the following, we provide a sufficient condition for the legal behavior equilibrium point.

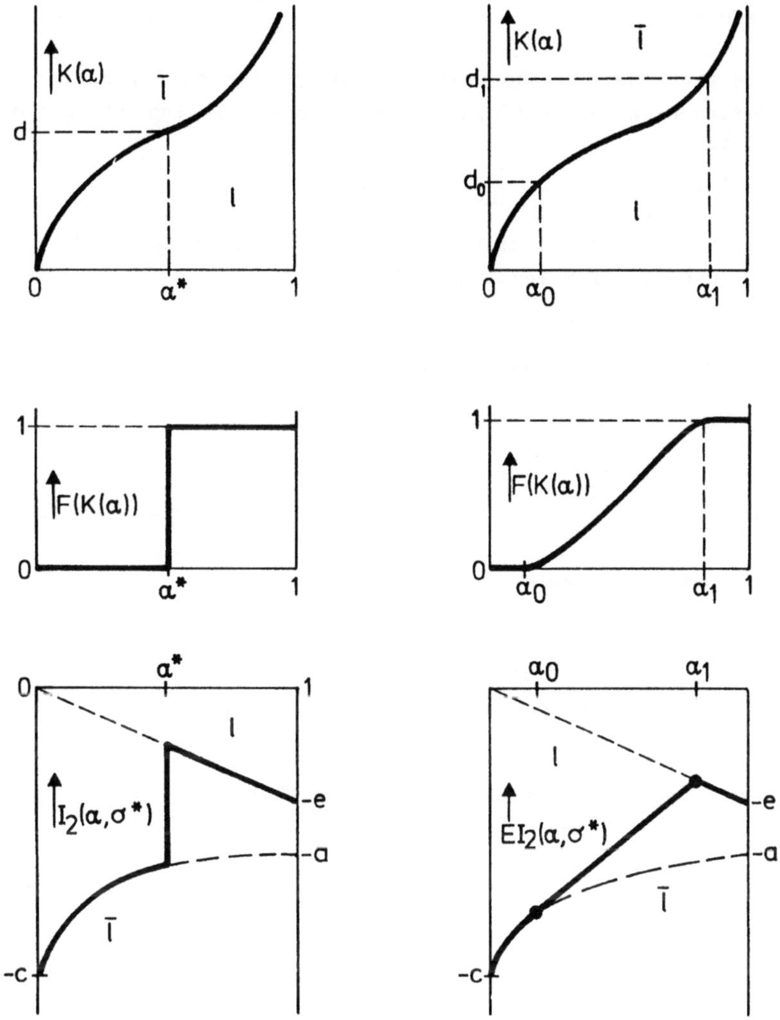

Figure 11: Critical value K(α) and probability F(K(α)) for legal behavior and inspectee's payoff as functions of α for complete (left hand side) and incomplete (right hand side) information; $EI_2(\alpha, \tau^*) = \max_{\sigma \in \Sigma_\alpha} EI_2(\sigma, \tau^*)$

Theorem 4

Assume that $F(x)$ is the uniform distribution on $[d_0, d_1]$ and that $K(\alpha)$ is differentiable in (d_0, d_1) where $K(\alpha_0) = d_0$ and $K(\alpha_1) = d_1$. If the condition

$$\frac{1}{d_1 - d_0} \cdot \left(\min_{\alpha_0 \le \alpha \le \alpha_1} \frac{dK(\alpha)}{d\alpha} \right) \ge \frac{e}{a-e} \tag{5-16}$$

holds, then the equilibrium false alarm probability α^* is $\alpha^* = \alpha_1$, and the probability for legal behavior is

$$F(K(\alpha^*)) = \int_{d_0}^{K(\alpha^*)} dF(t) = 1.$$

Proof

We can write $F(t)$ as

$$F(t) = \frac{t-d_0}{d_1-d_0}, \quad d_0 \le t \le d_1 .$$

Then the inspector's expected payoff is given by

$$G(\alpha) \equiv -e \cdot \alpha \cdot \frac{K(\alpha)-d_0}{d_1-d_0} + (-a \cdot (1-\beta(\alpha)) - c \cdot \beta(\alpha)) \cdot \frac{d_1-K(\alpha)}{d_1-d_0} .$$

It can be seen without much difficulty that

$$\frac{dG(\alpha)}{d\alpha} = -e \cdot \frac{K(\alpha)-d_0}{d_1-d_0} - (c-a) \cdot \frac{d\beta(\alpha)}{d\alpha} \cdot \frac{d_1-K(\alpha)}{d_1-d_0} +$$

$$+ (a \cdot (1-\beta(\alpha)) + c \cdot \beta(\alpha) - e \cdot \alpha) \cdot \frac{dK(\alpha)/d\alpha}{d_1-d_0} .$$

Therefore, we have

$$\frac{dG(\alpha)/d\alpha}{d_1-d_0} > -e + (a \cdot (1-\beta(\alpha)) + c \cdot \beta(\alpha) - e \cdot \alpha) \cdot \frac{dK(\alpha)/d\alpha}{d_1-d_0}$$

for any α, $\alpha_0 \le \alpha \le \alpha_1$. Since

$$a \cdot (1-\beta(\alpha)) + c \cdot \beta(\alpha) - e \cdot \alpha$$

is monotonically decreasing in α, we obtain

$$\frac{dG(\alpha)}{d\alpha} > -e + (a-e) \cdot \frac{dK(\alpha)/d\alpha}{d_1-d_0} \quad \text{for all} \quad \alpha : \alpha_0 \le \alpha \le \alpha_1 .$$

The condition (5-16) implies that

$$dG(\alpha)/d\alpha > 0 \quad \text{for all} \quad \alpha : \alpha_0 \le \alpha \le \alpha_1 .$$

Therefore, the boundary value $\alpha^* = \alpha_1$ is the solution of the maximization problem (5-8) in Theorem 3. ■

Using some properties of the function $K(\alpha)$ which are given in AVENHAUS and OKADA (1988), the condition (5-16) shows us two situations in which we can obtain the legal behavior equilibrium point:

(1) the inspector's cost e in case of false alarm is small, and

(2) the possible values of the inspectee's benefit t in case of no detection are high under the condition that the difference $d_1 - d_0$ is fixed.

We can interpret our results in both cases as follows: In case (1), the inspector does not worry very much about his loss in case of false alarm. Therefore, he can employ the test procedure with the high false alarm probability α_1. In case (2), the inspectee has a high incentive to behave illegally. Therefore, the inspector employs the test procedure with the high false alarm probability α_1 in order to prevent the inspectee from behaving illegally.

6. Applications

Even though we assumed that the inspector is incompletely informed about the inspectee's gain in case of undetected illegal behavior, in practice, it will be difficult if not impossible for him to get any information about the inspectee's payoff parameters. Therefore, it is very important that Theorem 3 gives us an advise for the construction of best test procedures which do not depend on the players' payoff parameters, once the value of the false alarm probability is given. In fact, this theorem establishes a bridge between the game theoretical approach which is necessary for the appropriate description of the inspection problem and the traditional statistical treatment where the error second kind probability is minimized for given value of the error first kind probability.

According to Theorem 3, the optimal decision scheme of the inspector for a given value of the false alarm probability α is the scheme guaranteeing the *maxmin value* of the auxiliary zero-sum game $G_\alpha = (\Sigma_\alpha, \Delta_M, 1-\beta)$ in which the inspector as player 1 chooses the test $\sigma \varepsilon \Sigma_\alpha$, the inspectee as player 2 chooses $\mu \varepsilon \ _M$, and the payoff from player 2 to player 1 is the detection probability $1 - \beta(\sigma,\mu)$. So, one has to find

$$\max_{\sigma} \min_{\mu} (1-\beta(\sigma,\mu)) \quad (6\text{-}1)$$

and the coresponding strategies. Operationally, it is much more convenient to compute

$$\min_{\mu} \max_{\sigma} (1-\beta(\sigma,\mu)) \quad (6\text{-}2)$$

which, of course, would be the same if G_α had a value. However, the existence of a value G_α does not follow from a standard minmax theorem, since the payoff function $1 - \beta(\sigma,\mu)$ does not satisfy the usual convexity requirements. Therefore, in applications one establishes the existence of this value by finding it as well as the optimal strategies.

There are many applications of this procedure (see AVENHAUS (1986)). As examples, we consider once more the two decision problems which we formulated in the second section. In the following, we consider only the auxiliary game G_α which is an important element of the solution of both games, with or without commitment power.

In doing so, we make full use of the multinormal distribution properties in conjunction with the Neyman-Pearson test. We will formulate a general theorem which covers both problems given in the second section and, thereafter, apply it to these problems, specifically.

Theorem 5

Given the multivariate normally distributed random vector X and the two alternative hypotheses

$$H_0(1) : X \sim N(0,\Sigma) \quad (6\text{-}3a)$$

$$H_1(\bar{1}) : X \sim N(\mu,\Sigma),\ \mu'\cdot e = M. \quad (6\text{-}3b)$$

(Do not confuse the covariance matrix Σ with the inspector's strategy set Σ_α now!) Consider the game $G_\alpha = (\Sigma_\alpha, \Delta_M, 1-\beta)$ in which the inspector as player 1 chooses the test $\sigma \in \Sigma_\alpha$ with a given false alarm probability α for a decision beween H_0 and H_1, player 2 chooses a diversion strategy $\mu \in \Delta_M$, and where the payoff from player 2 to player 1 is the detection probability $1 - \beta(\sigma,\mu)$.

Then the game G_α has a value v given by

$$v = 1 - \beta \; = \phi\left(\frac{M}{\sqrt{e'\cdot\Sigma\cdot e}} - U(1-\alpha)\right) \tag{6-4}$$

where ϕ is the cumulative standard normal distribution, and U its inverse. The optimal strategy $\sigma_\alpha{}^*$ of the inspector is given by the alarm set

$$\{x\,|\,x'\cdot e > \sqrt{e\cdot\Sigma\cdot e}\cdot U(1-\alpha)\}. \tag{6-5}$$

The optimal strategy of the operator is the (pure) diversion vector μ^* given by

$$\mu^* = \frac{M}{e'\cdot\Sigma\cdot e}\cdot\Sigma\cdot e. \tag{6-6}$$

Proof

We shall prove this theorem by showing that $(\sigma_\alpha{}^*,\mu^*)$ is in fact a saddle point for $1 - \beta(\sigma_\alpha,\mu)$ which means for all σ_α and μ

$$\beta(\sigma_\alpha{}^*,\mu) \le \beta(\sigma_\alpha{}^*,\mu^*) \le \beta(\sigma_\alpha,\mu^*). \tag{6-7}$$

We start by proving the right hand side inequality.

If the inspectee behaves legally, the density of the random variable X is according to (6-3a)

$$f_0(x) = (2\Pi)^{-n/2}\cdot|\Sigma|^{-1/2}\cdot\exp(-\frac{1}{2}\cdot x'\cdot\Sigma^{-1}\cdot x) \tag{6-8a}$$

while, if he makes a deviation μ, the density is according to (6-3b)

$$f_\mu(x) = (2\Pi)^{-n/2}\cdot|\Sigma|^{-1/2}\cdot\exp(-\frac{1}{2}\cdot(x-\mu)'\cdot\Sigma^{-1}\cdot(x-\mu)). \tag{6-8b}$$

As it is well known that, for any given μ, the most powerful test σ_μ for testing (the alternative hypothesis) f_μ versus (the null hypothesis) f_0 with type I error probability α is the Neyman-Pearson test the alarm set of which is given by:

$$\{x\,|\,\frac{f_\mu(x)}{f_0(x)} > K_\alpha''\} = \{x\,|\,x'\cdot\Sigma^{-1}\cdot\mu > K_\alpha'\}$$

where K_α'' and K_α' are constants determined by α (only). This means that for all μ and the corresponding Neyman-Pearson test σ_μ we have

$$\beta(\sigma_\mu,\mu) \le \beta(\sigma,\mu) \quad \text{for all} \quad \sigma.$$

For $\mu = \mu^*$ denote the corresponding σ_{μ^*} by σ^* whose alarm set is therefore:

$$\{x \mid \frac{M}{e'\cdot\Sigma\cdot e}\cdot x'\cdot e > K_\alpha'\} \quad \text{or} \quad \{x \mid x'\cdot e > K_\alpha\}$$

for some other constant K_α.

Now, as a linear combination of multinormal random variables, the random variable $x'\cdot e$ is normally distributed with expectation $E(x'\cdot e) = E(x')\cdot e = \mu'\cdot e = M$ under f_μ and $E(x'\cdot e) = 0$ under f_0. For both alternatives (and indpendently of μ) the variance of this random variable is $var(x'\cdot e) = e'\cdot\Sigma\cdot e$. Therefore, the false alarm probability α is given by

$$1 - \alpha = prob_0(x'\cdot e \le K_\alpha) = \phi\left(\frac{K_\alpha}{\sqrt{e'\cdot\Sigma\cdot e}}\right),$$

and the probability of not detecting the deviation μ is given by

$$\beta^* = prob_{\mu^*}(x'\cdot e \le K_\alpha) = \phi\left(\frac{K_\alpha - M}{\sqrt{e'\cdot\Sigma\cdot e}}\right).$$

Eliminating K_α from these two equations implies (6-4) and (6-5).

In order to complete the proof, it remains to establish the left hand side inequality of (6-7). This, in fact, follows readily, since for all μ with $\mu'\cdot e = M$

$$E(e'\cdot X) = e'\cdot\mu = M,$$

and the variance of $e'\cdot X$ is independent of μ, implying

$$\beta(\sigma_\alpha^*, \mu) = \beta(\sigma_\alpha^*, \mu^*) \quad \text{for all} \quad \mu.$$ ■

This result shows that in both examples, i.e., independently of the structure of the covariance matrix, the optimal strategy of the inspector is to just add all observed data and to perform a threshold test on the sum. Notice that this procedure is independent of the value of M, the order of magnitude of which the inspector might know, but not its precise value.

In case of the material accountancy example this result is especially interesting: With (5-4) we get

$$\sum_{i=1}^{n} X_i = \sum_{i=1}^{n} (I_{i-1} + R_i - S_i - I_i) = I_0 + \sum_{i=1}^{n} R_i - \sum_{i=1}^{n} S_i - I_n$$

which is the material balance for the whole sequence of inventory periods. Hence, this result means that all intermediate inventories $I_1, \ldots, I_{n-1}$ must not be taken into account in the optimal procedure.

Let us repeat that this surprising result has been derived under the assumption that "time is not valuable" as formulated in the second section: Obviously, one will have to use the intermediate inventories for closing intermediate balances if it is important for the inspector to detect *timely* any diversion of material.

In case of the data verification example the result shows that the so-called D-statistic

$$D := \sum_{i=1}^{n} x_i$$

which has been introduced many years ago also in the area of auditing (DUKE, NETER, LEITCH (1982)) is the best test statistic of the inspector.

Let us mention that also in the other extreme of minimal sample size (k=1) the trivial D-statistic D = x is the best test statistic, however, the optimal falsification strategy is more complicated (AVENHAUS, BATTENBERG, FALKOWSKI (1983)): If the total falsification is smaller than some critical value, then again the equal distribution is optimal, otherwise, only one datum has to be falsified. This is intuitive: For small total falsification, the falsification can be "hidden" in the measurement uncertainty, whereas for large total falsification this is no longer possible; therefore, here the inspectee has to play "vabanque": He has to put everything on one card and hope that this card is not drawn. For sample sizes between one and n, the situation gets very complicated: numerical calculations indicate that beyond some total falsification the D-statistic is no longer optimal, however, one can give upper limits of that total falsification below which the D-statistic is optimal.

These results are important for practitioners: For many years already the IAEA has used the D-statistic for the comparison of operators' and inspectors' data; originally, it has been justified by heuristic arguments (STEWART (1970)). The theory presented here shows its range of applicability and, furthermore, how the best test procedure would have to be determined in general.

7. References

R. AVENHAUS, Safeguards Systems Analysis - With Applications to Nuclear Safeguards and other Inspection Problems. Plenum Press, New York and London 1986

R. AVENHAUS, H.P. BATTENBERG, B.J. FALKOWSKI, Optimale Testverfahren bei der Datenverifikation (Optimal Data Verification Test Procedures). Methods of Operations Research 50, pp. 154-164, 1985

R. AVENHAUS and H. FRICK, Analyse von Fehlalarmen in Überwachungssystemen mit Hilfe von Zweipersonen-Nichtnullensummenspielen (Analysis of False Alarms in Safeguards Systems by Means of Two-Person Non-Zero-Sum Games). Operations Research Verfahren XXVI, pp. 629-639, 1977

R. AVENHAUS and A. OKADA, Inspector Leadership Games with Incomplete Information. Discussion Paper No. 17 of the Zentrum für Interdisziplinäre Forschung, Universität Bielefeld, July 1988

R. AVENHAUS and S. ZAMIR, Safeguards Games with Applications to Material Control. Discussion Paper No. 12 of the Zentrum für Interdisziplinäre Forschung, Universität Bielefeld, May 1988

H.P. BATTENBERG, Optimale Gegenstrategien bei Datenverifikationstests (Optimal Counterstrategies for Data Verification Tests). Ph.D. Dissertation of the Hochschule der Bundeswehr München, 1983

D. BIERLEIN, Direkte Überwachungssysteme (Direct Inspection Systems). Opeations Research Verfahren VI, pp. 57-68, 1969

D. BIERLEIN, Auf Bilanzen und Inventuren basierende Safeguards-Systeme (Safeguards Systems Based on Balances and Inventories). Operations Research Verfahren III, pp. 36-43, 1970

D. BIERLEIN, Game Theoretial Models of Safeguarding Different Types of Illegal Activities. Proceedings of the 4th Formator-Symposium on Mathematical Methods for the Analysis of Large-Scale Systems, Czechosl. Acad. Sc., Prague 1983

S.J. BRAMS, Superpower Games. Yale University Press, New Haven, London 1985

M. DRESHER, A Sampling Inspection Problem in Arms Control Agreements: A Game Theoretic Analysis. Memorandum RM-2972-ARPA, The Rand Corporation, Santa Barbara 1962

G.L. DUKE, J. NETER, P.A. LEITCH, Power Characteristics of Test Statistics in the Auditing Environment: An Empirical Study. Journal of Accounting Research, Vol. 20, No. 1, pp. 42-47, Spring 1982

R.A. DYE, Optimal Monitoring Policies in Agencies. Rand Journal of Economics, Vol. 17, No. 3, pp. 339-350, Autumn 1986

J. FICHTNER, Statistische Tests zur Abschreckung von Fehlverhalten (Statistical Tests for Deterring Illegal Behavior). Ph.D. Dissertation, Universität der Bundeswehr München 1985

H. FRICK, Spieltheoretische Behandlung mehrfacher Inventurprobleme (Game Theoretical Treatment of Multiple Inventory Problems). Ph.D. Dissertation, Universität Fridericiana Karlsruhe 1976

J. HARSANYI, Games with Incomplete Information Played by "Bayesian" Players. Parts I-III, Management Science 14, pp. 159-162, 320-334, and 486-502, 1967/68

E. HÖPFINGER, Zuverlässige Inspektionsstrategien (Reliable Inspection Strategies), Z. Wahrscheinlichkeitstheorie verw. Gebiete, pp. 35-46, 1974

International Atomic Energy Agency, The Structure and Content of Agreements between the Agency and States Required in Connection with the Treaty on the Non-Proliferation on Nuclear Weapons, IAEA Document, INF/CIRC/153, Vienna 1971

C.S. KANODIA, Stochastic and Moral Hazard. Journal of Accounting Research, Vol. 23, No. 1, pp. 175-193, Spring 1985

H.W. KUHN, Recursive Inspection Games, Applications of Statistical Metholology to Arms Control and Disarmament. A US ACDA Report under Contract ACDA/ST-3, pp. 169-182, 1963

E.L. LEHMANN, Testing Statistical Hypotheses. Wiley, New York 1959

M. MASCHLER, A Price Leadership Method for Solving the Inspector's Non-Constant-Sum Game. Nav. Res. Logistics Q. 13, pp. 11-33, 1966

M. MASCHLER, The Inspector's Non-Constant-Sume Game: Its Dependence on a System of Detectors. Nav. Res. Logistics Q. 14, pp. 275-290, 1967

R. SELTEN, A Simple Model of Imperfect Competition, Where 4 are Few and 6 are Many. Int. J. Game Theory 3, pp. 141-201, 1973

H. v.STACKELBERG, Marktform und Gleichgewicht (Market Form and Equilibrium). Julius Springer, Berlin 1934

K.B. STEWART, A Cost-Effectiveness Approach to Inventory Verification. Proceedings of the IAEA Symposium Safeguards Techniques in Karlsruhe, Vol. II, pp. 387-409, IAEA, Vienna 1971

E. Van DAMME, Stability and Perfection of Nash Equilibria, Springer-Verlag 1986

The International Journal of Game Theory is the leading international periodical devoted exclusively to game theoretical developments. Distinguished academics from around the world here present fundamental research contributions on all aspects of game theory.

Some of the interesting papers which appeared in 1996 are:

- [illegible]: Values for Two-Stage Games: Another View of the [illegible]
- [illegible]: Evolution [illegible] Dynamics [illegible]
- [illegible] Game Theory [illegible]
- [illegible] Repeated Games [illegible]
- [illegible]

[illegible]

MIX
Papier aus verantwortungsvollen Quellen
Paper from responsible sources
FSC® C105338

If you have any concerns about our products,
you can contact us on
ProductSafety@springernature.com

In case Publisher is established outside the EU,
the EU authorized representative is:
Springer Nature Customer Service Center GmbH
Europaplatz 3, 69115 Heidelberg, Germany

Printed by Libri Plureos GmbH
in Hamburg, Germany